Bioactive Compounds in Foods: New and Novel Sources, Characterization, Strategies, and Applications

Bioactive Compounds in Foods: New and Novel Sources, Characterization, Strategies, and Applications

Guest Editors

Ivana Generalić Mekinić
Vida Šimat

Basel • Beijing • Wuhan • Barcelona • Belgrade • Novi Sad • Cluj • Manchester

Guest Editors
Ivana Generalić Mekinić
Department of Food Technology and Biotechnology
University of Split
Split
Croatia

Vida Šimat
Department of Marine Studies
University of Split
Split
Croatia

Editorial Office
MDPI AG
Grosspeteranlage 5
4052 Basel, Switzerland

This is a reprint of the Special Issue, published open access by the journal *Foods* (ISSN 2304-8158), freely accessible at: https://www.mdpi.com/journal/foods/special_issues/8RR2A2Y7P1.

For citation purposes, cite each article independently as indicated on the article page online and as indicated below:

Lastname, A.A.; Lastname, B.B. Article Title. *Journal Name* **Year**, *Volume Number*, Page Range.

ISBN 978-3-7258-4731-0 (Hbk)
ISBN 978-3-7258-4732-7 (PDF)
https://doi.org/10.3390/books978-3-7258-4732-7

Cover image courtesy of Ivana Generalić Mekinić

Contents

About the Editors

Ivana Generalić Mekinić

Ivana Generalić Mekinić is an Associate Professor at the Department of Food Technology and Biotechnology at Faculty of Chemistry and Technology University of Split, Croatia. She earned her Ph.D. in Biotechnical Sciences (Food Technology) from the Faculty of Food Technology and Biotechnology of the University of Zagreb in 2011. Her research interests include the chemical composition of various plant materials and their related biological activities as well as isolation, identification and stability testing of isolated compounds and their application in food model systems. She has authored over 70 papers and participated in 15 scientific projects.

Vida Šimat

Vida Šimat is a full professor at the Department of Marine Studies of the University of Split, Croatia, in the field of food science and biotechnology. She holds a Ph.D. from the University of Zagreb, Croatia, and has been involved in numerous national and international research projects. She is currently coordinating an H2020 PRIMA project entitled Innovative Sustainable Solutions for Ready-to-Eat Traditional Mediterranean Products And Non-Conventional Healthy Foods (InnoSol4Med, ID 1836). She is also involved in industrial research activities specifically related to technology transfer to the food industry. In the last 5 years, she has authored and co-authored over 50 research papers and 8 book chapters as well as invited talks at international conferences on research topics related to the utilization of seaweed and microalgae, the valorization of agro-food industry by-products and the application of natural products to improve food quality, stability, and safety. According to Stanford University's new ranking for 2024, she is listed among the top 2% of most influential scientists in the world.

Preface

The global food industry is undergoing a transformative phase, driven by the imperative to develop sustainable, health-promoting, and functionally enriched food products. This reprint, derived from the Special Issue "*Bioactive Compounds in Foods: New and Novel Sources, Characterization, Strategies, and Applications*", addresses these challenges by compiling contemporary research focused on the exploration and application of bioactive compounds in food systems.

The scope of this Special Issue encompasses the identification of both traditional and emerging sources of bioactive compounds, their safety profiles, functional properties, nutritional significance, and health benefits. An emphasis is placed on innovative strategies for incorporating these compounds into food products, considering aspects such as bioaccessibility, bioavailability, and consumer acceptability. The contributions also delve into predictive analyses of future consumption trends and potential market opportunities for bioactive-enriched foods.

This compilation is intended for researchers, industry professionals, and policymakers engaged in food science, nutrition, biotechnology, and related disciplines. The collective efforts presented herein aim to foster advancements in the development of functional foods that cater to the evolving needs of global consumers.

We extend our sincere appreciation to all contributing authors for their valuable research and to the peer reviewers for their critical evaluations. Our gratitude also goes to the editorial team of *Foods* for their unwavering support throughout the publication process.

Ivana Generalić Mekinić and Vida Šimat
Guest Editors

foods

MDPI

Editorial

Bioactive Compounds in Foods: New and Novel Sources, Characterization, Strategies, and Applications

Ivana Generalić Mekinić [1,*] and Vida Šimat [2]

[1] Department of Food Technology and Biotechnology, Faculty of Chemistry and Technology, University of Split, R. Boškovića 35, 21000 Split, Croatia
[2] University Department of Marine Studies, University of Split, R. Boškovića 37, 21000 Split, Croatia; vida@unist.hr
* Correspondence: gene@ktf-split.hr

The current global population has reached almost 8 billion, and it is predicted that by 2050, this number will rise to 9.6 billion, representing a serious challenge for the food industry in the future. According to the UN, the demand for food will increase by almost 70% by that time, accentuating problems such as malnutrition, food insecurity and food shortages in certain regions of the world, as well as the rise of lifestyle- and diet-related diseases and an increasingly pronounced economic imbalance [1–3]. Global agricultural production is also adapting to climate changes, unsustainable agricultural practices, environmental degradation, technological challenges, biodiversity loss, depletion and poor management of water resources, among other concerns [4,5]. All of the above exert severe pressure on global food production, which, in addition to the mentioned challenges, needs to adapt to current food trends and ensure that the safety and quality of food products are maintained [6]. Only through innovations in the food sector, such as new preservation methods, the development of healthier and functional products, processing steps that can enhance and enrich the nutritional profile of products, finding novel sources food, natural preservatives, and the development of clean label products is it possible to overcome these challenges [7].

Bioactive compounds are highly valuable non-nutrient components that exert physiological effects, which are often protective and beneficial for human health. These include diverse phytochemicals, volatile metabolites, and even micronutrients and microbe-derived compounds. Epidemiological evidence links diets rich in bioactives (for example, the Mediterranean diet) with lower incidence of cardiovascular, metabolic, and neurodegenerative diseases, as well as cancer [8–10]. Over the last five years (2020–2025), research has rapidly advanced with regard to identifying these compounds, understanding their biological activities, and developing technologies for their integration into food products. This editorial synthesizes recent findings on food bioactive compounds, their sources, characterization strategies, food industry applications, and implications for human health. Well-established sources of bioactive molecules include fruits, vegetables, grains, legumes, herbs, and fermented foods, which are rich in flavonoids, phenolic acids, carotenoids, glucosinolates, alkaloids, vitamins, and probiotics [10]. Volatile organic compounds in foods are typically associated with aroma and flavor, but many also possess bioactivity. Essential oils and flavor components from herbs, spices, fruits, and other botanicals contain terpenes, aldehydes, and phenolics that exhibit antioxidant, anti-inflammatory, and antimicrobial properties [11]. Traditional nutrients like vitamins, minerals, and peptides can also be considered bioactive when they perform regulatory functions or confer protection beyond basic nutrition. Also, beneficial microbes used in food fermentations—especially lactic acid bacteria (LAB) and certain *Bacillus* or yeast strains—produce a variety of metabolites with bioactive properties [12]. Across all these categories, bioactive compounds often serve functional roles in food systems in addition to their biological activities. Many phytochemicals are pigments (such as anthocyanins and carotenoids) that impart appealing colors, or antioxidants that delay lipid oxidation and spoilage, thereby naturally preserving foods. Volatiles contribute to a product's aroma, but can also inhibit microbes. Meanwhile, organic acids produced during fermentation lower a product's pH, acting as preservatives and flavor enhancers. In essence, these compounds can improve the shelf-life, safety, and sensory qualities of foods while also providing health benefits. This dual functionality is highly attractive for researchers aiming to produce cleaner food products (e.g., those containing fewer synthetic additives) and aligns with consumer demand for natural health-promoting ingredients.

Recent research highlights alternative, underutilized and novel sources of food or components/isolates that can be used to create new, safe and functional products. These include agri-food byproducts, microalgae, seaweed, insect-derived food, fungi, medicinal plants, etc. [13–18]. These sources provide unique bioactive profiles and promote food sector sustainability. Extraction methods including ultrasound-assisted, supercritical CO_2, and microwave-assisted extraction are preferred for their efficiency and sustainability [19]; however, with the chemical diversity of bioactive compounds, advanced analytical strategies have become crucial for identifying and characterizing them. Structural characterization and activity screening through in vitro, in vivo, and in silico methods are key to validating functional properties. Additionally, scientists are actively investigating new processing techniques with a less negative impact on nutritive properties and valuable food constituents. Best practices include the following:

- Using whole natural sources that are high in bioactives (e.g., adding berry purees rich in anthocyanins and flavonols to beverages).
- Food supplementation and fortified foods (addition of isolated bioactive compounds or concentrated/dried extracts to food products to boost their health value).
- Applying measures to stabilize the bioactive compounds.
- Assessing the bioavailability of the compounds, as well as their controlled release in the body.
- Selecting specific protective strains or probiotic cultures to boost the content of certain vitamins (e.g., LAB that synthesize B-vitamins), release bound phenolics from plant fibers, or generate unique bioactive peptides.
- Exploring the synergy of combined bioactives, such as pairing probiotics with prebiotic fibers (synbiotics) or blending different plant extracts to target various health concerns through a single product, developing edible coatings and films infused with natural antimicrobials (like chitosan films with plant extracts), combining bioactives with synergistic nutrients, maintaining clean-label formulations, and ensuring regulatory compliance are essential to successful functional food development.

In addition, these natural, valuable and bioactive ingredients often provide antioxidant, anti-inflammatory, cardioprotective, neuroprotective, and gut-modulatory effects that are essential for promoting overall health and wellness [12,19]. They also play a role in reducing oxidative stress, enhancing metabolic function, supporting immunity, and potentially preventing chronic diseases. Postbiotics and polyphenol-rich diets have shown promise in improving gut and cognitive health, while fermented and fortified foods contribute to overall wellbeing [12,19,20]. It should be noted that bioactives often work synergistically, and that the combination of compounds available in a whole food may offer greater benefits than a single isolated compound.

Despite significant progress, there remain several challenges and frontiers in the field of food bioactive compounds. These include enhancing the bioavailability of the neutral compounds; clarifying precise molecular mechanisms for many effects; expanding the search for new bioactive sources in underutilized plants, marine organisms, and agro-industrial byproducts; ensuring the safety of concentrations of bioactives added to foods, as well as recommended intake levels; and bridging the gap between scientific advancements and consumer perspective and acceptance of the novel solutions. Research should focus on personalized nutrition, sustainable sourcing, and effective communication of health claims to maximize public health impact. The synergy between food science, biotechnology, and nutrition continues to shape the next generation of smarter functional foods. Such foods will not only nourish but also provide targeted benefits (from heart health to cognitive support), helping consumers take charge of their health through their diets.

The primary aim of this Special Issue (SI), "Bioactive Compounds in Foods: New and Novel Sources, Characterization, Strategies, and Applications", is to investigate both current and emerging sources of valuable natural bioactive compounds, and best practices for incorporating them into food products. The SI comprises ten original research papers and one review; together, these works address recent advances and current knowledge in the proposed field (please see the list of contributions for further details). We believe that further investigations in this scientific field will offer solutions to current global problems and meet the nutritional needs of future generations in a sustainable manner, protecting the planet's ecosystems.

Finally, we would like to thank all the authors and the reviewers who contributed their work, knowledge and suggestions to this SI, as well as MDPI and the *Foods* journal for their help and support.

List of Contributions

1. Costa, A.S.G.; Peixoto, J.A.B.; Machado, S.; Espírito Santo, L.; Soares, T.F.; Andrade, N.; Azevedo, R.; Almeida, A.; Costa, H.S.; Oliveira, M.B.P.P.; et al. Coffee Pulp from Azores: A Novel Phytochemical-Rich Food with Potential Anti-Diabetic Properties. *Foods* **2025**, *14*, 306. https://doi.org/10.3390/foods14020306.
2. Nikolić, V.; Žilić, S.; Simić, M.; Šavikin, K.; Stević, T.; Živković, J.; Sarić, B.; Milovanović, D.; Kandić Raftery, V. Characterization and Potential Food Applications of Oat Flour and Husks from Differently Colored Genotypes as Novel Nutritional Sources of Bioactive Compounds. *Foods* **2024**, *13*, 3853. https://doi.org/10.3390/foods13233853.
3. Coyago-Cruz, E.; Barrigas, A.; Guachamin, A.; Heredia-Moya, J.; Zuñiga-Miranda, J.; Vera, E. Bioactive Composition of Tropical Flowers and Their Antioxidant and Antimicrobial Properties. *Foods* **2024**, *13*, 3766. https://doi.org/10.3390/foods13233766.
4. Katsuki, T.; Ogi, K.; Kinno, A.; Kasamatsu, S.; Ihara, H.; Sumitani, H. Inhibition of Amyloid β Accumulation by Protease-Digested Whitebait (Shirasu) in a Murine Model of Alzheimer's Disease. *Foods* **2024**, *13*, 2858. https://doi.org/10.3390/foods13182858.
5. Murthy, H.N.; Yadav, G.G.; Joseph, K.S.; HS, S.K.; Magi, S.M.; Dewir, Y.H.; Mendler-Drienyovszki, N. Nutritional Value, Fatty Acid and Phytochemical Composition, and Antioxidant Properties of Mysore Fig (*Ficus drupacea* Thunb.) Fruits. *Foods* **2024**, *13*, 2845. https://doi.org/10.3390/foods13172845.

6. Higuchi, J.; Kurogochi, M.; Yamaguchi, T.; Fujio, N.; Mitsuduka, S.; Ishida, Y.; Fukudome, H.; Nonoyama, N.; Gota, M.; Mizuno, M.; et al. Qualitative and Quantitative Analyses of Sialyl O-Glycans in Milk-Derived Sialylglycopeptide Concentrate. *Foods* **2024**, *13*, 2792. https://doi.org/10.3390/foods13172792.
7. Bebek Markovinović, A.; Bosiljkov, T.; Janči, T.; Kostić, M.; Dedović, N.; Lučić, E.; Bavrka, K.; Pavlić, B.; Bursać Kovačević, D. Characterization of Antioxidant Bioactive Compounds and Rheological, Color and Sensory Properties in 3D-Printed Fruit Snacks. *Foods* **2024**, *13*, 1623. https://doi.org/10.3390/foods13111623.
8. Lemoine, C.; Rodrigues, M.J.; Dauvergne, X.; Cérantola, S.; Custódio, L.; Magné, C. A Characterization of Biological Activities and Bioactive Phenolics from the Non-Volatile Fraction of the Edible and Medicinal Halophyte Sea Fennel (*Crithmum maritimum* L.). *Foods* **2024**, *13*, 1294. https://doi.org/10.3390/foods13091294.
9. Politeo, O.; Ćurlin, P.; Brzović, P.; Auzende, K.; Magné, C.; Generalić Mekinić, I. Volatiles from French and Croatian Sea Fennel Ecotypes: Chemical Profiles and the Antioxidant, Antimicrobial and Antiageing Activity of Essential Oils and Hydrolates. *Foods* **2024**, *13*, 695. https://doi.org/10.3390/foods13050695.
10. Cardinali, F.; Belleggia, L.; Reale, A.; Cirlini, M.; Boscaino, F.; Di Renzo, T.; Del Vecchio, L.; Cavalca, N.; Milanović, V.; Garofalo, C.; et al. Exploitation of Black Olive (*Olea europaea* L. cv. Piantone di Mogliano) Pomace for the Production of High-Value Bread. *Foods* **2024**, *13*, 460. https://doi.org/10.3390/foods13030460.
11. Sayas-Barberá, E.; Paredes, C.; Salgado-Ramos, M.; Pallarés, N.; Ferrer, E.; Navarro-Rodríguez de Vera, C.; Pérez-Álvarez, J.Á. Approaches to Enhance Sugar Content in Foods: Is the Date Palm Fruit a Natural Alternative to Sweeteners? *Foods* **2024**, *13*, 129. https://doi.org/10.3390/foods13010129.

Author Contributions: Conceptualization, I.G.M. and V.Š.; investigation, I.G.M. and V.Š.; writing—original draft preparation, I.G.M.; writing—review and editing, V.Š. All authors have read and agreed to the published version of the manuscript.

Funding: This paper is supported by the PRIMA program under projects SEAFENNEL4MED (https://seafennel4med.com/) and InnoSol4Med (https://innosol4med.unist.hr/). The PRIMA program is supported by the European Union.

Conflicts of Interest: The authors declare no conflicts of interest.

References

1. United Nations. *World Population Prospects 2024: Summary of Results*; UN DESA/POP/2024/TR/NO. 9; United Nations: New York, NY, USA, 2024; Available online: https://population.un.org/wpp/assets/Files/WPP2024_Summary-of-Results.pdf (accessed on 8 April 2025).
2. Food and Agriculture Organization (FAO); International Fund for Agricultural Development (IFAD); United Nations Children's Fund (UNICEF); World Food Programme (WFP); World Health Organization (WHO). *The State of Food Security and Nutrition in the World 2023: Urbanization, Agrifood Systems Transformation and Healthy Diets across the Rural–Urban Continuum*; FAO: Rome, Italy, 2023. [CrossRef]
3. Alexandratos, N.; Bruinsma, J. *World Agriculture Towards 2030/2050: The 2012 Revision*; ESA Working Paper No. 12-03; FAO: Rome, Italy, 2012. [CrossRef]
4. Bakhtary, H.; Palmegiani, I.; Rodezno Ayestas, M.J.; Hahn, G. *Biodiversity and Climate Action in Agriculture and Food Systems: Opportunities for Building Synergies*; WWF: Gland, Switzerland, 2024; Available online: https://wwfint.awsassets.panda.org/downloads/biodiversity-and-climate-action-in-agriculture-and-food-systems.pdf (accessed on 8 April 2025).
5. Harwatt, H.; Wetterberg, K.; Giritharan, A.; Benton, T.G. *Aligning Food Systems with Climate and Biodiversity Targets: Assessing the Suitability of Policy Action over the Next Decade*; Research Paper; Royal Institute of International Affairs: London, UK, 2022. [CrossRef]

6. Hubert, B.; Rosegrant, M.; van Boekel, M.A.J.S.; Ortiz, R. The Future of Food: Scenarios for 2050. *Crop Sci.* **2010**, *50*, S-33–S-50. [CrossRef]
7. McKenzie, F.C.; Williams, J. Sustainable Food Production: Constraints, Challenges and Choices by 2050. *Food Sec.* **2015**, *7*, 221–233. [CrossRef]
8. Lisboa, H.M.; Pasquali, M.B.; dos Anjos, A.I.; Sarinho, A.M.; de Melo, E.D.; Andrade, R.; Batista, L.; Lima, J.; Diniz, Y.; Barros, A. Innovative and Sustainable Food Preservation Techniques: Enhancing Food Quality, Safety, and Environmental Sustainability. *Sustainability* **2024**, *16*, 8223. [CrossRef]
9. Sharif, Z.I.M.; Mustapha, F.A.; Jai, J.; Mohd Yusof, N.; Zaki, N.A.M. Review on Methods for Preservation and Natural Preservatives for Extending the Food Longevity. *Chem. Eng. Res. Bull.* **2017**, *19*, 145–153. [CrossRef]
10. Câmara, J.S.; Albuquerque, B.R.; Aguiar, J.; Corrêa, R.C.G.; Gonçalves, J.L.; Granato, D.; Pereira, J.A.M.; Barros, L.; Ferreira, I.C.F.R. Food Bioactive Compounds and Emerging Techniques for Their Extraction: Polyphenols as a Case Study. *Foods* **2021**, *10*, 37. [CrossRef] [PubMed]
11. Dahiya, D.; Terpou, A.; Dasenaki, M.; Nigam, P.S. Current Status and Future Prospects of Bioactive Molecules Delivered through Sustainable Encapsulation Techniques for Food Fortification. *Sustain. Food Technol.* **2023**, *1*, 500–510. [CrossRef]
12. Abdul Hakim, B.N.; Xuan, N.J.; Oslan, S.N.H. A Comprehensive Review of Bioactive Compounds from Lactic Acid Bacteria: Potential Functions as Functional Food in Dietetics and the Food Industry. *Foods* **2023**, *12*, 2850. [CrossRef] [PubMed]
13. Quintieri, L.; Nitride, C.; De Angelis, E.; Lamonaca, A.; Pilolli, R.; Russo, F.; Monaci, L. Alternative Protein Sources and Novel Foods: Benefits, Food Applications and Safety Issues. *Nutrients* **2023**, *15*, 1509. [CrossRef] [PubMed]
14. Maurya, N.; Kushwaha, R. Novel Protein Foods: Alternative Sources of Protein for Human Consumption. *Zenodo* **2019**. [CrossRef]
15. Zhang, X.; Zhang, Z.; Shen, A.; Zhang, T.; Jiang, L.; El-Seedi, H.; Zhang, G.; Sui, X. Legumes as an Alternative Protein Source in Plant-Based Foods: Applications, Challenges, and Strategies. *Curr. Res. Food Sci.* **2024**, *9*, 100876. [CrossRef] [PubMed]
16. Kanchan; Singh, A.; Parihar, V.; Sadhvi. Bioconversion of Food Wastes and By-products to Value-added Products: A Comprehensive Review. *Eur. J. Nutr. Food Saf.* **2024**, *16*, 297–305. [CrossRef]
17. Koptelova, A.; Ducrey, A.; Lunven, B.; Köller, L.; Nagornova, N.; Appenzeller, E.W.; Abitbol, T. Waste-to-Taste: Transforming Wet Byproducts of the Food Industry into New Nutritious Foods. *Chimia* **2024**, *78*, 824–830. [CrossRef] [PubMed]
18. Tsegay, Z.T.; Gebreegziabher, S.T.; Mulaw, G. Nutritional Qualities and Valorization Trends of Vegetable and Fruit Byproducts: A Comprehensive Review. *J. Food Qual.* **2024**, *2024*, 5518577. [CrossRef]
19. Hrelia, S.; Barbalace, M.C.; Angeloni, C. Agri-Food Wastes as Natural Source of Bioactive Antioxidants—Third Edition. *Antioxidants* **2025**, *14*, 198. [CrossRef] [PubMed]
20. Liu, C.; Liu, X. Innovative Application of Metabolomics on Bioactive Ingredients of Foods. *Foods* **2022**, *11*, 2974. [CrossRef] [PubMed]

foods

MDPI

Article

Coffee Pulp from Azores: A Novel Phytochemical-Rich Food with Potential Anti-Diabetic Properties

Anabela S. G. Costa [1,2], Juliana A. Barreto Peixoto [1], Susana Machado [1], Liliana Espírito Santo [1], Thiago F. Soares [1], Nelson Andrade [1,3], Rui Azevedo [1], Agostinho Almeida [1], Helena S. Costa [1,4], Maria Beatriz Prior Pinto Oliveira [1], Fátima Martel [3,5], Jesus Simal-Gandara [2] and Rita C. Alves [1,*]

[1] Network of Chemistry and Technology/Associated Laboratory for Green Chemistry (REQUIMTE/LAQV), Department of Chemical Sciences, Faculty of Pharmacy, University of Porto, 4050-313 Porto, Portugal; acosta@ff.up.pt (A.S.G.C.); jpeixoto@ff.up.pt (J.A.B.P.); smachado@ff.up.pt (S.M.); lsanto@ff.up.pt (L.E.S.); up201902664@edu.ff.up.pt (T.F.S.); nandrade@med.up.pt (N.A.); rmazevedo@ff.up.pt (R.A.); aalmeida@ff.up.pt (A.A.); helena.costa@insa.min-saude.pt (H.S.C.); beatoliv@ff.up.pt (M.B.P.P.O.)

[2] Nutrition and Bromatology Group, Department of Analytical Chemistry and Food Science, Faculty of Science, University of Vigo, E-32004 Ourense, Spain; jsimal@uvigo.es

[3] Unit of Biochemistry, Department of Biomedicine, Faculty of Medicine of Porto, University of Porto, 4200-319 Porto, Portugal; fmartel@med.up.pt

[4] Department of Food and Nutrition, National Institute of Health Dr. Ricardo Jorge, I.P., Av. Padre Cruz, 1649-016 Lisbon, Portugal

[5] Instituto de Investigação e Inovação em Saúde (i3S), University of Porto, 4200-135 Porto, Portugal

[*] Correspondence: rcalves@ff.up.pt

Abstract: Coffee pulp, a by-product of wet coffee processing, shows significant potential in the food and health domains, but its real applications remain underexplored. This work investigated the chemical composition and bioactive properties of coffee pulp from São Miguel Island (Azores, Portugal). The studied coffee pulp exhibited high fiber content (52% dw), mostly insoluble; notable mineral levels (10.6%), mainly K, Ca, and Mg; and 6% dw of total amino acids, with hydroxyproline, aspartic acid, glutamic acid, and leucine in higher amounts. Despite containing low fat (1.6% dw), mainly saturated, it also showed considerable amounts of polyunsaturated fatty acids with a favorable n6/n3 ratio (1.40) and vitamin E (α-, β-, and γ-tocopherols). Its antioxidant capacity can be partially explained by the chlorogenic acid content (9.2 mg/g dw), and caffeine (0.98%) was present in similar amounts to those observed in some arabica coffee beans. A decrease in glucose uptake in Caco-2 cells was found, but not in fructose, suggesting selective inhibition of SGLT1 and potential antidiabetic effects. These results show that Azorean coffee pulp has potential as a sustainable and bioactive ingredient for incorporation into functional foods or dietary supplements.

Keywords: coffee by-product; valorization; sustainability; chemical composition; bioactivity; Caco-2 cells

1. Introduction

A fresh coffee cherry contains approximately 35–45% of pulp [1,2]. With global coffee production reaching around ten million tons of raw beans annually [3], a substantial amount of this by-product is generated, underscoring the need for an effective means of valorization. Moreover, this by-product is frequently released into natural ecosystems, leading to harmful environmental effects due to the substantial accumulation of organic matter and phytotoxic compounds. However, coffee pulp holds significant potential as a functional ingredient for innovative applications in the food industry [1,2,4,5]. In fact, although the use of dried coffee pulp for food purposes is not usual in Portugal or other European countries, significant progress was made in 2022 when the European Union (EU) approved the commercialization of dried coffee pulp from *Coffea arabica* L. and *Coffea canephora* Pierre ex A. Froehner, along with its infusion, as a traditional food from third countries [6]. This decision was based on the long-standing history of safe consumption of coffee pulp and its beverage in countries such as Yemen, Ethiopia, and Bolivia, where they have been traditionally used [6]. As a result, the market for coffee pulp is expected to grow, highlighting the need for a comprehensive understanding of its potential health effects when consumed.

Overall, coffee pulp is recognized as a good source of antioxidants, particularly phenolics, such as chlorogenic acids, caffeic acid, ferulic acid, and coumaric acid [7]. Its chemical composition includes ~50% carbohydrates, 20% dietary fiber, 10% protein, 2.5% lipids, and 1.3% caffeine [4], making coffee pulp highly versatile. In addition to potential applications in bioethanol production [8] or as composting material [1,2], its anthocyanin content also makes it suitable as a natural food colorant [9]. It can also be processed into flour, containing 18% dietary fiber [4,10], or be used in functional infusions [1]. Coffee pulp can also undergo hydrolysis to recover sugars from the matrix, namely xylose, arabinose, fructose, glucose, sucrose, and maltose [4]. Several studies have also demonstrated a range of bioactive properties, including antioxidant, anti-inflammatory, hypolipidemic, hepatoprotective, antidiabetic, and antiobesity, as well as the capacity to modulate gut microbiota [11–14].

In this study, we conducted, for the first time, a comprehensive analysis of the chemical composition and bioactive potential of coffee cherry pulp from S. Miguel, Azores (Portugal), focusing on its antioxidant and antidiabetic properties. Although not yet well known in the international market or even among the general population, the Azores archipelago boasts a small but promising coffee production. These volcanic islands provide unique microclimatic conditions, such as mild temperatures (15 °C in winter to 27 °C in summer) and high humidity (76%), which contribute to optimal coffee plant growth [15]. These distinct environmental factors bring novelty and significance to this study, as they may influence the chemical and bioactive profiles of this by-product. To the best of our knowledge, this is the first detailed report on the nutritional profile of Azorean coffee pulp, covering its mineral composition, amino acid and fatty acid profiles, vitamin E content, free sugars, chlorogenic acids, and caffeine contents. Additionally, we explored its antioxidant and antidiabetic potential, including the first evaluation of sugar uptake inhibition by coffee pulp extracts in an intestinal cell line. Our findings can offer new insights into the unique qualities of Azorean coffee pulp and its potential applications.

2. Materials and Methods

2.1. Reagents, Standards and Materials

Kjeldahl tablets were purchased from Merck (Darmstadt, Germany). The dietary fiber assay kit and celite were provided by Sigma-Aldrich (St. Louis, MO, USA).

For mineral analysis, nitric acid (≥69% (*w*/*w*) Trace Metal™ was purchased from Fisher Scientific (Leicestershire, UK), hydrogen peroxide (30% *w*/*w*) Suprapur® was provided by Merck (Darmstadt, Germany), and hydrochloric acid (>30% *w*/*w*) TraceSELECT™ was supplied by Honeywell Fluka™ (Seelze, Germany). Certified reference materials (BCR-679, ERM-BB422, ERM-BC382) were provided by the European Commission, Joint Research Centre (Brussels, Belgium). A 10 µg/mL internal standard (Mix1-SCP-IS7) and single-element stock solutions (1000 μg/mL) of gallium (Ga), rhodium (Rh), and calcium (Ca) were from SCP SCIENCE (Quebec, QC, Canada); single-element stock solutions (1000 μg/mL) of iron (Fe), magnesium (mg), sodium (Na), and potassium (K) for Atomic Absorption Spectroscopy were from Fluka (Seelze, Germany); Sigma-Aldrich (Buchs, Switzerland) provided phosphorus (P) and mercury (Hg) single-element standards (1000 μg/mL); and the ICP multi-element standard solution (21 elements) was obtained from Supelco (Darmstadt, Germany).

For chromatographic analyses, the following standards/reagents were used: standards of amino acids (98.0%–≥99.5%), tocopherols (α-, β, λ-, and δ-), fructose, glucose, and sucrose, as well as the Supelco 37 Component FAME Mix, were all provided by Sigma-Aldrich (St. Louis, MO, USA); Larodan (Solna, Sweden) provided tocotrienols (α-, β-, λ-, and δ-) and tocol; L-norvaline was purchased from Sigma (Deisenhofen, Germany); caffeine and chlorogenic acids (3-, 4-, and 5-caffeoylquinic acids) were supplied by Honeywell Riedel-de Haën™ (Seetze, Germany). The derivatization reagents (9-fluorenylmethyl chloroformate and o-phthalaldehyde/3-mercaptopropionic acid) were purchased from Agilent Technologies (Palo Alto, CA, USA). For eluent preparation, 1,4-dioxane was obtained from Sigma-Aldrich (St. Louis, MO, USA); HPLC-grade methanol, acetonitrile, and n-hexane were acquired from Honeywell (Düsseldorf, Germany).

For spectrophotometric analyses, Merck (Darmstadt, Germany) provided the Folin–Ciocalteu reagent. Chlorogenic acid (5-caffeoylquinic acid), (±)-catechin hydrate, iron (II) sulfate heptahydrate, 2,4,6-tris(2-pyridyl)-s-triazine (TPTZ), 2,2-diphenyl-1-picrylhydrazyl (DPPH), (±)-6-hydroxy-2,5,7,8-tetramethylchromane-2-carboxylic acid (Trolox), and iron (III) chloride hexahydrate were all supplied by Sigma-Aldrich (St. Louis, MO, USA).

For cellular assays, minimum essential medium (MEM), HEPES, trypsin–EDTA, antibiotic/antimycotic solution, NADH, SRB, sodium pyruvate, and trichloroaceric acid were all purchased from Sigma (St. Louis, MO, USA); fetal calf serum was supplied by Invitrogen Corporation (Carlsbad, CA, USA); Triton X-100 was purchased from Merck (Darmstadt, Germany); [1,2-^{3}H(N)]-deoxy-D-glucose (^{3}H-DG; 60 mCi/mmol) and ^{14}C-D-fructose (^{14}C-FRU; 250–360 mCi/mmol) were purchased from American Radiolabeled Chemicals (St. Louis, MO, USA).

All other reagents were of analytical grade. Ultrapure water was obtained from a Direct-Pure UP Ultrapure & RO Lab Water System (RephiLe Bioscience Ltd., Boston, MA, USA).

2.2. Sample and Sample Preparation

Dried coffee pulp from *Coffea arabica* (caturra vermelha variety) was kindly provided by a local Azorean producer (São Miguel, Azores, Portugal; 37°48′24.4″ N 25°37′29.5″ W), corresponding to the 2020 harvest (June). Coffee pulp was collected right after harvesting the fruits using a de-pulper (initial step of the wet processing method). It was then carefully spread in a thin layer on a net, in the open air, inside a greenhouse to be protected from direct sunlight and climatic conditions, being frequently turned and moved to dry evenly, until a moisture content lower than 10% was obtained. After being dried, the sample (~2–3 kg) was shipped to the laboratory. There, the ground coffee pulp was stored under vacuum and kept protected from light at room temperature, till analyses.

2.3. Nutritional Composition

Ash (AOAC 923.03), crude protein (AOAC 984.13), total fat (AOAC 991.36), and total and insoluble dietary fiber (AOAC 985.29 and 991.42, respectively) were determined using standard normalized methods [16]. Soluble fiber and available carbohydrate contents were estimated by calculation.

2.4. Mineral Composition

Sample mineralization was performed in an ETHOS™ EASY microwave oven (Milestone, Sorisole, Italy) equipped with an SK-15 EasyTEMP high-pressure rotor [17,18]. Briefly, 400 mg of sample were mixed with 9 mL HNO_3 (69% *w*/*w*). After pre-digestion, 1 mL H_2O_2 (30% *w*/*w*) was added. The vessels were then tightly closed and placed in a microwave oven (program: temperature increase in 20 min up to 210 °C; 15 min at 210 °C). After cooling down, 0.5 mL of HCl (>30% *w*/*w*) was diluted with ultrapure water. A digestion blank (with reagents only and no sample) was also prepared. To ensure the quality of the digestion process, 400 mg of three different reference materials—BCR-679 (white cabbage), ERM-BB422 (fish muscle), and ERM-BC382 (wheat flour)—were digested under the same conditions as the sample.

For the elemental analysis of the digested sample, an ICP-MS iCAP™ Q (Thermo Fisher Scientific, Bremen, Germany) containing a Meinhard® TQ+ quartz concentric nebulizer (Golden, CO, USA), a high-purity quartz cyclonic nebulization chamber, and a detachable quartz torch with a 2.5 mm internal diameter injector, and an interface of two nickel cones (sampler and skimmer), was used. High-purity argon (99.9997%; Gasin, Leça da Palmeira, Portugal) was used both as nebulization gas and for plasma formation. The operating conditions were as follows: 1.14 L/min of nebulization gas; 0.79 L/min of auxiliary gas; 13.9 L/min of plasma gas; 1550 W of radiofrequency power; and a dwell time ranging from 1 to 50 ms. The internal standard was prepared by proper dilution of single-element Ga and Rh solutions and a multi-element solution (Internal Standard Mix 1–SCP-IS7).

Calibration curves were prepared by properly diluting the stock standard solutions in the following ranges: Mn, Co, Ni, Cu, Zn, As, Se, Rb, Sr, Mo, Cd, Cs, Ba, and Pb (0.5–50 μg/L); Fe (100–1000 μg/L); Mg and Na (100–10,000 μg/L); K, Ca, and P (1000–50,000 μg/L); Hg (1–10 μg/L). The isotopes ^{23}Na, ^{25}Mg, ^{31}P, ^{39}K, ^{43}Ca, ^{55}Mn, ^{57}Fe, ^{59}Co, ^{60}Ni, ^{65}Cu, ^{66}Zn, ^{75}As, ^{82}Se, ^{87}Rb, ^{88}Sr, ^{98}Mo, ^{111}Cd, ^{133}Cs, ^{137}Ba, ^{202}Hg, ^{206}Pb, ^{207}Pb, and ^{208}Pb were analyzed for sample elemental content, and the isotopes ^{6}Li, ^{45}Sc, ^{71}Ga, ^{90}Y, ^{103}Rh, ^{115}In, and ^{209}Bi were monitored as internal standards. The reference materials were analyzed at the beginning and end of the analytical series.

2.5. Amino Acid Profile

For the determination of total amino acids, the method outlined by Machado et al. [19] was followed. Very briefly, the sample (~150 mg) was mixed with 6M HCl to undergo an acid hydrolysis (110 °C, 24 h). In parallel, alkaline hydrolysis using 3 mL KOH (4 M) for 6 h was conducted specifically to determine tryptophan content. The hydrolyzed samples were then centrifuged, and a supernatant aliquot (50 μL) was collected and neutralized.

For free amino acids, the protocol described by Machado et al. [19] was also followed. In this case, free amino acids were extracted from the sample (0.5 g) with deionized water (10 mL) by magnetic stirring (40 °C, 30 min), followed by re-extraction (5 mL; 40 °C, 15 min).

The internal standard (norvaline, 2 mg/L) was added to hydrolysates/extracts prior to online automatic derivatization, as described by Machado et al. [19].

Total and free amino acids were quantified via reversed phase-HPLC using an integrated system from Jasco (Jasco, Tokyo, Japan), composed of an LC-NetII/ADC hardware interface, two PU-980 pumps, an AS-4150 automatic derivatizer/injector, an FP-2020 Plus detector, and a CO-4061 oven. Separation of amino acids was carried out on a ZORBAX

Eclipse Plus C_{18} column (4.6 × 250 mm, 5 μm; Agilent Technologies, Santa Clara, CA, USA) at 50 °C, using the same elution conditions as those described by Machado et al. [19].

2.6. Vitamin E Profile Analysis

Vitamin E was determined as described by Alves et al. [20]. Very briefly, 150 mg of sample was mixed for 30 min with BHT (75 μL, 0.1%), tocol (50 μL, 0.1 mg/mL) and absolute ethanol (1 mL). Then, n-hexane (4 mL) was added, and another vortexing cycle of 30 min was employed. Subsequently, 2 mL of NaCl (1%, *w/v*) were added. After vortexing, the organic phase was collected, and the residue was re-extracted with n-hexane. The organic phases were mixed and anhydrous Na_2SO_4 was used to ensure that no water was left in the extract. The final solution was concentrated under a N_2 stream and analyzed using an HPLC system (Jasco, Tokyo, Japan) equipped with an autosampler (AS-4050), a pump (PV-4180), a Supelcosil™ LC-SI column (7.5 cm × 3 mm, 3 μm, Supelco, Bellefonte, PA, USA), a diode-array detector (MD-4015), and a fluorescence detector (FP-4025). The system was operated according to the same conditions described by Alves et al. [20].

2.7. Fatty Acid Profile Analysis

The same lipid extraction protocol described in Section 2.5 Vitamin E profile analysis was employed. Fatty acids were then derivatized into methyl esters, according to ISO 12966-2: 2011 [21], using a procedure based on both transesterification and methylation of the free fatty acids. The resulting fatty acid methyl esters (FAMEs) were then analyzed in a GC-FID system (Shimadzu, Tokyo, Japan), equipped with an AOC-20i autosampler, a split/splitless auto-injector (250 °C), and a flame ionization detector. FAME separation was achieved on a CP-Sil 88 silica capillary column (50 m × 0.25 mm, 0.2 μm) from Varian (Middelburg, The Netherlands). The carrier gas was helium (3.0 mL/min), and the temperature program used was 120 °C, 5 min; 2 °C/min till 160 °C; 160 °C, 2 min; 2 °C/min till 220; 220 °C for 10 min. FAMEs were identified by comparing the respective retention times with those of the Supelco 37 Component FAME Mix.

2.8. Free Sugar Determination

To extract free sugars, the sample (200 mg) was mixed with deionized water (10 mL) for 20 min. The samples were then centrifuged, and the supernatant was analyzed on an HPLC system (Jasco, Tokyo, Japan) equipped with an autosampler (AS-4050), a pump (PU-4180), a column oven (CO-4061), and a low-temperature evaporative light scattering detector (LT-ELSD Sedex 80, Sedere, Alfortville, France). Fructose, glucose, and sucrose were quantified after chromatographic separation on a Shodex column (Asahipak NH2P-50 4E, 4.6 mm ID × 250 mm) at 30 °C, using water/acetonitrile (1:3) as eluent (1 mL/min) for 20 min [22].

2.9. Antioxidant Profile

To obtain antioxidants, 100 mg of sample were extracted with 40 mL water/ethanol (1:1). The extractions were carried out for 1 h at 40 °C with constant agitation. The obtained extracts were filtered and stored at −20 °C for the following analyses.

The ferric-reducing antioxidant power was determined according to Rufino et al. [23] with minor modifications. The extract (35 μL) was mixed with 265 μL of FRAP reagent [23] at 37 °C for 30 min. Absorbance was measured at 595 nm. A calibration curve was prepared with ferrous sulfate (50–600 μmol/L). The $DPPH^{\bullet}$ scavenging activity assay was carried out according to Silveira et al. [24] with minor adjustments. In brief, 30 μL of extract were mixed with 270 μL of an ethanolic $DPPH^{\bullet}$ solution ($6{\cdot}10^{-5}$ M) at room temperature, protected from light, for 20 min. Absorbance was read at 525 nm. A calibration curve was prepared with Trolox (5–150 mg/L).

Total phenolic content was estimated according to Zhang et al. [25] with some minor alterations. The extract (30 μL) was mixed with 150 μL of Folin–Ciocalteu reagent (1:10) and 120 μL of sodium carbonate (7.5% *w/v*) at 45 °C for 15 min, followed by 30 min at room temperature. Absorbance was measured at 765. Chlorogenic acid (5–160 mg/L) was used to prepare the calibration curve. In turn, total flavonoid content was estimated according to Zou et al. [26] using minor adjustments. The extract (1 mL) was diluted with water (1:5) and mixed with 300 μL of 5% (*w/v*) sodium nitrite. After 5 min, 300 μL of 10% aluminium chloride was added. After one more minute, 2 mL of 1 M sodium hydroxide and 2.5 mL of water were also added and the absorbance was read at 510 nm. Catechin (5–100 mg/L) was used to prepare the calibration curve.

2.10. Chlorogenic Acids Profile and Caffeine Content by HPLC

The extracts prepared in Section 2.9. Antioxidant profile were further analyzed to quantify chlorogenic acids and caffeine. An HPLC integrated system from Jasco (Tokyo, Japan) equipped with an automatic sampler (AS-2057 Plus), a pump (PU-2089 Plus), a column oven (CO-2060 Plus, 28 °C), a Zorbax-SB-C18 (5 μm, 250 mm × 4.6; Agilent Technologies, Santa Clara, CA, USA), and a diode-array detector (MD-2018 Plus) was used for analysis, using exactly the same conditions as those described by Machado et al. [27].

2.11. Antidiabetic Potential

The Caco-2 cell line (human colorectal adenocarcinoma; passage numbers 14–39) was obtained from ATCC (Manassas, VA, USA). The cells were grown in MEM (containing glucose and supplemented with fetal calf serum, HEPES, and antibiotics), sub-cultured, and prepared, as described by Peixoto et al. [28].

The extracts were prepared as described in Section 2.9. Antioxidant profile. Ethanol was then evaporated under a N_2 stream and the wet residue was freeze-dried. An amount (~10 mg) of this freeze-dried powder was dissolved in distilled water (100 mg/mL). For each of the following experiments, decreasing concentrations (2, 1, and 0.5 mg/mL) were prepared. Cells were exposed to those concentrations for 24 h in FCS-free culture medium. Distilled water was used as control, and did not influence the measured parameters.

^{3}H-DG and ^{14}C-FRU uptake studies were performed exactly as described by Peixoto et al. [28]. Cell radioactivity was measured by liquid scintillation counting (LKB Wallac 1209 Rackbeta, Turku, Finland). When tested, the extracts were present for 24 h before the uptake assays (in FCS-free culture medium), and also during preincubation and incubation periods. In controls, the extract was replaced by distilled water.

At the end of 24 h of exposure to the extract, cell viability was also assessed using two distinct methods, the lactate dehydrogenase (LDH) activity assay (which measures the activity of LDH released from cell-membrane-damaged cells) and the sulforhodamine B (SRB) assay (which quantifies total cell proteins) [28]. The protein content of cell monolayers was determined as described by Bradford [29], using human serum albumin as standard.

2.12. Statistical Analysis

Data were expressed as average ± standard (n = 3) for chemical parameters. For cell experiments, data were presented as average ± standard error of the average (n = 9). Student's t-test at $p < 0.05$ was used to evaluate statistical differences between two groups (GraphPad Prism version 7.0 software, San Diego, CA, USA).

3. Results and Discussion

Upcycling by-products offers significant opportunities for sustainable advancement across social, environmental, and economic dimensions, contributing to the circular economy requirements within the coffee chain. This approach entails not only prioritizing coffee beans but also recognizing the potential of by-products like cherry pulp, which is frequently discarded despite its valuable properties. The present study aimed to evaluate the chemical composition and the bioactive potential of coffee cherry pulp from São Miguel Island, in the Azores archipelago. Given the distinctive microclimatic conditions of the Azores, we sought to investigate whether the coffee cherry pulp from this region might exhibit specific characteristics that could enhance its bioactive properties.

Table 1 shows the proximate composition (in dry weight) of the dried coffee pulp analyzed in this study, which is essential for determining its potential as a food ingredient. The results showed a diverse and rich nutritional profile, with a high dietary fiber content (52%). The results also showed that the coffee pulp has substantial amounts of total minerals (11%) and total protein (10%), together with a low-fat content (<2%). These results are in the same range as those reported by authors who studied coffee pulp from different geographical origins [27,30–32], although some small differences can be noticed. In Table 1, it is possible to compare the nutritional composition of coffee pulps from different species (arabica and robusta) and geographical origins (data extracted from the literature [27,30–32]). The results are all expressed as percentage of dry weight (% dw), enabling a cross-comparison of the nutrient levels.

Table 1. Nutritional composition of coffee pulp (% dry weight) from different species/varieties and geographical origins.

	Current Study *	Machado et al. [27]	Gil-Ramirez et al. [30]	Ameca et al. [31]	Phuong et al. [32]
Coffee species (variety)	Arabica (caturra vermelha)	Arabica	Arabica (caturra)	Arabica	Robusta
Geographical origin	Azores, Portugal	Colombia	Nicaragua	Mexico	Vietnam
Composition:					
Total ash	10.6 ± 0.3	10.7 ± 0.2	7.6 ± 0.1	7.4 ± <0.1	6.3
Crude protein	10.1 ± 0.1	10.2 ± 0.1	12.9 ± 0.0	10.9 ± 0.2	9.5
Total lipids	1.6 ± <0.1	1.7 ± <0.1	2.8 ± 0.2	1.2 ± 0.1	1.2
Total dietary fiber	52.0 ± 0.7	46.1 ± <0.1	49.8 ± 1.7	49.3 ± 2.6	53.9
Insoluble	44.4 ± 0.1	37.0 ± 0.1	38.6 ± 0.5		49.6
Soluble	7.6 ± 0.7	9.1 ± 0.1	11.2 ± 1.2		4.4
Available carbohydrates *	25.6 ± 0.7	31.2 ± 0.3	26.9 ± 3.7		
Free sugars	5.0 ± 0.1				9.2
Frutose	2.9 ± <0.1		7.6 ± 0.3		
Glucose	2.0 ± <0.1		1.6 ± 0.8		
Saccharose	n.d.				
Arabinose			1.6 ± 0.1		

The results within the first column are expressed as mean ± standard deviation (n = 3). * Available carbohydrates were estimated by subtracting the ash, protein, fat, and total dietary fiber contents to 100%. n.d., not detected.

3.1. Carbohydrate Profile

Total dietary fiber varied slightly among the samples, with the highest content observed in robusta pulp from Vietnam (54%) and the lowest in Colombian arabica pulp (46%). The similarity between the total fiber content of our sample and that from Vietnam (Table 1) suggests a minor influence of the coffee species in this parameter. However, when looking at the different fiber profiles, it seems that pulp from arabica species is significantly richer in soluble fiber (~2-fold). This is particularly relevant as different fiber types have distinct physiological functions, and the fiber profile will directly influence the suitability of coffee pulp for specific dietary and functional applications. For example, insoluble dietary fiber is recognized for supporting bowel regularity, adding bulk to stool, facilitating intestinal transit, and helping to relieve constipation, thereby improving overall digestive health. In turn, soluble fiber can slow glucose absorption and plays an essential role in gut health through its fermentability, providing a substrate for beneficial gut bacteria and supporting immune function. Additionally, soluble fiber has been linked to an improved lipid profile and glycemic control, which can be particularly beneficial for people managing metabolic conditions such as diabetes [33,34].

In what concerns free sugars, Nicaraguan arabica pulp contained 2.6-fold higher fructose contents and lower contents of glucose compared to our sample, which could be due to differences in growing conditions, agricultural practices, post-harvest processing, and particularly the ripeness of the fruit at the moment of harvest [35–37]. These differences will certainly influence the sweetness of the fruit.

3.2. Mineral Profile

Regarding the ash content, indicative of the total mineral composition, this showed slight variations among the species and regions. In the current study, we found an ash content of 10.6%, which aligned closely with the Colombian sample (10.7%) but exceeded the values found for Vietnamese robusta pulp (6.3%) [32] and Mexican (7.4%) [31] and Nicaraguan arabica (7.6%) [30] pulps (Table 1). This suggests that geographical factors, including the soil mineral content and cultivation methods, can influence the mineral concentration in coffee pulp. To better understand the mineral profile of the coffee pulp considered in this study, our sample was subjected to further analysis. Table 2 depicts the results obtained by ICP-MS, showing that our Azorean coffee pulp contains various macroelements, essential trace elements, and non-essential or potentially toxic elements, each with unique roles and impacts on human health.

Potassium was the most abundant macroelement (42.5 mg/g), followed by calcium (3.55 mg/g), magnesium (1.84 mg/g), and phosphorus (1.46 mg/g). These results are in accordance with those of Hurtado and Abarca [38], who found a similar macroelement profile (although the potassium content in our sample was slightly higher) in arabica coffee pulp from the Typica variety produced in Ecuador (results in dry basis: 3.1% of potassium, 0.46% of calcium, 0.14% of magnesium, and 0.13% of phosphorus [38]).

The same profile was also reported by Gil-Ramirez et al. [30] for Nicaraguan arabica pulp, although lower concentrations were observed in their study: 24.2 mg/g of potassium, 2.4 mg/g of calcium, 1.2 mg/g of phosphorous, and 0.9 mg/g of magnesium (results in dry weight). This lower amount of macroelements reflects the lower ash content also described by the authors [30] in comparison to our Azorean coffee pulp (Table 1). In addition, in our study, we also found sodium at a concentration of 1.00 mg/g (Table 2). These macroelements are physiologically important in dietary sources since they play critical roles in numerous functions within the human body. For example, potassium is crucial for cellular function and supports numerous physiological processes, such as muscle function, glucose metabolism, and blood pressure, while calcium plays an essential role in blood clotting, muscle contraction, nerve transmission, and the formation of bones and teeth. Magnesium, in turn, is involved in essential enzymatic reactions, including the activation of amino acids, DNA synthesis, and neurotransmission, supporting also the immune function; and sodium works alongside potassium to regulate the distribution of body water and blood pressure, being essential for maintaining acid–base balance and facilitating the transmission of nerve impulses [39]. Meanwhile, phosphorus primarily serves as the main storage form of metabolic energy and acts as a co-factor for various enzymes [39]. In addition to these five macroelements, we also identified essential trace elements in Azoren coffee pulp, such as iron (27 μg/g), copper (13.4 μg/g), zinc (9.3 μg/g), and manganese (8.6 μg/g), followed by molybdenum, selenium, and cobalt in the ng/g range (Table 2), which collectively offer numerous health benefits to human metabolism [39]. Contrary to what was observed for macroelements, for microelements, Gil-Ramirez et al. [30] reported significantly higher values compared to those found in this study: iron (77 μg/g), manganese (62 μg/g), cooper (18 μg/g), and zinc (17 μg/g). The authors also reported the presence of boron and silicon. These differences highly suggest that environmental factors, such as soil composition, climate, and agricultural practices, have a significant impact on the microelement composition of coffee pulp.

Table 2. Mineral profile of Azorean coffee pulp.

Macroelements	
K (mg/g)	42.50 ± 1.20
Ca (mg/g)	3.55 ± 0.08
Mg (mg/g)	1.84 ± 0.03
P (mg/g)	1.46 ± 0.04
Na (mg/g)	1.00 ± 0.02
Essential trace elements	
Fe (μg/g)	27.20 ± 1.30
Cu (μg/g)	13.50 ± 0.40
Zn (μg/g)	9.26 ± 0.49
Mn (μg/g)	8.63 ± 0.24
Mo (ng/g)	663.03 ± 21.02
Se (ng/g)	64.20 ± 4.80
Co (ng/g)	28.0 ± 1.30
Non-essential and toxic trace elements	
Rb (μg/g)	135.22 ± 2.61
Sr (μg/g)	36.67 ± 0.43
Ba (μg/g)	7.05 ± 0.22
Al (μg/g)	6.80 ± 1.05
Ni (μg/g)	0.17 ± 0.01
Cs (ng/g)	76.80 ± 2.20
Li (ng/g)	20.70 ± 2.10
Pb (ng/g)	15.80 ± 1.50
Cd (ng/g)	8.87 ± 0.61
Be (ng/g)	<LOD (2)
As (ng/g)	<LOD (40)
Sb (ng/g)	<LOD (40)
Hg (ng/g)	<LOD (10)
Tl (ng/g)	<LOD (3)

The results are expressed as mean ± standard deviation (n = 3), in dry weight. LOD, limit of detection.

Finally, in terms of non-essential and toxic trace elements, several were found in the Azorean coffee pulp (Table 2), with values ranging from 9 ng/g (cadmium) to 135 μg/g (rubidium). Although we reported the presence of more toxic elements (Table 2) in comparison to the study of Gil-Ramirez et al. [30], the Azorean coffee pulp seems to have a lower level of contamination regarding some elements (based on the concentrations found). For example, in the Nicaraguan pulp, Gil-Ramirez et al. [30] found 25 ng/g of cadmium, 1.05 μg/g of nickel, and 0.18 μg/g of arsenic. In turn, lead was absent, while we found it in our samples (15.8 ng/g). Despite these differences, the concentrations of these elements in coffee pulp samples remain generally low. Indeed, the European Food Safety Authority

(EFSA) technical report [6] on the notification of dried cherry pulp as a traditional food from a third country provides data on the heavy metal content of six batches of dried coffee cherry pulp (lead: 18–123 ng/g, cadmium: <5–20 ng/g; arsenic: <10–40 ng/g; mercury: <10 ng/g; nickel: <0.05–0.45 µg/g), values that are within the same range or slightly higher than ours (Table 2), which were considered safe and not concerning for market placement.

3.3. Protein Profile

The crude protein content was relatively consistent across all samples, averaging around 10%, with the highest value in the Mexican arabica pulp (10.85%) [31] and the lowest in Vietnamese robusta pulp (9.52%) [32]. These values indicate that this by-product can be an interesting and alternative source of protein. To better understand the composition and the quality of the protein fraction, we analyzed the total (mg/g) and free (µg/g) amino acid profiles of the Azorean coffee pulp (Figure 1). The analysis of these profiles gives valuable and additional information to comprehend the nutritional value and the functional properties of the sample, as each amino acid plays a distinct role in physiological processes [40].

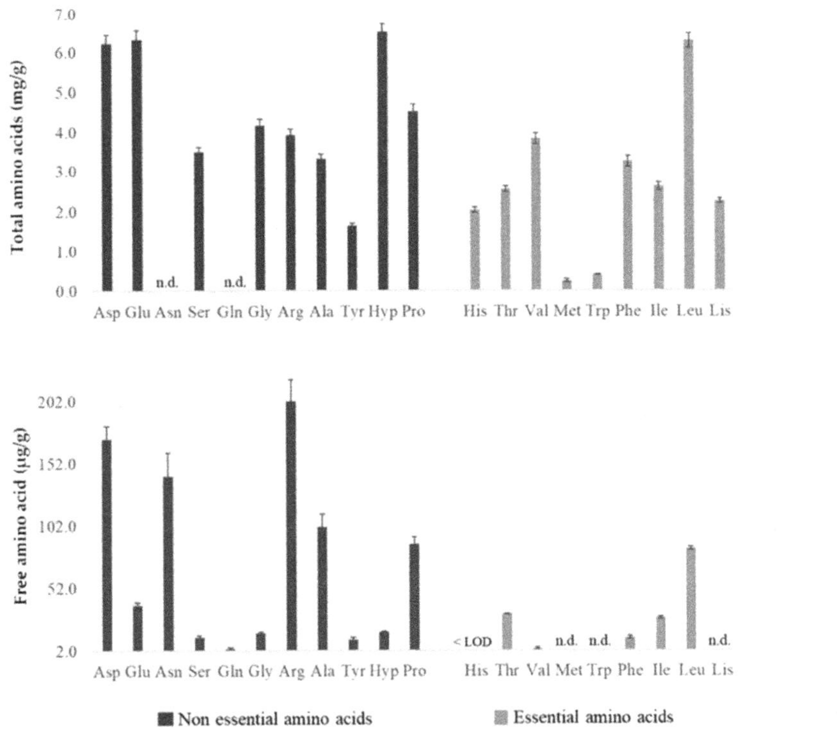

Figure 1. Total and free amino acid profiles of Azorean coffee pulp. The results are expressed as mean ± standard deviation ($n = 3$), in dry weight. n.d., not detected (For total amino acids, conversion of asparagine and glutamine into aspartic acid and glutamic acid occurs during acid hydrolysis, respectively).

The total amino acid content of the Azorean coffee pulp was 63.5 mg/g. This value is significantly lower than that described for crude protein in Table 1. Some hypotheses can explain this discrepancy. Although the Kjeldahl method is a classic and widely accepted approach for determining total nitrogen, which is then converted to an estimated protein content using an established factor, the determined total nitrogen encompasses all nitrogenous compounds, not just proteins. Indeed, non-protein nitrogen sources, such as nitrates or alkaloids like caffeine or related compounds, will raise the estimated crude protein values. In contrast, HPLC-FLD allows the direct measurement of amino acids, avoiding interference from non-protein nitrogen compounds. However, this chromatographic analysis requires hydrolysis to break down proteins into individual amino acids, which can lead to some degradation or loss of the most sensitive amino acids (e.g., methionine, serine), contributing to the underestimation of total amino acid content. In addition, in some cases, peptide bonds may only be partially broken down, depending on their type and structural characteristics. For example, those involving isoleucine and valine, such as isoleucine/isoleucine, valine/valine, valine/isoleucine, and isoleucine/valine, are not always fully hydrolyzed and may remain partially intact [19,41].

As depicted in Figure 1, the most abundant amino acids in the Azorean coffee pulp were hydroxyproline (6.5 mg/g), leucine (6.3 mg/g), glutamic acid (6.3 mg/g), and aspartic acid (6.2 mg/g). It is important to emphasize that these two last values also take into account the conversion of asparagine and glutamine into aspartic acid and glutamic acid, respectively, under the acidic conditions necessary for hydrolysis. Nonetheless, this deamination process, resulting in the corresponding carboxylic acids, does not impact the quantification, as the molecular weights within each amine/acid pair are almost identical [41]. Other amino acids found in high amounts were proline (4.5 mg/g), glycine (4.2 mg/g), arginine (3.9 mg/g), and valine (3.8 mg/g). All essential amino acids were detected (histidine, threonine, valine, methionine, tryptophan, phenylalanine, isoleucine, leucine, and lysine), in an amount that represents about 37% of the total amino acids (23.4 mg/g); however, the relatively low amount of some of them, in particular methionine and tryptophan, impairs protein quality. To enhance its nutritional value, especially for applications such as the production of coffee pulp-derived flour, it would be advisable to complement its consumption with other protein sources rich in methionine and tryptophan. For example, legumes such as chickpeas and lentils are particularly rich in methionine [42], while the cereal grains spelt and oat are rich in tryptophan [43]. These combinations are perfectly suitable and could effectively address the deficiencies in methionine and tryptophan found in coffee pulp, significantly enhancing the overall protein quality. Despite that limitation, coffee pulp can be seen as a very good source of amino acids with important physiological roles that can be used for specific food enrichment. For example, glycine, proline, and hydroxyproline account for 57% of the total amino acid composition in collagen, which is crucial for maintaining the integrity and strength of connective tissues (e.g., skin, bones, cartilage, and blood vessels) [44]. In turn, glutamic acid, aspartic acid, arginine, and proline, are key contributors to cognitive functions, while the branched-chain amino acids (leucine, isoleucine, and valine), also present in substantial amounts, have a relevant role in muscle growth [40].

The total amino acid profile obtained in the present study is quite different from that described by Gil-Ramirez et al. [30]. Although the authors described 12.9% protein content in Nicaraguan arabica coffee pulp (higher than that found in this study), they reported a total amino acid content of only 21 mg/g (~2%). This difference could be explained by the fact that different amino acids were quantified in both studies, probably due to standards availability. For example, we quantified hydroxyproline (a major amino acid in Azorean coffee pulp) and tryptophan (performing an extra basic hydrolysis), while they quantified cysteine (a minor amino acid). However, as this does not explain all the differences observed, the method employed to quantify amino acids should also be considered. For example, Gil-Ramirez et al. [30] used a different hydrolysis protocol and did not refer to a neutralization step before storage to avoid eventual amino acid degradation that may occur even when the sample is frozen at −20 °C. In addition, they used a different derivatization procedure as well as a distinct chromatographic separation method, and that could also influence the results. Overall, the amino acid contents reported in the present study are approximately 2- to 4-fold higher than those reported by Gil-Ramirez et al. [30], with some slight variations.

In Table 3, it is also possible to observe the free amino acid profile of Azorean coffee pulp. Free amino acids, which are not bound in protein structures, are available for rapid absorption and metabolism. They are essential to regulate both exocrine and endocrine secretions and influence protein digestion, metabolic processes, and nutrient absorption, while also contributing to the maintenance of the integrity and protective functions of the gastrointestinal mucosa [45]. In addition, their contribution to the sensorial characteristics of food has long been recognized [46]. The total free amino acid content of the sample analyzed in this study was 0.96 mg/g. Arginine (202 µg/g) and aspartic acid (172 µg/g) were the most abundant free amino acids, followed by asparagine (142 µg/g) and alanine (101 µg/g). In terms of flavor, these four amino acids can be responsible for sweetness (alanine), sourness (aspartic acid), sourness or sweetness (asparagine), and bitterness (arginine). Aspartic acid and alanine, together with glutamic acid (38 µg/g), threonine (31 µg/g), glycine (16 µg/g), serine (12.5 µg/g), and glutamine (3.6 µg/g), are also involved in umami taste [46]. In turn, some essential amino acids, such as histidine, lysine, methionine, and tryptophan were not detected in Azorean coffee pulp. This free profile is also different from that reported by Gil-Ramirez et al. [30]. The authors highlighted proline as the main free amino acid, followed by serine, aspartic acid, phenylalanine, and arginine, in this order; they also reported the presence of free histidine, lysine, and methionine, although in low contents. Several factors could have influenced the free amino acid profile of the samples analyzed in the different studies, leading to these variations, such as the growing conditions, the maturity at harvest, and post-harvest handling. In fact, even for the coffee bean, there are several studies that reveal the influence of these same factors on the free amino acid profile [47–49].

Table 3. Fatty acids profile of Azorean coffee pulp.

Fatty Acids		
Myristic	C14:0	1.43 ± 0.13
Palmitic	C16:0	40.92 ± 0.28
Palmitoleic	C16:1	1.45 ± 0.10
Stearic	C18:0	8.88 ± 0.29
Oleic	C18:1*n*9*c*	5.33 ± 0.21
Linoleic [1]	C18:2*n*6*c*	22.30 ± 0.05
Arachidic	C20:0	3.78 ± 0.29
α-Linolenic [1]	C18:3*n*3	15.91 ± 0.22
n6/n3		1.40 ± 0.02
n9/n6		0.24 ± 0.01
ΣSFA		55.01 ± 0.30
ΣMUFA		6.78 ± 0.21
ΣPUFA		38.21 ± 0.22

The results are expressed as relative % of total fatty acids (mean ± standard deviation, $n = 3$). [1] Essential fatty acid. ΣSFA, sum of saturated fatty acids (C14:0 + C16:0 + C18:0 + C20:0); ΣMUFA, sum of monounsaturated fatty acids (C16:1 + C18:1*n*9*c*); ΣPUFA, sum of polyunsaturated fatty acids (C18:2*n*6*c* + C18:3*n*3).

3.4. Lipid Profile

Although coffee pulp is a good source of amino acids and minerals, it is very low in fat. However, some differences can be highlighted among samples collected from distinct geographical origins and species (Table 1). For example, the Nicaraguan arabica pulp contained the highest total fat content (2.8%) [30], while the Vietnamese robusta pulp had the lowest (1.2%) [31].

Despite this low fat content, coffee pulp seems to be a good source of vitamin E, which is a liposoluble antioxidant comprised of eight naturally occurring compounds that share similar structures (vitamers). Their structures contain a chromanol ring with different methyl group arrangements (α-, β-, γ-, and δ-), and a 16-carbon phytyl side chain fully saturated in tocopherols and unsaturated in tocotrienols [20]. The eight forms show different levels of biological activity, being α-tocopherol the most active due to its selective recognition by the α-tocopherol transfer protein that plays a crucial role in maintaining plasma α-tocopherol levels [50]. Although vitamin E's biological effect is mostly derived from its antioxidant properties, by protecting cells from peroxidation and contributing to membrane integrity, it is also involved in the primary intracellular defense system and has been associated with the prevention of several diseases [51].

In a previous work published by our group [20], we found that from the eight possible vitamers, coffee beans contained only α- and β-tocopherols, with β-tocopherol present in approximately twice the amount of α-tocopherol in arabica coffee beans (both raw and roasted). In the case of the Azorean arabica pulp (total vitamin E: 72.67 ± 1.01 μg/g dw), besides α- and β-tocopherols, (51.18 ± 1.26 and 2.12 ± 0.07 μg/g dw, respectively) we also identified γ-tocopherol (19.37 ± 0.77 μg/g dw). However, in this case, α-tocopherol was the major vitamer (approximately 2.6- and 25-fold higher than γ- and β-tocopherols, respectively), being this a more desirable profile based on α-tocopherol higher biological activity. These results are in accordance with those reported by Tavares et al. [52] for coffee husks (the main by-product obtained from the dry post-harvest coffee processing), which consists of dried skin, pulp, mucilage, and parchment. The authors found a vitamin E profile similar to that reported above for our sample, with α-tocopherol being the major vitamer (3.7–7.1 mg/100 g), followed by γ-tocopherol (1.2–2.1 mg/100 g) and β-tocopherol (0.4–0.9 mg/100 g) [52].

In terms of the fatty acid profile of Azorean coffee pulp (Table 3), palmitic acid was the major fatty acid found (41%), contributing significantly to the total saturated fatty acids (SFAs) content (55%). Consequently, the other SFAs—stearic acid (8.9%), arachidic acid (3.8%), and myristic acid (1.4%)—were present in lower amounts. Although the lipid profile showed this predominance in SFAs, polyunsaturated fatty acids (PUFAs) were also detected in substantial amounts (38%). The essential linoleic and α-linolenic acids were the main PUFAs (22 and 16%, respectively), resulting in an n6/n3 ratio of 1.4. This falls perfectly within the recommended range for a balanced intake of omega-6 and omega-3 fatty acids, since it has long been recognized that a low n6/n3 ratio (ideally near 1) is crucial for suppressing inflammation and reducing the risk of developing cancer, cardiovascular diseases, and autoimmune disorders [53]. Finally, two monounsaturated fatty acids (MUFAs) were also detected, but not in significant amounts (oleic acid: 5%; palmitoleic acid: 1.5%) when compared to the above-mentioned fatty acids.

The fatty acid profile found in this study for the Azorean coffee pulp is similar to that reported by Rios et al. [54] for dried arabica coffee pulp (Tabi variety), obtained through wet processing in Colombia, suggesting that the variety and/or geographical origin do not have a major influence the lipid profile of the fruits. In their study, the authors reported 47% SFAs, 40% PUFAs, and 12% MUFAs, with the major fatty acids being palmitic acid

(36%), linoleic acid (22%), and α-linolenic acid (17%) [54]—very similar values to those presented in Table 3 for Azorean coffee pulp.

3.5. Antioxidant and Phytochemical Profile

Azorean coffee pulp presented a high antioxidant capacity (Table 4), evidenced by its ferric-reducing antioxidant power (487.5 μmol FSE/g) and DPPH• scavenging activity (21.5 mg TE/g). These two methods were used in this study because they present complementary mechanisms of action. The FRAP assay involves the reduction of the complex formed between Fe^{3+} and 2,4,6-tripyridyl-s-triazine (TPTZ), which turns blue and is then quantified spectrophotometrically. This method is effective for compounds with redox potentials below 0.7 V, corresponding to the redox potential of Fe^{3+}-TPTZ, being a useful in vitro indicator of a compound's capacity to maintain redox balance in biological systems. In this way, the FRAP assay operates exclusively through an electron transfer mechanism, which means that it cannot detect antioxidants that neutralize radicals via hydrogen atom transfer. In contrast, the DPPH• scavenging assay evaluates antioxidants capable of neutralizing the DPPH• radical either by electron transfer or hydrogen atom transfer. This method measures the antioxidant-reducing capacity by observing the decrease in absorbance of the DPPH• solution, reflecting the reduction or quenching of the radical [55].

Table 4. Antioxidant activity and phytochemical composition of dried Azorean coffee pulp.

Antioxidant Activity	
Ferric reducing antioxidant power (μmol FSE/g dw)	487.47 ± 7.34
DPPH•-SA (mg TE/g dw)	21.49 ± 1.62
Phytochemicals	
Total flavonoids content (mg CE/g dw)	21.72 ± 0.11
Total phenolics content (mg CGAE/g dw)	45.87 ± 2.00
3-caffeoylquinic acid (mg/g dw)	0.40 ± 0.03
4-caffeoylquinic acid (mg/g dw)	0.83 ± 0.13
5-caffeoylquinic acid (mg/g dw)	7.97 ± 0.89
Caffeine (mg/g dw)	9.82 ± 0.58

The results are expressed as mean ± standard deviation (n = 9, for spectrophotometric assays; n = 3, for HPLC analyses). FSE, ferrous sulfate equivalents; DPPH•-SA, 2,2 diphenyl-1picrylhydrazyl radical scavenging activity; TE, trolox equivalents; CE, catechin equivalents; CGAE, chlorogenic acid equivalents.

The FRAP value found in the present study was slightly lower than those reported by Machado et al. [27] for Colombian arabica pulp (8.58 g FSE/100 g or, after conversion, 565 μmol FSE/g) using a similar protocol, while the DPPH• scavenging activity in our sample was higher (7.7 mg TE/g in the Colombian one). These differences suggest differences in phytochemical composition that could result, as mentioned above, from growing, ripening, or post-harvest processing conditions. To explore that, total flavonoids, total phenolics, chlorogenic acid profile, and caffeine contents were also studied. We found differences in the phenolic profile that could influence the results in the antioxidant assays. In the present study, Azorean coffee pulp presented higher values for total flavonoid (21.7 mg CE/g) and total phenolic contents (45.9 mg CGAE/g) than those described for Colombian coffee pulp (12.3 mg CE/g and 23.7 mg CGAE/g) [27]. In addition, our Azorean sample was also richer in chlorogenic acids (5-caffeoylquinic acid: 8.0 mg/g vs. 2.2 mg/g for Colombian coffee pulp [28]; 4-caffeoylquinic acid: 0.8 mg/g vs. 0.1 mg/g; 3-caffeoylquinic acid: 0.40 vs. 0.06 mg/g, respectively). Notwithstanding, although total flavonoids and other phenolic compounds have an established role as primary contributors to the antioxidant capacity of coffee pulp, other components (such as vitamin E, described in Section 3.4) also play a significant role in this property. In addition, our sample was also slightly richer in caffeine: 9.8 mg/g, against 0.85 mg/g reported by Machado et al. [27] for Colombian pulp.

This richness in phytochemical compounds, such as caffeine and chlorogenic acids, suggests potential beneficial physiological effects when coffee pulp is consumed. By comparison with coffee beans, particularly arabica beans, that contain caffeine in a similar range (~1%) [56], we can infer similar physiological actions. Caffeine is well-documented for its psychoactive effects, including improvements in cognitive and psychomotor performance, with minimal effective doses around 12.5–50 mg [57,58]. Based on the caffeine content in coffee pulp, approximately 1–5 g of coffee pulp flour would be required to reach this threshold. Based on this, coffee pulp could be suggested as a natural source of caffeine to enrich specific foods aimed at enhancing mental and physical performance. However, it is important to consider potential limitations to its consumption. Specific populations, such as children, pregnant or breastfeeding women, and those sensitive to caffeine may avoid or limit intake. Furthermore, excessive consumption should be avoided, as consuming more than 20 g of coffee pulp flour could exceed the safe caffeine limit for a single dose (200 mg) established by the EFSA [59]. In addition, the cumulative caffeine intake from all sources, including coffee beverages, food supplements, energy drinks, medications, and other natural caffeine-containing foods, should also be taken into account to prevent exceeding the recommended daily limits of caffeine ingestion (400 mg) [59].

Chlorogenic acids (esters of hydroxycinnamic acids with quinic acid) have also been associated with several health benefits, including a significant reduction in the risk of developing type II diabetes. Indeed, in recent years, their role in the regulation of sugar and lipid metabolism has been studied, with reports of relevant antioxidant, antidiabetic, and anti-inflammatory effects [60].

3.6. Antidiabetic Potential

Although some studies have already reported antidiabetic effects of coffee pulp, namely an inhibitory effect on α-amylase enzyme activity in vitro [61] and the ability to decrease plasma glucose levels and insulin resistance in diabetic rats [12], as far as we know, there are no studies reporting the effects of coffee pulp on intestinal glucose and fructose uptake. Considering that the intestine is an organ of primary importance for the absorption of dietary sugar, the inhibition of this process might be a useful therapeutic strategy to modulate postprandial glycemic control, sugar metabolism, and metabolic health. In this context, radiolabeled glucose (^{3}H-deoxyglucose, ^{3}H-DG) and fructose (^{14}C-fructose, ^{14}C-FRU) were used as tracers to study the transport dynamics in Caco-2 cells.

The effects of Azorean coffee pulp on the glucose (^{3}H-DG) and fructose (^{14}C-FRU) uptake by Caco-2 cells are depicted in Figure 2.

As can be observed, the Azorean coffee pulp extract was able to significantly inhibit ^{3}H-DG uptake at all concentrations tested, while no significant effects ($p > 0.05$) were found on ^{14}C-FRU uptake (Figure 2). Of note, the reductions in ^{3}H-DG uptake caused by the extract were very expressive (from 54 to 64%) and concentration-dependent.

In the intestine, the absorption of glucose in the apical membrane of small intestinal cells is mediated by the sodium-dependent glucose co-transporter (SGLT1) and by the facilitative glucose transporter 2 (GLUT2), while the absorption of fructose is mediated by the facilitative glucose transporter 5 (GLUT5) and by GLUT2 [62]. Considering the differential effects of the extract on ^{3}H-DG and ^{14}C-FRU uptake, it might be suggested that SGLT1 is the main target of the extract's inhibitory effect.

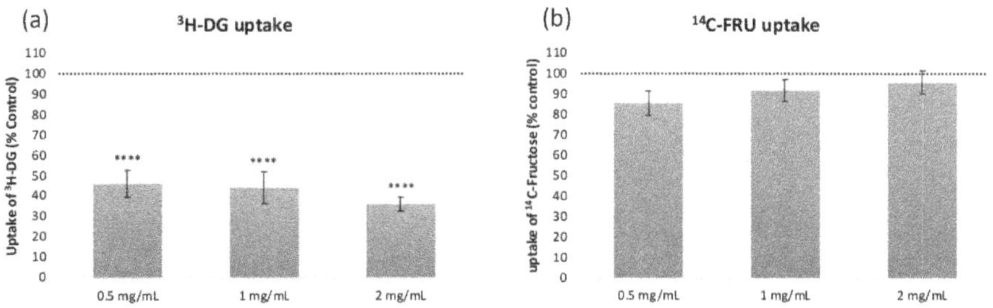

Figure 2. Effect of coffee pulp extracts on (**a**) ^{3}H-DG and (**b**) ^{14}C-FRU uptake by Caco-2 cells in comparison with control (100%, horizontal dash line). The results are expressed as average ± SEM (n = 9); ****, $p < 0.0001$ significantly different from control by Student's t-test.

These results are quite encouraging and might be intimately related to the richness of coffee pulp extract in caffeine and chlorogenic acids CGA (Table 4). In fact, in a previous study conducted in our laboratory with extracts of coffee silverskin (a coffee by-product obtained during bean roasting), we found a synergistic effect between 5-caffeoylquinic acid and caffeine in inhibiting both ^{3}H-DG and ^{14}C-FRU uptake. Regardless, the extracts presented even higher reductions compared to the mixture of both standards at the concentrations present in the extracts [28], highlighting the potential contribution of other compounds besides 5-caffeoylquinic acid and caffeine to the effects found.

To confirm that the effects found on ^{3}H-DG and ^{14}C-FRU uptake were not related to a cytotoxic effect of the coffee pulp extract on Caco-2 cells, LDH and SRB assays were further performed. Among the concentrations tested, the coffee pulp extract presented a significant cytotoxic effect at 2 mg/mL, observable with both the LDH (Figure 3a) and SRB (Figure 3b) assays. However, the extract presented a protective effect in the LDH test at the lowest concentration tested (0.5 mg/mL) (Figure 2a). Therefore, up to 1 mg/mL, the extract is able to decrease ^{3}H-DG uptake without interfering with cell viability, which is important in the context of further studies using this coffee pulp extract.

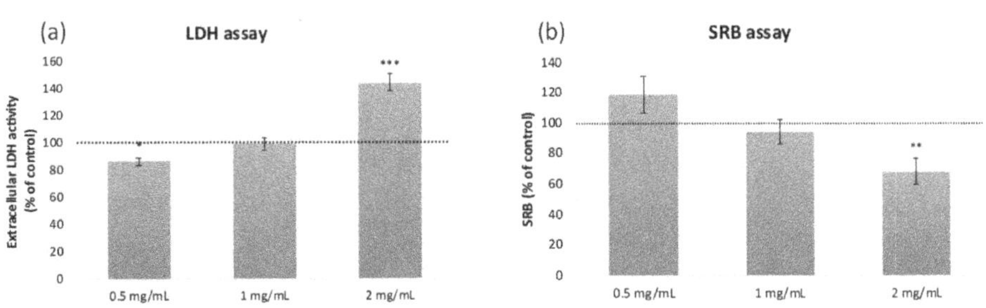

Figure 3. Effect of coffee pulp extracts on (**a**) Caco-2 cell viability (extracellular LDH activity), and (**b**) culture mass (SRB assay), in comparison with control (100%, horizontal dash line). The results are expressed as average ± SEM (n = 9). *, $p < 0.05$; **, $p < 0.01$; ***, $p < 0.001$ significantly different from control by Student's t-test.

4. Conclusions

In this work, a comprehensive analysis of the chemical composition and bioactive potential of coffee cherry pulp from S. Miguel, Azores (Portugal) was carried out. Azorean coffee pulp proved to be a valuable natural resource, particularly from a circular economy perspective, as it constitutes a commonly discarded by-product that can be upcycled into functional ingredients or nutraceuticals.

Chemical analyses revealed that Azorean coffee pulp is a good source of fiber (predominantly insoluble), minerals (mainly K, Ca, and Mg), and protein (with hydroxyproline, aspartic acid, glutamic acid, and leucine being the major amino acids). Although its fat content is low and mostly saturated, the pulp also provides considerable amounts of polyunsaturated fatty acids with a favorable n6/n3 ratio (1.40), as well as vitamin E (mainly α-tocopherol). Its antioxidant capacity can be partially attributed to chlorogenic acids (3-, 4-, and 5-caffeoylquinic acids), and its caffeine content is comparable to that found in arabica coffee beans. In addition to this rich phytochemical profile, a decrease in glucose uptake (but not in fructose uptake) was observed in Caco-2 cells, suggesting selective inhibition of the SGLT1 transporter and a potential antidiabetic effect. These results show that Azorean coffee pulp has potential as a sustainable and bioactive ingredient for incorporation into innovative functional foods or dietary supplements with health-promoting properties.

From a sustainability point of view, upcycling coffee pulp can offer several advantages, since coffee producers can reduce the environmental impact associated with pulp disposal, addressing waste management challenges, while generating additional revenues. Overall, coffee pulp upcycling can contribute to a more resilient and resource-efficient coffee supply chain, promoting local economic development and aligning with global sustainability goals.

Author Contributions: Conceptualization, A.S.G.C. and R.C.A.; methodology, A.S.G.C., S.M., L.E.S. T.F.S., N.A. and R.A.; validation, F.M. and R.C.A.; formal analysis, A.S.G.C., J.A.B.P., L.E.S. and N.A.; investigation, A.S.G.C., J.A.B.P. and N.A.; data curation, A.S.G.C., J.A.B.P., S.M., L.E.S., T.F.S. and N.A.; resources, A.S.G.C., A.A., H.S.C., M.B.P.P.O., F.M. and R.C.A.; writing—original draft preparation, A.S.G.C., J.A.B.P. and R.C.A.; writing—review and editing, A.S.G.C., N.A., H.S.C., M.B.P.P.O., F.M., J.S.-G. and R.C.A.; visualization, A.S.G.C. and N.A.; supervision, R.C.A. and J.S.-G.; project administration, R.C.A. and H.S.C.; funding acquisition, R.C.A., H.S.C., M.B.P.P.O. and F.M. All authors have read and agreed to the published version of the manuscript.

Funding: This work was funded by FCT/MCTES (Fundação para a Ciência e Tecnologia and Ministério da Ciência, Tecnologia e Ensino Superior) through the project PTDC/SAU-NUT/2165/2021 (COBY4HEALTH—Can coffee by-products decrease the risk of metabolic syndrome? A comprehensive approach to reduce waste and valorize health benefits).

Institutional Review Board Statement: Not applicable.

Informed Consent Statement: Not applicable.

Data Availability Statement: The original contributions presented in this study are included in the article. Further inquiries can be directed to the corresponding author.

Acknowledgments: The authors acknowledge FCT/MCTES for the national funds received through the projects UIDB/50006/2020 (DOI 10.54499/UIDB/50006/2020), UIDP/50006/2020 (DOI 10.54499/UIDP/50006/2020), and LA/P/0008/2020 (DOI 10.54499/LA/P/0008/2020). J.A.B.P. gives thanks to FCT/MCTES and ESF (European Social Fund) through NORTE 2020 (Programa Operacional Região Norte) for her PhD grant, ref. SFRH/BD/07329/2021. L.E.S. is grateful to LAQV-Tecnologias e Processos Limpos-UIDB/50006/2020 for her grant (REQUIMTE 2023-49). N.A. and R.C.A. thanks FCT for funding through the Scientific Employment Stimulus—Individual Call (CEECIND/08492/2022 and CEECIND/01120/2017, respectively). The authors thank Quinta do Avô João (Azores, Portugal) for kindly providing the sample for the study.

Conflicts of Interest: The authors declare no conflicts of interest.

References

1. Heeger, A.; Kosińska-Cagnazzo, A.; Cantergiani, E.; Andlauer, W. Bioactives of coffee cherry pulp and its utilisation for production of cascara beverage. *Food Chem.* **2017**, *221*, 969–975. [CrossRef] [PubMed]

2. Magoni, C.; Bruni, I.; Guzzetti, L.; Dell'Agli, M.; Sangiovanni, E.; Piazza, S.; Regonesi, M.E.; Maldini, M.; Spezzano, R.; Caruso, D. Valorizing coffee pulp by-products as anti-inflammatory ingredient of food supplements acting on IL-8 release. *Food Res. Int.* **2018**, *112*, 129–135. [CrossRef] [PubMed]
3. International Coffee Organization (ICO). Coffee Report and Outlook, April 2023. *Int. Coffee Organ.* 2023. Available online: https://icocoffee.org/documents/cy2022-23/Coffee_Report_and_Outlook_April_2023_-_ICO.pdf (accessed on 9 January 2025).
4. Pandey, A.; Soccol, C.R.; Nigam, P.; Brand, D.; Mohan, R.; Roussos, S. Biotechnological potential of coffee pulp and coffee husk for bioprocesses. *Biochem. Eng. J.* **2000**, *6*, 153–162. [CrossRef] [PubMed]
5. Murthy, P.S.; Naidu, M.M. Sustainable management of coffee industry by-products and value addition—A review. *Resour. Conserv. Recycl.* **2012**, *66*, 45–58. [CrossRef]
6. European Food Safety Authority (EFSA). Technical Report on the notification of dried cherry pulp from *Coffea arabica* L. and *Coffea canephora* Pierre ex A. Froehner as a traditional food from a third country pursuant to Article 14 of Regulation (EU) 2015/2283. *EFSA Support. Publ.* **2021**, *18*, 6808E.
7. Rodriguez-Duran, L.V.; Ramírez-Coronel, M.A.; Aranda-Delgado, E.; Nampoothiri, K.M.; Favela-Torres, E.; Aguilar, C.N.; Saucedo-Castañeda, G. Soluble and bound hydroxycinnamates in coffee pulp (*Coffea arabica*) from seven cultivars at three ripening stages. *J. Agric. Food Chem.* **2014**, *62*, 7869–7876. [CrossRef] [PubMed]
8. Hu, S.; Gil-Ramírez, A.; Martín-Trueba, M.; Benítez, V.; Aguilera, Y.; Martín-Cabrejas, M.A. Valorization of coffee pulp as bioactive food ingredient by sustainable extraction methodologies. *Curr. Res. Food Sci.* **2023**, *6*, 100475. [CrossRef] [PubMed]
9. Prata, E.R.; Oliveira, L.S. Fresh coffee husks as potential sources of anthocyanins. *LWT-Food Sci. Technol.* **2007**, *40*, 1555–1560. [CrossRef]
10. Velez, A.R.; Lopez, J.C.J. Process for Obtaining Honey and/or Flour of Coffee From the Pulp or Husk and the Mucilage of the Coffee Bean. U.S. Patent 20150017270A1, 2 May 2017.
11. Bhandarkar, N.S.; Mouatt, P.; Majzoub, M.E.; Thomas, T.; Brown, L.; Panchal, S.K. Coffee pulp, a by-product of coffee production, modulates gut microbiota and improves metabolic syndrome in high-carbohydrate, high-fat diet-fed rats. *Pathogens* **2021**, *10*, 1369. [CrossRef] [PubMed]
12. Boonphang, O.; Ontawong, A.; Pasachan, T.; Phatsara, M.; Duangjai, A.; Amornlerdpison, D.; Jinakote, M.; Srimaroeng, C. Antidiabetic and renoprotective effects of *Coffea arabica* pulp aqueous extract through preserving organic cation transport system mediated oxidative stress pathway in experimental type 2 diabetic rats. *Molecules* **2021**, *26*, 1907. [CrossRef]
13. Braojos, C.; Gila-Díaz, A.; Rodríguez-Rodríguez, P.; Monedero-Cobeta, I.; Morales, M.D.; Ruvira, S.; Ramiro-Cortijo, D.; Benítez, V.; Martín-Cabrejas, M.A.; Arribas, S.M. Effect of supplementation with coffee and cocoa by-products to ameliorate metabolic syndrome alterations induced by high-fat diet in female mice. *Foods* **2023**, *12*, 2708. [CrossRef]
14. Sales, A.L.; Iriondo-DeHond, A.; DePaula, J.; Ribeiro, M.; Ferreira, I.M.; Miguel, M.A.L.; Del Castillo, M.D.; Farah, A. Intracellular antioxidant and anti-inflammatory effects and bioactive profiles of coffee cascara and black tea kombucha beverages. *Foods* **2023**, *12*, 1905. [CrossRef] [PubMed]
15. Azores Islands: Location, History, Climate and More. Available online: https://azores.com/azores (accessed on 26 December 2024).
16. AOAC. *Official Methods of Analysis of AOAC International*, 21st ed.; AOAC International: Rockville, MD, USA, 2019.
17. EPA. *Method 3052: Microwave Assisted Acid Digestion of Siliceous and Organically Based Matrices*; United States Environmental Protection Agency: Washington, DC, USA, 1996. Available online: https://www.epa.gov/sites/default/files/2015-12/documents/3052.pdf (accessed on 26 December 2024).
18. Azevedo, R.; Oliveira, A.R.; Almeida, A.; Gomes, L.R. Determination by ICP-MS of essential and toxic trace elements in gums and carrageenans used as food additives commercially available in the Portuguese market. *Foods* **2023**, *12*, 1408. [CrossRef] [PubMed]
19. Machado, S.; Costa, A.S.G.; Pimentel, F.B.; Oliveira, M.B.P.P.; Alves, R.C. A study on the protein fraction of coffee silverskin: Protein/non-protein nitrogen and free and total amino acid profiles. *Food Chem.* **2020**, *326*, 126940. [CrossRef]
20. Alves, R.C.; Casal, S.; Oliveira, M.B.P.P. Determination of vitamin E in coffee beans by HPLC using a micro-extraction method. *Food Sci. Technol. Int.* **2009**, *15*, 57–63. [CrossRef]
21. *ISO 12966-2: 2011*; Animal and Vegetable Fats and Oils-Gas Chromatography of Fatty Acid Methyl Esters-Part 2: Preparation of Methyl Esters of Fatty Acids. International Organization for Standardization: Geneva, Switzerland, 2011.
22. Montesano, D.; Cossignani, L.; Giua, L.; Urbani, E.; Simonetti, M.S.; Blasi, F. A Simple HPLC-ELSD Method for Sugar Analysis in Goji Berry. *J. Chem.* **2016**, *2016*, 6271808. [CrossRef]
23. Rufino, M.S.M.; Alves, R.E.; Brito, E.S.; Morais, S.M.; Sampaio, C.G.; Pérez-Jiménez, J.; Saura-Calixto, F.D. *Metodologia Científica: Determinação da Atividade Antioxidante Total em Frutas Pelo Método De Redução do Ferro (FRAP)*; Embrapa Agroindústria Tropical: Fortaleza, Brazil, 2006.

24. da Silveira, A.C.; Kassuia, Y.S.; Domahovski, R.C.; Lazzarotto, M. *Método de DPPH Adaptado: Uma Ferramenta Para Analisar Atividade Antioxidante de Polpa de Frutos da Erva-Mate de Forma Rápida e Reprodutível*; Comunicado Técnico No. 421; Embrapa Florestas: Colombo, Brazil, 2018.
25. Zhang, Q.; Zhang, J.; Shen, J.; Silva, A.; Dennis, D.A.; Barrow, C.J. A Simple 96-Well Microplate Method for Estimation of Total Polyphenol Content in Seaweeds. *J. Appl. Phycol.* **2006**, *18*, 445–450. [CrossRef]
26. Zou, Y.; Lu, Y.; Wei, D. Antioxidant Activity of a Flavonoid-Rich Extract of *Hypericum perforatum* L. in Vitro. *J. Agric. Food Chem.* **2004**, *52*, 5032–5039. [CrossRef]
27. Machado, M.; Espírito Santo, L.; Machado, S.; Lobo, J.C.; Costa, A.S.G.; Oliveira, M.B.P.P.; Ferreira, H.; Alves, R.C. Bioactive Potential and Chemical Composition of Coffee By-Products: From Pulp to Silverskin. *Foods* **2023**, *12*, 2354. [CrossRef] [PubMed]
28. Peixoto, J.A.B.; Andrade, N.; Machado, S.; Costa, A.S.G.; Puga, H.; Oliveira, M.B.P.P.; Martel, F.; Alves, R.C. Valorizing Coffee Silverskin Based on Its Phytochemicals and Antidiabetic Potential: From Lab to a Pilot Scale. *Foods* **2022**, *11*, 1671. [CrossRef] [PubMed]
29. Bradford, M.M. A rapid and sensitive method for the quantitation of microgram quantities of protein utilizing the principle of protein-dye binding. *Anal. Biochem.* **1976**, *72*, 248–254. [CrossRef] [PubMed]
30. Gil-Ramírez, A.; Rebollo-Hernanz, M.; Cañas, S.; Monedero Cobeta, I.; Rodríguez-Rodríguez, P.; Gila-Díaz, A.; Benítez, V.; Arribas, S.M.; Aguilera, Y.; Martín-Cabrejas, M.A. Unveiling the Nutritional Profile and Safety of Coffee Pulp as a First Step in Its Valorization Strategy. *Foods* **2024**, *13*, 3006. [CrossRef]
31. Ameca, G.M.; Cerrilla, M.E.O.; Córdoba, P.Z.; Cruz, A.D.; Hernández, M.S.; Haro, J.H. Chemical composition and antioxidant capacity of coffee pulp. *Ciênc. Agrotec.* **2018**, *42*, 307–313. [CrossRef]
32. Phuong, D.V.; Quoc, L.P.T.; Tan, P.V.; Duy, L.N.D. Production of bioethanol from Robusta coffee pulp (*Coffea robusta* L.) in Vietnam. *Foods Raw Mat.* **2019**, *7*, 10–17. [CrossRef]
33. McRorie, J.W.; McKeown, N.M. Understanding the Physics of Functional Fibers in the Gastrointestinal Tract: An Evidence-Based Approach to Resolving Enduring Misconceptions about Insoluble and Soluble Fiber. *J. Acad. Nutr. Diet.* **2017**, *117*, 251–264. [CrossRef] [PubMed]
34. Guan, Z.-W.; Yu, E.-Z.; Feng, Q. Soluble Dietary Fiber, One of the Most Important Nutrients for the Gut Microbiota. *Molecules* **2021**, *26*, 6802. [CrossRef]
35. Redgwell, R.; Fischer, M. Coffee carbohydrates. *Braz. J. Plant Physiol.* **2006**, *18*, 165–174. [CrossRef]
36. Pua, A.; Choo, W.X.D.; Goh, R.M.V.; Liu, S.Q.; Cornuz, M.; Ee, K.-H.; Sun, J.; Lassbliere, B.; Yu, B. A systematic study of key odourants, non-volatile compounds, and antioxidant capacity of cascara (dried *Coffea arabica* pulp). *LWT-Food Sci. Technol.* **2021**, *138*, 110630. [CrossRef]
37. Osorio Pérez, V.; Matallana Pérez, L.G.; Fernandez-Alduenda, M.R.; Alvarez Barreto, C.I.; Gallego Agudelo, C.P.; Montoya Restrepo, E.C. Chemical Composition and Sensory Quality of Coffee Fruits at Different Stages of Maturity. *Agronomy* **2023**, *13*, 341. [CrossRef]
38. Hurtado, J.G.F.; Abarca, J.M. Quantification of K, Ca, Mg y P minerals in coffee (*Coffea arabica* L. var. Typica) pulp and parchment. *Rev. Venez. Cienc. Tecnol. Aliment.* **2010**, *1*, 221–230.
39. Stein, A.J. Global impacts of human mineral malnutrition. *Plant Soil.* **2010**, *335*, 133–154. [CrossRef]
40. Wu, G. Amino acids: Metabolism, functions, and nutrition. *Amino Acids* **2009**, *37*, 1–17. [CrossRef] [PubMed]
41. European Pharmacopoeia. *2. Methods of Analysis. 2.2.56. Amino Acid Analysis. European Pharmacopoeia 5.0*; Council of Europe: Strasbourg, France, 2005; pp. 86–92.
42. Multescu, M.; Culetu, A.; Susman, I.E. Screening of the Nutritional Properties, Bioactive Components, and Antioxidant Properties in Legumes. *Foods* **2024**, *13*, 3528. [CrossRef]
43. Comai, S.; Bertazzo, A.; Bailoni, L.; Zancato, M.; Costa, C.V.L.; Allegri, G. The content of proteic and nonproteic (free and protein-bound) tryptophan in quinoa and cereal flours. *Food Chem.* **2007**, *100*, 1350–1355. [CrossRef]
44. Li, P.; Wu, G. Roles of dietary glycine, proline, and hydroxyproline in collagen synthesis and animal growth. *Amino Acids* **2018**, *50*, 29–38. [CrossRef] [PubMed]
45. Gabriel, A.S.; Uneyama, H. Amino acid sensing in the gastrointestinal tract. *Amino Acids* **2013**, *45*, 451–461. [CrossRef]
46. Kawai, M.; Sekine-Hayakawa, Y.; Okiyama, A.; Ninomiya, Y. Gustatory sensation of L- and D-amino acids in humans. *Amino Acids* **2012**, *43*, 2349–2358. [CrossRef]
47. Abreu, H.M.C.; Nobile, P.M.; Shimizu, M.M.; Yamamoto, P.Y.; Silva, E.A.; Colombo, C.A.; Mazzafera, P. Influence of air temperature on proteinase activity and beverage quality in *Coffea arabica*. *Braz. J. Bot.* **2012**, *35*, 357–376. [CrossRef]
48. Dias, E.C.; Borém, F.M.; Pereira, R.G.F.A.; Guerreiro, M.C. Amino acid profiles in unripe Arabica coffee fruits processed using wet and dry methods. *Eur. Food Res. Technol.* **2012**, *234*, 25–32. [CrossRef]

49. Bastian, F.; Hutabarat, O.S.; Dirpan, A.; Nainu, F.; Harapan, H.; Emran, T.B.; Simal-Gandara, J. From Plantation to Cup: Changes in Bioactive Compounds during Coffee Processing. *Foods* **2021**, *10*, 2827. [CrossRef]
50. Arai, H.; Kono, N. α-Tocopherol transfer protein (α-TTP). *Free Radic. Biol. Med.* **2021**, *176*, 162–175. [CrossRef] [PubMed]
51. Szewczyk, K.; Chojnacka, A.; Górnicka, M. Tocopherols and Tocotrienols—Bioactive Dietary Compounds; What Is Certain, What Is Doubt? *Int. J. Mol. Sci.* **2021**, *22*, 6222. [CrossRef] [PubMed]
52. Tavares, K.M.; Lima, A.R.; Nunes, C.A.; Silva, V.A.; Mendes, E.; Casal, S.; Pereira, R.G.F.A. Free tocopherols as chemical markers for Arabica coffee adulteration with maize and coffee by-products. *Food Control* **2016**, *70*, 318–324. [CrossRef]
53. Simopoulos, A.P. The importance of the ratio of omega-6/omega-3 essential fatty acids. *Biomed. Pharmacoter.* **2002**, *56*, 365–379. [CrossRef]
54. Rios, M.B.; Iriondo-DeHond, A.; Iriondo-DeHond, M.; Herrera, T.; Velasco, D.; Gómez-Alonso, S.; Callejo, M.J.; del Castillo, M.D. Effect of Coffee Cascara Dietary Fiber on the Physicochemical, Nutritional and Sensory Properties of a Gluten-Free Bread Formulation. *Molecules* **2020**, *25*, 1358. [CrossRef]
55. Prior, R.L.; Wu, X.; Schaich, K. Standardized methods for the determination of antioxidant capacity and phenolics in foods and dietary supplements. *J. Agric. Food Chem.* **2005**, *53*, 4290–4302. [CrossRef]
56. Gebeyehu, B.T.; Bikila, S.L. Determination of Caffeine Content and Antioxidant Activity of Coffee. *Am. J. Appl. Chem.* **2015**, *3*, 69–76. [CrossRef]
57. Smit, H.J.; Rogers, P.J. Effects of Low Doses of Caffeine on Cognitive Performance, Mood and Thirst in Low and Higher Caffeine Consumers. *Psychopharmacology* **2000**, *152*, 167–173. [CrossRef] [PubMed]
58. Childs, E.; de Wit, H. Subjective, Behavioral, and Physiological Effects of Acute Caffeine in Light, Nondependent Caffeine Users. *Psychopharmacology* **2006**, *185*, 514–523. [CrossRef] [PubMed]
59. FSA NDA Panel (EFSA Panel on Dietetic Products, Nutrition and Allergies). Scientific Opinion on the safety of caffeine. *EFSA J.* **2015**, *13*, 4102.
60. Tajik, N.; Tajik, M.; Mack, I.; Enck, P. The potential effects of chlorogenic acid, the main phenolic components in coffee, on health: A comprehensive review of the literature. *Eur. J. Nutr.* **2017**, *56*, 2215–2244. [CrossRef] [PubMed]
61. Cañas, S.; Rebollo-Hernanz, M.; Cano-Muñoz, P.; Aguilera, Y.; Benítez, V.; Braojos, C.; Gila-Díaz, A.; Rodríguez-Rodríguez, P.; Cobeta, I.M.; Pablo, Á.L.L.d.; et al. Critical Evaluation of Coffee Pulp as an Innovative Antioxidant Dietary Fiber Ingredient: Nutritional Value, Functional Properties, and Acute and Sub-Chronic Toxicity. *Proceedings* **2021**, *70*, 65.
62. Loureiro, G.; Martel, F. The effect of dietary polyphenols on intestinal absorption of glucose and fructose: Relation with obesity and type 2 diabetes. *Food Rev. Int.* **2019**, *35*, 390–406. [CrossRef]

foods

MDPI

Article

Characterization and Potential Food Applications of Oat Flour and Husks from Differently Colored Genotypes as Novel Nutritional Sources of Bioactive Compounds

Valentina Nikolić [1,*], Slađana Žilić [1], Marijana Simić [1], Katarina Šavikin [2], Tatjana Stević [2], Jelena Živković [2], Beka Sarić [1], Danka Milovanović [1] and Vesna Kandić Raftery [3]

[1] Research Department, Maize Research Institute, Zemun Polje, Slobodana Bajića 1, 11080 Belgrade, Serbia; szilic@mrizp.rs (S.Ž.); marijana.simic@mrizp.rs (M.S.); bsaric@mrizp.rs (B.S.); dmilovanovic@mrizp.rs (D.M.)
[2] Institute of Medicinal Plants Research, Dr. Josif Pančić, Tadeuša Koščuška 1, 11000 Belgrade, Serbia; ksavikin@mocbilja.rs (K.Š.); tstevic@mocbilja.rs (T.S.); jzivkovic@mocbilja.rs (J.Ž.)
[3] Breeding Department, Maize Research Institute, Zemun Polje, Slobodana Bajića 1, 11080 Belgrade, Serbia; vkandic@mrizp.rs
* Correspondence: valentinas@mrizp.rs

Abstract: Oats are gluten-free cereals rich in dietary fiber, β-glucans, phenolic acids, flavonoids, carotenoids, vitamin E, and phytosterols. They have been used in traditional medicine for centuries to treat hyperacidity, acute pancreatitis, burns, and skin inflammation. This study assessed the nutritional and phenolic profile of oat flour (OF) and ground oat husks (OHs) from white, brown, and black hulled oat genotypes, as well as the antioxidant and antimicrobial activity of their extracts. The extracts were tested on six strains of gastrointestinal tract pathogens. OF samples had, on average, a high protein content (15.83%), fat content (6.27%), and β-glucan content (4.69%), while OH samples were rich in dietary fiber. OHs had significantly higher average total phenolic compounds compared to OF and had twice as high antioxidant capacity. Ferulic acid was predominant in all samples, followed by *p*-coumaric, isoferulic, vanillic, and syringic acid. The traditionally prepared OH extracts manifested the best bactericidal activity against *Listeria monocytogenes*, *Escherichia coli*, and *Staphylococcus haemolyticus*, while *Salmonella typhimurium* was the least sensitive to the bactericidal effect of all the investigated samples. Both OF and powdered OHs have potential applications in the functional food industry and pharmacy due to their bioactive compounds, their biological activity, as well as their overall nutritional profile.

Keywords: oats; chemical composition; phenolic compounds; antioxidant capacity; antimicrobial activity; human gut microbiota

1. Introduction

Hippocrates' maxim, "Let food be thy medicine and medicine be thy food," which dates back two millennia, was nearly forgotten in the 19th century due to the development of modern medicine. However, a renewed focus on the role of nutrition in promoting health and preventing disease by utilizing plants with medicinal properties has regained popularity in the past several decades [1]. Epidemiological studies have shown that the consumption of whole-grain cereals is associated with a decreased risk of some chronic diseases and conditions such as diabetes [2] cardiovascular disease [3], and obesity [4]. Their beneficial activity is mostly connected to the presence of dietary fiber, vitamins, essential fatty acids, and phenolic compounds [5].

Oat (*Avena sativa* L.) is the seventh most extensively grown and commercially significant cereal in the world. It was first cultivated for its medicinal properties, before being employed as a nutrient-dense food source for humans and animals [6]. Oats exhibit a broad range of biological activity, which suggests their potential therapeutic value [7,8]. Namely, oats have been found to promote physiological benefits such as reducing hyperglycemia, hyperinsulinemia, hypercholesterolemia, hypertension, and cancer [9]. Oats have been used in European, Chinese, and Middle Eastern traditional medicine for centuries to treat mental and physical ailments [6,10,11], as well as obesity, constipation, loss of appetite, and headaches [12–14]. In Serbian traditional medicine, oat grains are used as a remedy for hyperacidity and acute pancreatitis, and as a treatment for burns and skin inflammation. Furthermore, oat straw is frequently used to treat frostbite, sciatica, and rheumatism [15].

Oats are naturally gluten-free grains suitable for persons with gluten-related disorders, including celiac disease [16]. They are predominantly composed of starch, which makes up more than 60% of the grain weight and is crucial in determining oats' physicochemical and structural characteristics [17]. Whole-grain oats are rich in beneficial macronutrients, such as lipids with a high degree of unsaturated fatty acids, which make up about 40 and 36% of total fatty acids, respectively, and proteins with a well-balanced essential amino acid composition [18]. Oats contain dietary fiber with a high level of β-glucan (2–8.5% *w/w* of oat grain), a soluble form of fiber that has been shown to have a variety of physiological effects, including controlling postprandial blood glucose levels and lowering serum cholesterol levels [19,20]. Nevertheless, oat grains are abundant in bioactive compounds such as avenanthramides, polyphenols, phenolic acids, carotenoids, vitamin E, and phytosterols [6,8,21].

The inedible outer husks of the oat grain, typically removed during harvest and processing, are the main byproduct of oat production and can make up as much as 35% of the grain [22]. Oat husks (OHs) are frequently underestimated as agro-industrial wastes and are commonly fed directly to animals or disposed of in landfills [23]. OHs also have a range of biologically important components, primarily proteins (1–7%, dry weight basis), cellulose (16–26%), hemicellulose (24–35%), and lignin (13–25%) [22,23]. OHs are a rich source of bioactive compounds, including phenolic acids and avenanthramides, which have many health-promoting qualities. The most prevalent phenolic components in oats are phenolic acids, mostly present in the three-layer bran (peel, seed coat, and aleurone) [24]. The consumption of phenolic acids has been associated with lowering the risk of cardiovascular disease [25]. In addition to their potential health benefits, phenolic acids have been demonstrated to improve gut health by promoting the growth of beneficial bacteria and suppressing pathogens, which supports overall digestive health [26,27]. Only a small portion of the phenolic compounds are immediately absorbed by the small intestine; the gut microbiota will metabolize up to 90% of these compounds in the colon [28]. Accordingly, the breakdown and metabolism of phenolic compounds are greatly influenced by the gut microbiota [29]. However, aside from using them as a source of dietary fiber, the OH utilization level is still considerably low [23].

The prebiotic impact of dietary phenolic compounds on gut microbiota has been demonstrated in numerous in vitro and in vivo investigations in recent years [30,31]. For instance, adding bound phenolic compounds from rice bran to human fecal homogenate in vitro typically suppresses the growth of harmful bacteria [32]. According to Gong et al. [30], higher intakes of phenolic acids from whole wheat were linked to increased abundances of *Bifidobacterium* and *Lactobacillus* and lower abundances of *Escherichia coli, Clostridiaceae,* and *Clostridium perfringens*. Furthermore, oat β-glucan is a prebiotic that modulates intestinal flora and aids in the treatment of diarrhea and related conditions by preserving the energy balance thanks to several significant physicochemical properties, including solubility, viscosity, and gelation [19]. Human endogenous digestive enzymes and the gastric acid environment scarcely hydrolyze oat β-glucan; instead, it enters the large intestine and is broken down by gut bacteria [19].

The objective of this study was to assess the nutritional and phenolic profile of oat flour and ground oat husk obtained from black, brown, and white hulled oat genotypes, as well as the antioxidant and antimicrobial activity of their extracts rich in bioactive compounds in order to evaluate their potential for the production of functional foods and pharmacological application.

2. Materials and Methods

2.1. Chemicals and Consumables

The HPLC grade chemicals–6-Hydroxy-2,5,7,8-tetramethylchroman-2-carboxylic acid (Trolox), 2,2′-azino-bis/3-ethylbenzothiazoline-6-sulfonic acid (ABTS), and phenolic acid standards: gallic acid, 3,4-dihydroxybenzoic acid, chlorogenic acid, vanillic acid, caffeic acid, syringic acid, *p*-coumaric acid, sinapic acid, ferulic acid, and isoferulic acid–were purchased from Sigma-Aldrich (Steinheim, Germany). Formic acid (98%) and methanol were purchased from J.T. Baker (Deventer, The Netherlands). The p.a. grade Folin-Ciocalteu reagent was procured from Sigma-Aldrich (Steinheim, Germany). The p.a. grade chemical potassium persulfate (dipotassium peroxdisulfate) was bought from Fluka Chemie AG (Buchs, Switzerland). Sodium hydroxide, sulfuric acid, ethyl acetate, diethyl ether, and ethanol were purchased from Merck (Darmstad, Germany). Sodium tetraborate-10-hydrate and boric acid were bought from Sigma-Aldrich (St. Louis, MO, USA). Assay kits for the determination of resistant starch (K-RSTAR) and β-glucan (K-BGLU) contents were purchased from Megazyme (Wicklow, Ireland). The enzymes used for the in vitro digestion protocol were all procured from Sigma-Aldrich, Merck, namely: pepsin from porcine gastric mucosa (P7000-25G), bile extract porcine (B8631-100G), pancreatin from porcine pancreas (P1750-25G), protease from *Streptomyces griseus* (P5147-1G), and viscozyme L cellulolytic enzyme mixture (V2010-50ML). Syringe filters (nylon, 0.45 mm) and Hypersil GOLD aQ C18 column (150 mm × 4.6 mm, i.d., 3 µm) were supplied by Thermo Fisher Scientific (Waltham, MA, USA). Ultrapure water was used throughout the experiments (LaboStar Pro system, Evoqua, Pittsburgh, PA, USA).

2.2. Plant Material

The experimental material comprised three oat (*Avena sativa* L.) genotypes obtained from the Maize Research Institute, Zemun Polje (MRIZP) gene bank. The genotypes were chosen based on differences in agronomic traits such as yield and its components, as well as the grain husk color. Single genotypes of yellow, brown, and black oats were cultivated at the location of MRIZP Zemun Polje, Serbia (44°52′ N, 20°19′ E, 82 m a.s.l.), sown in the growing season of 2023/2024. Standard cropping practices were applied to provide adequate nutrition and to keep the plots disease- and weed-free. After harvesting, the broken and damaged grains, as well as extraneous matter, were removed from the samples. The oat grains were hulled using a scalpel to separate the OHs and groats manually. The separated OHs and groats were ground on a Perten 120 lab mill (Perten Instruments AB, Hägersten, Sweden) to obtain fine powder (particle size < 500 µm) OF and OHs for the analyses. The oat grain is referred to as OF (oat flour) throughout the manuscript for simplicity reasons. All the prepared samples were stored at −70 °C before analysis.

2.3. Chemical Procedures

2.3.1. Analysis of Basic Chemical Composition

The dry matter content of the samples was assessed by using the conventional drying method in an oven at 105 °C to a constant mass. The total starch content was determined on UniPol L 2020 polarimeter (Schmidt + Haensch GmbH and Co., Berlin, Germany) according to the Ewers polarimetric method [33]. The total protein content was determined by the standard micro- Kjeldahl method (AOAC 920.87) (AutoKjeldahl distillation unit K-350 and speed digester K-439, BÜCHI Labortechnik, Flawil, Switzerland) as the total nitrogen multiplied by 5.7 [34]. The fat content was determined according to the standard Soxhlet method (AOAC 920.39) [35] on a FatExtractor E-500 (Büchi Labortechnik, Flawil, Switzerland). The ash content was analyzed by the slow combustion of the sample at 550 °C in a muffle furnace (L47, 1200 °C, Naber Industrieofenbau, Lilienthal, Germany) by following the AOAC 923.03 method [35]. All the results are given as means ± standard deviation of three repetitions and expressed as a percentage per dry matter (d.m.).

2.3.2. Analysis of the Alcohol-Soluble Protein Fraction

The ethanol-soluble protein fraction was obtained by successive extractions of OF and ground OHs with a series of solvents (in a ratio of 1:10 *w/v*) according to the Osborne procedure, as described by Lookhart and Bean [36], with some modifications. Distilled water, 0.5 M NaCl, and 70% ethanol were used to extract albumin, globulin, and prolamin fractions. Except for the albumin fraction, all the others were extracted from the water-washed pellet remaining after the previous extraction. The extraction of each protein fraction was performed by repeated stirring three times for 30 min at room temperature, followed by centrifugation at 10,000 rpm for 5 min at 4 °C. The final volume of each protein extract was 50 mL. The protein extracts (10 mL) were evaporated for 12 h at 100 °C. The protein content in the ethanol-soluble fraction was analyzed after the micro Kjeldahl method [35] on the BÜCHI Kjeldahl System (Auto Kjeldahl Distillation Unit K-350 and Speed Digester K-439, BÜCHI Labortechnik, Flawil, Switzerland) and calculated from the nitrogen content determined, using 5.7 as the conversion factor. The results are given as the percentage of the d.m. and the percentage of the total protein (protein solubility index).

2.3.3. Analysis of Dietary Fibers

The contents of hemicellulose, cellulose, neutral detergent fibers (NDF), acid detergent fibers (ADF), and lignin (ADL) were determined by the Van Soest detergent method using the Fibertec system 2010 (Foss, Hillerød, Denmark). The procedure is described in detail in a previous paper [37]. The method is based on the fibers' solubility in neutral, acid, and alkali reagents. NDF practically represents total insoluble fibers (not soluble in water); ADF mainly consists of cellulose and lignin; and ADL is pure lignin. The hemicellulose content was obtained as a difference between the NDF and ADF contents, while the cellulose content was calculated as the difference between the ADF and lignin contents. All the results are given as the percent per d.m.

2.3.4. Analysis of β-Glucan

The content of β-glucan was determined using the Megazyme protocol [38]. Briefly, samples (0.5 g) were suspended and hydrated in ethanol (1 mL, 50% *v/v*) and sodium phosphate buffer solution (4 mL, 20 mM, pH 6.5), stirred, and then incubated at 40 °C for 1 h with purified lichenase enzyme (200 μL, 10 U). After the addition of 5 mL of 200 mM sodium acetate buffer pH 4.0 and centrifugation at 5000 rpm for 10 min, an aliquot (100 μL) was then hydrolyzed to completion with purified β-glucosidase (100 μL, 0.2 U) during incubation at 50 °C for 10 min. A total of 50 mM acetate buffer pH 4.0 was used as the reaction blank. The D-glucose produced was assayed using a glucose oxidase/peroxidase reagent (GOPOD) (3 mL). The absorbance was measured at 510 nm (Agilent 8453 UV-visible spectroscopy system, Agilent Technologies, Inc, Santa Clara, CA, USA). The β-glucan content was expressed as a percentage of d.m.

2.3.5. Analysis of Resistant Starch

The resistant starch content was determined according to the Megazyme protocol [39]. Summarily, 4.0 mL of pancreatic α-amylase (10 mg/mL) containing amyloglucosidase (AMG) (3 U/mL) are added to 100 mg oat samples and incubated for 16 h at 37 °C in a shaking water bath. During this time, the combined action of the two enzymes hydrolyzes non-resistant starch to D-glucose. Adding 4.0 mL of 99% ethanol stops the reaction, and centrifugation recovers the RS as a pellet. After that, the pellets are centrifuged twice (1.500× g, 10 min) and washed with 2 mL of ethanol (50% *v/v*). Decantation is used to eliminate free liquid. RS in the pellet is dissolved in 2 mL of 2 M KOH by agitating the mixture strongly over a magnetic stirrer in an ice-water bath. AMG is used to quantitatively hydrolyze the starch to glucose after neutralization with 8 mL of 1.2 M (pH 3.8) acetate buffer. GOPOD reagent (3 mL) is used to measure D-glucose, which is a marker for the sample's RS concentration. By combining the original supernatant and the washings, increasing the volume to 100 mL, and using GOPOD to measure the D-glucose level, non-resistant starch can be identified. The absorbance of each solution was measured at 510 nm against the reagent blank. The resistant starch content of the samples was calculated by using the Megazyme Mega-Calc™ spreadsheet, and the results were expressed as a percentage per dry matter (d.m.).

2.3.6. Extraction of Total Phenolic Compounds

Alkaline hydrolysis was applied at room temperature for 4 min using 10 mL of 4 M NaOH to release phenolic compounds from 500 mg of OF and ground OHs. From the hydrolyzate, phenolic compounds were extracted according to the procedure described by Žilić et al. [40]. The extracts were evaporated to dryness under the N_2 stream at 25 °C (Reacti-Therm nitrogen evaporator system 18821, Thermo Fisher Scientific Inc., Waltham, MA, USA) and the residues were redissolved in methanol. Such prepared methanolic solutions were used for the analyses of total phenolic compounds and phenolic acids. All the extractions were performed in triplicate for each sample and kept at −70 °C before analyses.

2.3.7. Analysis of Total Phenolic Compounds (TPCs)

The total phenolic content was determined by the Folin-Ciocalteu assay as described by Singleton et al. [41]. Briefly, 300 and 100 μL of the OF extracts and OH extracts were transferred into test tubes, and their volume was filled up to 500 μL with distilled water. After the addition of Folin-Ciocalteu reagent (250 μL) and 20% aqueous Na_2CO_3 solution (1.25 mL), the tubes were vortexed and the absorbance of the mixture was measured at 750 nm after 40 min and centrifugation at 8000 rpm. The content of total phenolics was expressed as μg of gallic acid equivalent (GAE) per g of dry matter (d.m.).

2.3.8. Analysis of Phenolic Acids

To determine the phenolic acids, the clear supernatants were filtered through the 0.45 μm nylon filter, and the pure extracts were analyzed using the HPLC-DAD system (Thermo Scientific Ultimate 3000). The chromatographic separation was performed on the Thermo Scientific Hypersil GOLD aQ C18 column (150 mm × 4.6 mm, i.d., 3 μm) at 25 °C using a gradient mixture of 1% formic acid in water (solvent A) and 100% methanol (solvent B) at a flow rate of 0.8 mL/min and run time of 30 min. The solvent gradient was programmed as previously described by Žilić et al. [42]. The chromatograms were recorded at 280 nm by monitoring the spectra within the wavelength range of 190–400 nm. Standards of gallic acid, 3,4-dihydroxybenzoic acid, chlorogenic acid, vanillic acid, caffeic acid, syringic acid, *p*-coumaric acid, sinapic acid, ferulic acid, and isoferulic acid were used (10, 20, 40, 50 and 100 μg/g). The identified phenolic acids' peaks were confirmed and quantified by data acquisition and spectral evaluation using the Thermo Scientific Dionex Chromeleon 7.2. Chromatography software. The content of phenolic acids was expressed as μg per g of d.m.

2.3.9. Analysis of Total Antioxidant Capacity (TAC)

The antioxidant capacities of OH fine powder and OF were measured according to the QUENCHER method described by Serpen et al. [43], using a 7 mM aqueous solution of ABTS (2,2-azino-bis/3-ethil-benothiazoline-6-sulphonic acid) with 2.45 mM $K_2O_8S_2$ as the stock solution. The working solution of $ABTS^{\bullet+}$ was obtained by diluting the stock solution in water/ethanol (50:50, *v/v*). Depending on the sample, 2 and 9 mg of ground OHs and OF, respectively, were mixed with 20 mL of $ABTS^{\bullet+}$ working solution, and the mixture was rigorously shaken for 25 min. Afterwards, the centrifugation absorbance was measured at 734 nm. The total antioxidant capacity was expressed as the Trolox equivalent antioxidant capacity (TEAC) in mmol of Trolox per kg of d.m.

2.4. Baking Functionality Properties

2.4.1. Gelling Properties

The water solubility index (WSI), water absorption index (WAI), and swelling power (SP) were determined according to the method described by Cornejo and Rosell [44]. Powdered OHs and OF (1 g) were weighed into centrifuge tubes and 20 mL of distilled water was added. The tubes were shaken in a water bath for 15 min at 90 °C and then centrifuged at 3000× *g* at 4 °C for 10 min. The WAI was calculated as the quotient of the mass of the sediment and the initial mass of the sample. Dry matter of the supernatant evaporated in the ventilation oven for 12 h at 110 °C was used for the WSI calculation. The SP was calculated as the quotient of the sediment mass and the difference between the initial mass of the sample and the dry mass of the evaporated supernatant. Each sample was analyzed in three replicates, and the WSI and SP values were expressed in g.

2.4.2. Solvent Retention Capacity (SRC)

Solvent retention capacity (SRC), a test that indicates the ability of flour to retain individual diagnostic solvents (distilled water, 50% sucrose, 5% sodium carbonate, and 5% lactic acid water solutions) based on the swelling behavior of polymer networks in flour, was determined according to the American Association of Cereal Chemists (AACC) method 56–11 adapted by Haynes et al. [45]. Each flour polymer network is associated with the corresponding diagnostic solvent. Considering the analyzed samples, OH powder and OF, two solvents were individually used to determine the SRC values: 50% sucrose in water and 5% lactic acid in water. Five grams of sample (OH powder and OF) were weighed and vortexed in 25 mL of an appropriate solvent for a total of 25 min at one-minute intervals every 5 min to allow the samples to solvate and swell. After centrifugation at 3000 rpm for 10 min, the SRC values were calculated and expressed as percentages of the mass of flour gel after exposure to the solvent in relation to the original flour weight.

2.5. In Vitro Multistep Enzymatic Digestion Protocol

To determine the potential OF and ground OH digestibility for human consumption an in vitro multistep digestion procedure was applied. The method, consisting of oral, gastric, duodenal, and colon phases, was proposed by Papillo et al. [46] and modified according to Hamzalioğlu and Gökmen [47]. Digestion fluids simulating the saliva (simulated salivary fluid, SSF), gastric juice (simulated gastric fluid, SGF), and duodenal juice (simulated duodenal fluid, SDF), as well as pepsin solution in 0.1 M HCl, bile salts, pancreatin solution in distilled water, protease water solution, and viscozyme L, were used and prepared according to the procedure described by Hamzalioğlu and Gökmen [47]. The multistep digestion conditions are described in detail by the same authors. Samples obtained after in vitro digestion were filtered through qualitative filter paper, air-dried in a ventilated oven (Memmert UF 55; Memmert GmbH + Co. KG) for 2 h, and then dried to constant mass at 105 °C for 4 h. After weighing the samples, their digestibility was calculated according to the following equation:

$$\text{Digestibility} = ((m_0 - m_d)/m_0) \times 100 \quad (1)$$

where m_0 is the mass of the absolutely dry sample prior to digestion, and m_d is the remaining (undigested) mass of the absolutely dry sample. The digestibility of dry matter was expressed as a percentage.

2.6. Antimicrobial Activity Evaluation

2.6.1. Extract Preparation

The samples were prepared according to the recipe from the traditional medicine of Serbia [9]. Briefly, OF and powdered OH samples (10 g) were poured into 150 mL of water and boiled until the volume was reduced to 100 mL. Samples were filtered and the supernatant was left to cool. Afterward, the extract prepared after alkaline hydrolysis, as described by Žilić et al. [40] and evaporated to dryness under the N_2 stream at 25 °C, was dissolved in water and used for the antimicrobial assay.

2.6.2. Antimicrobial Assay

The antimicrobial effect of oat samples was tested on six important pathogens of the gastrointestinal tract, namely: *Listeria monocytogenes*, *Staphylococcus haemolyticus*, *Salmonella typhimurium*, *Enterococcus faecalis*, *Escherichia coli*, and *Shigella flexneri*. An overnight culture of bacteria was produced by inoculating several colonies in MH (Muller Hinton) broth and incubating in a thermostat at 37 °C for up to 24 h. The cultures thus prepared were brought to a cell concentration of 1×10^8 CFU/mL with a McFarland tube densitometer (DEN-1, BIOSAN, Riga, Latvia). The densitometers are designed and factory-calibrated to measure the turbidity of cell suspensions in a variety of life science applications. Dilutions were made from the obtained bacterial suspensions so that the final concentration of bacteria was 10^6 CFU/mL. First, an attempt was made to determine the MIC (minimum inhibitory concentration) values of the tested water extracts in microtiter dilution plates with 96 wells. As the wells in the plates were defined to hold 200 mL of total volume (test substance, liquid broth, and bacterial suspension), and water extracts for inhibiting the growth of the tested bacteria required higher concentrations, we used a modified method. Different volumes of tested extracts were added to Petri dishes with 1 mL of overnight culture of tested bacteria with a final concentration of 10^6 CFU/mL. A medium containing the MH agar with a melting point below 45 °C was poured over it. Petri dishes prepared this way were incubated in a thermostat at 37 °C for 48–72 h after hardening the agar. Each concentration was determined in triplicate. After incubation, grown colonies were counted and compared with control Petri dishes in which only bacterial suspensions and medium for bacterial growth were added.

2.7. Statistical Analysis

The data were reported as a mean ± standard deviation of three independent repetitions per sample. Statistical analyses were performed using Minitab19 Statistical Software. The one-way ANOVA analysis of variance with Tukey's test was. Pearson's correlation coefficients (r) were calculated to evaluate the relationship between the individual parameters. Differences between the means with probability $p < 0.05$ were accepted as statistically significant.

3. Results and Discussion

3.1. Nutritional Profile of OF and OH

3.1.1. Basic Chemical Composition

The results of the basic chemical composition of the investigated OF samples and powdered OHs are shown in Table 1.

The starch content in the OF of all three oat genotypes was expectedly high, ranging from 55.71% in the yellow oat genotype to 57.58% in the black oat genotype. On the other hand, starch was not determined in the OHs of all three genotypes. Furthermore, oat starch is the main grain ingredient predominant in the endosperm, and differs in content between 51 and 65% due to variations caused by environmental factors during cultivation and plant genotype [48]. Oat starch exhibits several distinct structural characteristics in contrast to other cereal starches, including a clustered granular structure, smaller granule size, lower relative crystallinity, and higher concentration of amylose-lipid complexes [17]. These properties allow starch to be used as a food ingredient or additive for the improvement of texture, emulsion stability, and moisture retention. Oats are suitable for human consumption and have various applications, including oat flour and oatmeal, biscuits, noodles, bread, bars, and yogurt [49].

Table 1. Basic chemical composition of OF and OHs.

Sample	Dry Matter	Total Starch	Total Protein	Alcohol-Soluble Proteins		Fat	Ash
	(%)	(% d.m.)	(% d.m.)	(% d.m.)	(% t.p.)	(% d.m.)	(% d.m.)
OF							
Yellow	90.60 e	55.71 ± 0.01 b	17.38 ± 0.10 a	2.04 ± 0.05 b	11.73	5.45 ± 0.16 c	2.44 ± 0.01 d
Brown	90.92 d	57.38 ± 0.54 a	16.67 ± 0.09 b	2.31 ± 0.11 a	13.86	7.20 ± 0.14 a	2.30 ± 0.02 e
Black	90.99 d	57.58 ± 1.15 a	13.43 ± 0.01 c	1.60 ± 0.01 c	11.91	6.15 ± 0.21 b	2.19 ± 0.04 f
OHs							
Yellow	96.06 a	n.d.	4.79 ± 0.06 d	0.64 ± 0.09 d	13.36	1.23 ± 0.18 d	5.85 ± 0.02 a
Brown	94.69 b	n.d.	4.54 ± 0.02 e	0.64 ± 0.10 d	14.10	1.02 ± 0.05 d	4.64 ± 0.00 b
Black	94.51 c	n.d.	3.09 ± 0.02 f	0.56 ± 0.01 d	18.12	1.32 ± 0.06 d	5.00 ± 0.00 b

OF—oat flour; OHs—oat husks; t.p.—total protein. Values are means of three determinations ± standard deviation. Means followed by the same letter within the same column are not significantly different ($p < 0.05$).

The highest fat content determined in our samples (7.20%) was in the brown OF, and while it amounted to around 1% in the OHs it did not differ significantly among the oat genotypes. Oats are a unique cereal, with 2–18% fat, which accumulates mainly in the endosperm, unlike other oily seeds which typically store fats in the embryo [50]. Oat lipids consist of 23% saturated (mostly palmitic acid), 34% monounsaturated (primarily oleic acid), and 43% polyunsaturated (primarily linoleic acid) fats [51]. This composition affects food oxidation, where the amount of unsaturated fatty acids is especially significant, and hence the flavor of oat products. The results of our study follow those of Ibrahim et al. [52], who reported fat contents ranging from 6.16% to 6.67%, and ash contents varying between 3.90% and 6.02% among the five oat cultivars cultivated in Pakistan. The ash content determined in our study was the highest in yellow OHs (5.85%) and the lowest in yellow OF (2.19%). According to Usman et al. [53], oat grains had ash and fat contents of 3.7% and 4.5%, respectively. Neitzel et al. [54] reported that OHs contained 6.27% ash. Zhu et al. [55] found that whole oats comprised 14.88% crude protein, 8.16% crude fat, 1.57% ash, and 60.15% total starch. Varietal variances, climatic circumstances, soil composition, and cultural practices could all represent contributing factors to the discrepancies in the study's conclusions [56].

After removing the OHs, the remainder of the oat grain—groat generally contains 15–25% protein. The protein content of oat grains rises from the center to the outside [57]. The starchy endosperm has about 12% protein, while the bran (pericarp, testa, nucellus, aleurone, and some subaleurone) has 18–26% and the germ has 29–38% [58]. The total protein content detected in our study was higher in the investigated OF, ranging from 13.43% in black oats to 17.38% in yellow oats, than in OHs where it varied between 3.09% in black oats and 4.79% in yellow oats. Ibrahim et al. [52] reported protein contents ranging from 8.13% to 12.69% among the five oat cultivars cultivated in Pakistan, while Usman et al. [53], reported average protein levels of 13.5% in oat grains. Zhou et al. [23] reported 5.3% of total protein in the ground OHs. Kouřimská et al. [51] analyzed hulled, dehulled, and nude oat grains (without OHs) and found the range of protein content was between 16.75 and 17.78% (d.m. basis), while the fat content was between 3.16% and 5.82%. Oat grain is the only cereal crop that contains avenalin, salt-soluble legumin-like globulin nutritional quality to soybean proteins, as the major storage protein (70–80%) [6,58]. In contrast to the predominant alcohol-soluble prolamins, commonly referred to as gluten proteins, which comprise 60–80% of the total protein found in the *Triticeae* cereals: wheat (gliadins and glutenins), barley (hordeins), and rye (secalins) avenins make up about 10–15% of the total protein in oats [59–61]. As long as there is no cross-contamination with other gluten-containing cereals and the food's gluten content is less than 20 mg/kg, oats, oat products can be regarded as gluten-free under EU law ((EU) No 828/2014) [62]. The content of alcohol-soluble avenin prolamin detected in our study ranged from 11.73% (yellow oats) to 13.86% (brown oats) in the OF, and from 13.36% (yellow oats) to 18.12%

(black oats) of total protein found in the oat OHs. In terms of percentage of dry matter, these values are well below the benchmark of 20 mg/kg of oats, which means that the OF and OHs are safe for human consumption as gluten-free. To sum up, the primary distinctions between the prolamins of different cereals are found in their physicochemical properties, including their source, solubility, molecular weight, disulfide bonds, and amino acid makeup [63]. Despite a lack of research on avenin, the nutritional qualities of oats and the health benefits of avenins have received more attention in recent years. Compared to other prolamins, avenin is more hydrophilic; its maximum solubility was found in 45% (*w/w*) ethanol [61].

3.1.2. Dietary Fibers

Dietary fibers are crucial nutrients that possess beneficial properties such as regulating various physiological processes, ranging from bowl regulation to treating chronic illnesses [52]. According to Zhu [64], the health benefits of dietary fibers include antioxidative properties, possible anticancer effects, control of body weight and glycemic levels, neuroprotective qualities, protection of retinal health, hypolipidemic effects, hepatoprotective properties, and potential anti-aging effects. Some studies indicate that the antioxidant effect of dietary fiber is the result of the phenolic compounds bound to polysaccharide complexes that comprise the dietary fiber [65].

The results of the investigated dietary fibers are presented in Table 2.

Table 2. Content of dietary fibers in OF and OHs (% d.m.).

Sample	NDF	ADF	ADL	Hemicellulose	Cellulose	*β*-Glucan	Resistant Starch
OF							
Yellow	15.83 ± 0.31 ^{f}	3.11 ± 0.08 ^{d}	0.89 ± 0.20 ^{d}	12.72 ± 0.40 ^{e}	2.22 ± 0.28 ^{d}	5.33 ± 0.01 ^{a}	2.95 ± 0.39 ^{b}
Brown	32.36 ± 0.25 ^{d}	3.65 ± 0.12 ^{d}	1.29 ± 0.47 ^{cd}	28.71 ± 0.35 ^{c}	2.36 ± 0.38 ^{d}	4.07 ± 0.16 ^{c}	2.43 ± 1.03 ^{a}
Black	17.22 ± 0.04 ^{e}	3.34 ± 0.26 ^{d}	0.89 ± 0.17 ^{d}	13.88 ± 0.23 ^{d}	2.45 ± 0.09 ^{d}	4.66 ± 0.00 ^{b}	2.91 ± 0.50 ^{b}
OHs							
Yellow	80.27 ± 0.08 ^{b}	38.96 ± 0.54 ^{a}	6.00 ± 0.11 ^{a}	41.31 ± 0.62 ^{b}	32.96 ± 0.43 ^{c}	0.06 ± 0.00 ^{d}	n.d.
Brown	77.91 ± 0.24 ^{c}	36.04 ± 0.69 ^{c}	1.86 ± 0.28 ^{b,c}	41.87 ± 0.45 ^{b}	34.18 ± 0.42 ^{b}	0.06 ± 0.00 ^{d}	n.d.
Black	83.82 ± 0.22 ^{a}	37.71 ± 0.08 ^{b}	2.39 ± 0.26 ^{b}	46.11 ± 0.13 ^{a}	35.32 ± 0.18 ^{a}	0.03 ± 0.01 ^{d}	n.d.

OF—oat flour; OHs—oat husks; NDF—neutral detergent fiber, ADF—acid detergent fiber, ADL—acid detergent lignin. Values are means of three determinations ± standard deviation. Means followed by the same letter within the same column are not significantly different ($p < 0.05$).

Significant differences in the fiber composition of the studied oat samples were detected, namely between the powdered OH which had higher lignocellulosic fibers, NDF (80.27–83.82%), ADF (36.04–38.96%), ADL (1.86–6.00%), hemicellulose (41.31–46.11%) and cellulose (32.96–35.32%) contents, compared to those in OF (15.83–32.36%, 3.11–3.65%, 0.89–1.29%, 12.72–28.71%, 2.22–2.45%, respectively), which can be explained by the fact that OHs consist mainly of lignocellulosic fiber.

These results are in accordance with the previous research of de Oliveira et al. [66] where OHs contained 40.1% cellulose, 25.1% hemicellulose, 26.1% lignin, and 8.7% ash. Furthermore, Neitzel et al. [54] reported that OHs contained 66.19% holocellulose, 29.80% α-cellulose, and 25.44% lignin. Neitzel et al. [54] reported that OHs contained 66.19% holocellulose (total polysaccharide fraction after removal of extractives and lignin), 29.80% α-cellulose, and 25.44% lignin. A previous study by Žilić et al. [67] found that hull-less oat grain contained on average 13.11% hemicellulose, 1.41% cellulose, 15.16% NDF, 2.04% ADF, and 0.98% lignin, which follows the results of our study. Cell wall components such as cellulose, hemicellulose, lignin, and silica make up the NDF fraction. Since lignin is completely indigestible and its presence decreases the availability of the plant material's cellulose and hemicellulose components, it is significant from a nutritional aspect [68].

Because OH contains considerable amounts of fiber, most research on OH valorization has been on using lignocellulose for the manufacture of biofuel or animal feed [23].

However, recent studies have demonstrated that OHs can find different applications in the food industry. For example, given that micronized OHs are a food byproduct rich in dietary fiber and polyphenols with antioxidant qualities, they can be used as an additive in bread making to improve the nutritional and textural properties of gluten-free bread [69].

The OF samples had a higher β-glucan content (from 4.07% to 5.33%) than those in OHs (from 0.03 to 0.06%). Majumdar et al. [70] reported levels of β-glucan in three oat lines ranging from 4.43% to 6.46%, which is in accordance with our results. Oat β-glucans are soluble dietary fibers that can be consumed by gut microbiota in the colon, leading to the production of short-chain fatty acid (SCFA) metabolites [71]. The European Commission approved a health claim for oat β-glucan in 2011 based on the European Food Safety Authority's scientific conclusion that consuming oats lowers postprandial glycemia [72].

Resistant starch is not digested in the small intestine but can instead be utilized in the colon by gut microbiota [73]. Resistant starch in oats, accounting for 29.31% of the starch content in raw granular form, can modulate blood glucose and contribute to the food glycemic index value [71]. The resistant starch in OFs ranged between 2.43% in brown oat grains and 2.95% in yellow oat grains, while its presence was not detected in the OH samples. However, Xia et al. [74] reported a slightly higher content in the untreated whole-grain oats, which amounted to 5.31%. According to Zhu et al. [55], whole oats samples comprised 4.72% resistant starch, and 13.53% total dietary fiber, while 5.08% of the whole-grain oats were β-glucan.

3.2. Phenolic Compounds and Antioxidant Capacity

The contents of phenolic compounds, phenolic acids, and antioxidant capacities of the investigated oat samples are given in Table 3. According to our study, the content of total phenolic compounds determined in the OF samples of the oat genotypes grown in Serbia was close to the content measured in grains of oat genotypes grown in the Czech Republic, as reported by Alemayehu et al. [75]. These authors studied the content of total phenolics in dehulled oat grains of genotypes originating from nine world countries and reported that the content was the highest (1688.0–2016.0 µg GAE/g) in the grains of oat genotypes originating from India. In addition, according to the results obtained in our study, OHs had 13 to 25 times higher contents of total insoluble-bound phenolic compounds (12,086.76 to 24,352.48 µg GAE/g d.m.) compared to OF (841.89–982.08 µg GAE/g d.m.). A previous study by Žilić et al. [21] reported a low content of soluble free phenolic compounds in the grain of the four analyzed standard yellow-colored hull-less oat genotypes. Varga et al. [76] also reported that OHs of twenty differently colored oat genotypes had higher both soluble free and insoluble-bound phenolic compounds than groats, as well as that the amount of bound phenolics was ten times higher than that of free phenolics. On the other hand, Emmons et al. [77] reported that the average content of total phenolic compounds in OF and OHs was not significantly different. There have been reports of various cereal species hierarchies based on their levels of total phenolic and antioxidant activity [21,78]. In addition to genotype, the antioxidant properties were found to be influenced by location and genotype x location interactions [76]. Conversely, Emmons et al. [79] reported that there were no significant differences in the content of total free phenolic compounds, *p*-coumaric, and ferulic acid in three oat genotypes grown in seven different locations, unlike the total antioxidant capacity and content of other phenolic compounds, which differed significantly. Five phenolic acids were detected in oat samples, *p*-coumaric, ferulic, isoferulic, vanillic, and syringic acid. In the study by Žilić et al. [21], a high content of soluble free caffeic acid was measured in the grains of hull-less oat genotypes. According to the findings of Varga et al. [76], the OHs and grain of the tested genotypes contained a total of 28 soluble phenolics, including phenolic aldehydes, benzoic acids, hydroxycinnamic acids, mono- and dihydroxycinnamoyl glycerol esters, and avenanthramides. The most prevalent phenolic aldehyde in the OHs and grain reported by Emmons et al. [77] was vanillin. In our study, the content of vanillic acid, an oxidized form of vanillin, ranged from 16.66 μg/g in black OF to 23.05 μg/g in yellow OF, and from 101.44 μg/g in yellow OHs to 619.38 μg/g in black OHs. In general, the phenolic acids content was higher in the OHs, especially of the brown and black genotypes, except for the *p*-coumaric acid, where the content was significantly higher in the yellow oat genotype's OHs. The *p*-coumaric acid content in the OHs was about 74, 34, and 33 times higher in the yellow, brown, and black OHs than in the OF, respectively. Ferulic acid was predominant, in both the OF (395.88 to 589.14 μg/g d.m.) and OHs (4987.02 to 13,794.82 μg/g d.m.). Ferulic acid is one of the phenolic acids that play a crucial part in anti-inflammatory processes through cyclooxygenase inhibition, both tumor necrosis factor-alpha and prostaglandin E2, as shown by the improved arterial endothelial function seen in both in vitro [80] and in vivo. Additionally, ferulic acid possesses gut-modulating qualities supporting its overall ability to promote health [26]. Compared to *p*-coumaric acid, especially ferulic acid, the content of isoferulic acid and syringic acid was low in all the tested samples.

Table 3. Content of phenolic compounds and antioxidant capacity of OF and OHs.

Sample	TPC (μg GAE/g)	*p*-Coumaric Acid (μg/g)	Ferulic Acid (μg/g)	Isoferulic Acid (μg/g)	Vanillic Acid (μg/g)	Syringic Acid (μg/g)	TAC (mmol Trolox Eq/kg)
OF							
Yellow	875.92 ± 11.73 ^{d}	90.96 ± 8.98 ^{c}	395.88 ± 13.72 ^{d}	n.d.	23.05 ± 0.48 ^{d}	18.14 ± 0.83 ^{c}	23.85 ± 1.46 ^{cd}
Brown	841.89 ± 54.16 ^{d}	82.1 ± 8.67 ^{c}	448.72 ± 20.65 ^{d}	8.35 ± 1.07 ^{a}	22.74 ± 0.95 ^{d}	19.76 ± 1.34 ^{c}	24.28 ± 1.36 ^{c}
Black	982.08 ± 43.69 ^{d}	84.40 ± 5.10 ^{c}	589.14 ± 13.01 ^{d}	10.96 ± 0.67 ^{a}	16.66 ± 0.72 ^{d}	19.47 ± 0.79 ^{c}	22.05 ± 1.01 ^{d}
OHs							
Yellow	12,086.76 ± 259.41 ^{b}	6732.36 ± 325.83 ^{a}	4987.02 ± 110.36 ^{c}	29.11 ± 0.94 ^{a}	101.44 ± 5.53 ^{c}	65.54 ± 0.73 ^{b}	37.32 ± 2.03 ^{b}
Brown	21,971.52 ± 890.64 ^{a}	2749.63 ± 30.12 ^{b}	13,794.82 ± 122.57 ^{a}	49.85 ± 57.29 ^{a}	442.62 ± 0.97 ^{b}	92.79 ± 5.36 ^{a}	48.19 ± 0.11 ^{a}
Black	24,352.48 ± 528.41 ^{c}	2726.76 ± 60.14 ^{b}	13,271.2 ± 59.1 ^{b}	70.59 ± 6.55 ^{a}	619.38 ± 8.26 ^{a}	63.25 ± 6.85 ^{b}	46.64 ± 1.01 ^{a}

OF—oat flour; OHs—oat husks; Means followed by the same letter within the same column are not significantly different ($p < 0.05$). TPC—total phenolic compounds; TAC—total antioxidant capacity. Values are means of three determinations ± standard deviation. Means followed by the same letter within the same column are not significantly different ($p < 0.05$).

The antioxidant capacity of the samples investigated in our study was higher in OHs, ranging from 42.31 mmol Trolox/kg d.m. in yellow OHs to 53.16 mmol Trolox/kg d.m. in brown OHs, and from 22.61 mmol Trolox/kg d.m. in black OF to 25.06 mmol Trolox/kg d.m. in brown OF (Table 3). The total antioxidant capacities of oat grains determined by different methods and expressed in Trolox equivalents reported by various sources varied between 1.7–3.0 μmol Trolox/g and 32.9–117.9 μmol Trolox/g [75]. Furthermore, in our study, unlike the phenolic compounds, the antioxidant capacity of OHs was 1.6 to 2.5 times higher compared to that of OF. This may indicate the antagonistic effect of the compounds present in the oat.

3.3. Baking Functionality

The properties of oat OF and powdered OHs that impact baking functionality, i.e., gelling properties WSI (water solubility index), WAI (water absorption index), SP (swelling power), and solvent retention capacities (SRC), are presented in Figure 1.

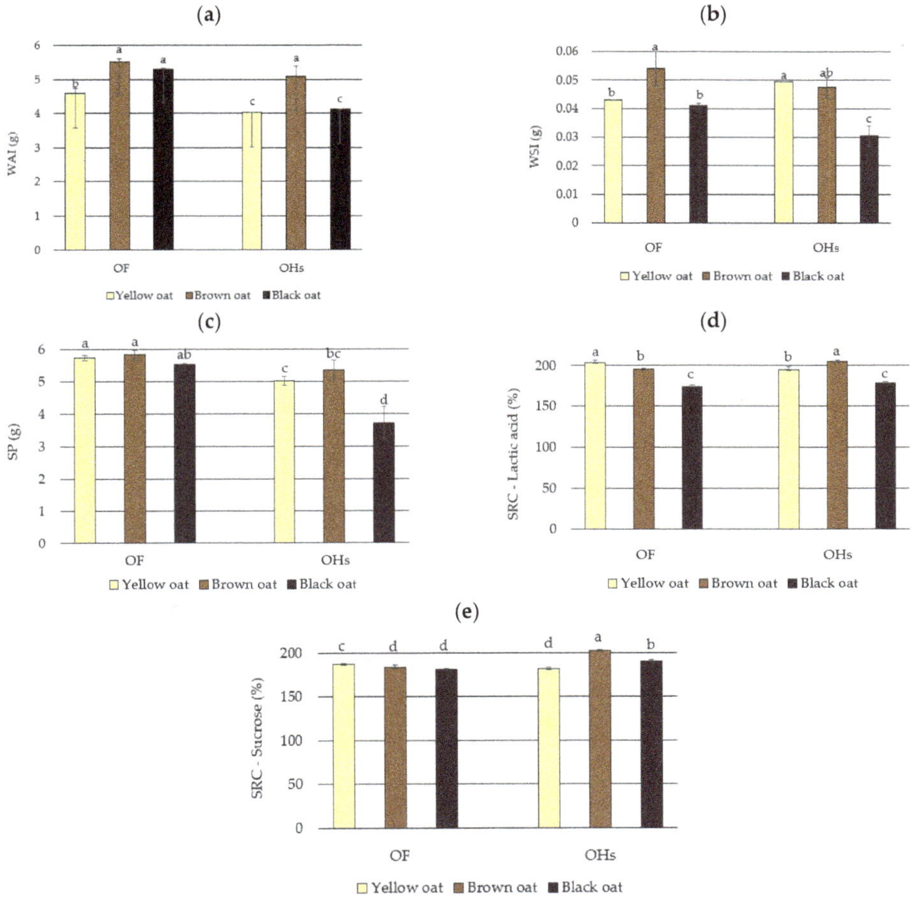

Figure 1. Gelling properties: (**a**) water absorption index; (**b**) water solubility index; (**c**) swelling power, and solvent retention capacity (**d**) lactic acid SRC; (**e**) sucrose SRC of oat flour (OF) and oat husks (OHs). Values are means of three determinations ± standard deviation. Means followed by the same letter within the same column are not significantly different ($p < 0.05$).

The WAI, WSI, and SP varied significantly. The WAI values were respectively lower in OHs than in OF. Except for the yellow OF, which had a lower WSI than the OHs, brown and black OF had a higher WSI than the respective OH. The SP was higher in all the OF than in the OHs. However, all of these gelling properties can be considered exceptional, given that the starting material before water treatment weighed 1 g, and the results indicate that the starting material absorbed 3–5 times more water than its initial mass. A significant correlation between the WAI and starch content (r = 0.646**) can be attributed to the fact that the degradation of the crystalline structure of the starch and the increase in amylose content leaching represent the cause of the maximum water absorption capacity. According to Ibrahim et al. [52], the OF with a higher percentage of polysaccharides has a far higher potential to absorb water. Compared to wheat flour, OF has a better water absorption ratio and dough-softening qualities because of its increased β-glucan content [81]. A positive correlation between the WAI and total protein content (r = 0.591*) follows the statement from the research of Ibrahim et al. [52] that the flour's ability to absorb water is determined by its polar amino acid content. The WAI showed positive correlations with β-glucan (r = 0.556*), and resistant starch contents (r = 0.588*), as well as a negative correlation with the NDF content (r = −0.587*). Furthermore, Ibrahim et al. [52] reported that the five investigated cultivars' water absorption capacities varied greatly, ranging from 173.66 to 188.33%, while Smuda et al. [82] reported a slightly lower water absorption index of oat bran of 115.2%. Varietal differences, variations in soil composition, variations in climate, and variations in agronomic techniques could be the reason why the cultivars differ considerably in terms of water absorption capacity in the study by Ibrahim 2020 [52].

The swelling power manifested positive correlations with the protein content (0.769**), β-glucan (0.695**), starch (0.705**), and resistant starch contents (0.691**), and a strong negative correlation with the NDF fiber content (r = 0.718**). According to a previous study on wheat genotypes conducted by Nikolić et al. [83], the presence of fiber-rich bran in whole-wheat flours had a significant impact on their baking functionality. For instance, sucrose SRC had the highest correlation with the lignin content, whereas the starch content improved absorption capacities for 50% sucrose and 5% lactic acid, respectively.

Since they are a rich source of dietary fiber with high water absorption capacity, OHs are considered a promising component for gluten-free baking [66]. Furthermore, thanks to their high water absorption capacity, utilizing the cellulose fibers from OHs to produce hydrogel and cellulose nanocrystals for food packaging is encouraging because, in addition to the recycling of the cellulose fibers and by-products of grain processing, they can be used to create valuable products that offer distinct attributes [66].

SRC tests have become increasingly popular for assessing the quality of wheat products by analyzing the functionality of flour with four diagnostic SRC solvents. However, their application for the evaluation of oat flour has been scarce until recently. The results of our study showed that the sucrose SRC ranged from 181.06% in black OF to 203.00% in brown OHs, while the lactic acid SRC ranged from 173.88% in black OF to 204.78% in brown OHs. In a study in which 30 oat varieties were analyzed, Zhang et al. [32] found that lactic acid SRC values varied between 76.20% and 154.00% and for sucrose SRC between 68.35% and 90.25%. They also reported that there were correlations between the β-glucan content and lactic acid SRC (0.373*), as well as between the ash content and lactic acid SRC (0.481*), while sucrose SRC did not correlate to any of the OF components. Conversely, the results of our study indicate that the lactic acid showed no significant correlations to the chemical constituents of the oat grain and OH. However, the sucrose SRC was negatively correlated with the starch (r = −0.529*) and resistant starch content (r = −0.511*), β-glucan (r = −0.493*), and total protein (r = −0.467*) content, and positively correlated with the NDF fiber content (r = 0.483*). The discrepancies between the correlations determined by different authors can be attributed to the genotypic differences, as well as the number of samples used in the study. Lactic acid is considered a measure of gluten strength, and is associated with protein quality, and therefore linked to flour's capacity to absorb water [32]. Research by Hammed et al. [84], and Pasha et al. [85] showed that the sucrose SRC ranged from 68.35 to 90.25% and averaged 82.49%, which was less than that of wheat flours (121.9–140.6% and 125.0–163.0%, respectively). The explanation could be that the pentosan content of oats is substantially lower than that of wheat, and sucrose SRC is correlated with pentosan content [32].

3.4. In Vitro Digestibility

The results of the multistep in vitro enzymatic digestibility protocol utilized to determine the potential digestibility of the OF and powdered OH samples are depicted in Figure 2.

The digestive properties of foods derived from oats have been receiving a lot of attention lately. Research suggests that oats, which are inherently slow-digesting cereals, can help control postprandial blood glucose levels with consistent daily consumption [71]. It is evident from the results depicted in Figure 2 that the OF samples had significantly higher digestibility (from 44.54% to 48.24% in yellow and brown OF, respectively) than the powdered OH samples (from 12.02% to 16.69% in black and brown OH, respectively). Lower digestibility was anticipated because of the OH's significantly higher fiber content, as supported by the strong negative correlation (r = −0.966**) between digestibility and NDF content. Furthermore, Schefer et al. [86] reported that phenolic compounds interact with proteins, lipids, and carbohydrate constituents, which further inhibit the digestive enzymes via chemical interactions, causing them to precipitate and reduce their activity in the degradation of carbohydrates. A strong negative correlation between the digestibility and total phenolic compounds determined in our study (r = −0.943) is in accordance with the abovementioned statement. On the other hand, Zhang et al. [87] found that bound phenolics from rice bran dietary fiber were released during simulated gastrointestinal digestion. However, over a quarter were released during colonic fermentation by microbiota stimulating the growth of certain beneficial bacteria during colonic fermentation.

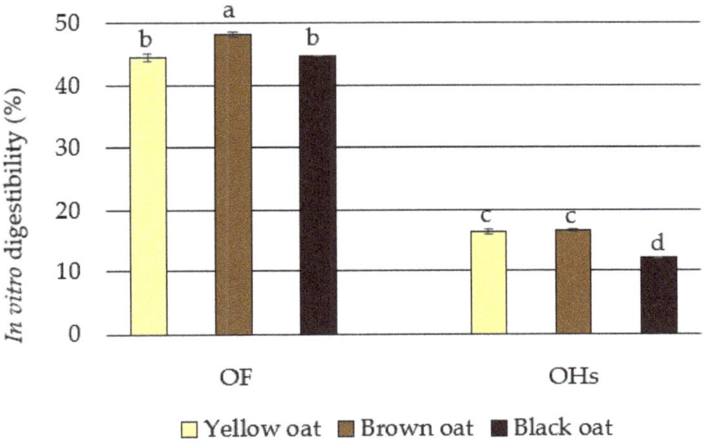

Figure 2. In vitro digestibility of oat flour (OF) and oat husks (OHs). Values are means of three determinations ± standard deviation. Means followed by the same letter within the same column are not significantly different ($p < 0.05$).

Oat β-glucans have a high viscosity, which slows gastric emptying and delays food digestion, reducing glucose release due to starch hydrolysis. This increases chyme viscosity, preventing stomach emptying and affecting glucose transfer to enterocytes [71]. The hypoglycemic effect of β-glucans begins in the stomach's stage of food digestion, where they form extremely viscous solutions in an acidic environment. This slows the digestion of carbohydrates and gastric contents into the duodenum, lowering glucose release and reducing postprandial glycemia and insulin requirement [88]. In our study, a significant negative correlation between the digestibility and the β-glucan (r = −0.869**) supports these statements.

3.5. *Antimicrobial Activity of the Oat Extracts*

Traditional Serbian medicine uses oat decoction to treat various digestive tract issues, including diarrhea, pancreatitis, and excess stomach acid [15]. The abovementioned conditions could represent symptoms or possible complications of bacterial infections. For this reason, the antimicrobial effect of the oat extract samples was investigated on six important pathogens of the gastrointestinal tract (*Listeria monocytogenes*, *Staphylococcus haemolyticus*, *Salmonella typhimurium*, *Enterococcus faecalis*, *Escherichia coli*, and *Shigella flexneri*). Extracts obtained after hydrolysis and samples prepared according to the traditional recipe were tested (Table 4).

Table 4. Minimal bactericidal concentration (MBC) of hydrolyzed and traditionally prepared oat extracts expressed in microliters of the preparation.

Strain	Yellow		Brown		Black	
	OF	OHs	OF	OHs	OF	OHs
Hydrolyzed samples						
Listeria monocytogenes	<140	<140	200	160	180	140
Staphylococcus haemolyticus	140	140	>200	>200	>200	>200
Salmonella typhimurium	>200	180	>200	200	>200	>200
Enterococcus faecalis	140	140	>200	180	180	>200
Escherichia coli	140	140	200	140	140	140
Shigella flexneri	160	150	>200	180	>200	>200
Traditionally prepared samples						
Listeria monocytogenes	>200	>200	>200	160	220	150
Staphylococcus haemolyticus	180	150	>200	150	>200	150
Salmonella typhimurium	>200	>200	>200	200	>200	200
Enterococcus faecalis	>200	200	>200	180	>200	180
Escherichia coli	180	180	220	160	180	160
Shigella flexneri	220	180	>200	180	>200	160

OF—oat flour; OHs—oat husks.

The samples obtained by hydrolysis showed a higher bactericidal activity, which can be associated with a higher content of polyphenols. The extracts obtained from the yellow OF and OHs showed the highest antimicrobial activity against the majority of the investigated bacterial strains. The most sensitive bacteria were *L. monocytogenes* and *E. coli*, which were suppressed by almost all samples in amounts less than 200 μL/mL. Moreover, OF and OH extracts of the yellow oat samples obtained after hydrolysis were also active in inhibiting the growth of *E. faecalis* (140 μL/mL for both samples) and *Shigella flexneri* (150 and 160 μL/mL for OH and OF samples, respectively). The least sensitive to the activity of all extracts obtained by hydrolyzation was *S. typhimurium* (Table 4).

As for the samples obtained by the traditional recipe, the strongest bactericidal activity was manifested by brown OF extract and black OH extract, especially against *L. monocytogenes, E. coli,* and *S. haemolyticus*. As in the case of hydrolyzed samples, *S. typhimurium* was the least sensitive bacterial strain (Table 4).

Alrahmany et al. [89] concluded that there is a direct connection between antimicrobial activity and the presence of phenolics in the bran extracts by showing that oat brans treated with enzymes cellulase, amyloglucosidase, and viscozyme reduced the growth of *E. coli* by 37% compared to controls. A study reported that *Escherichia coli* was susceptible to the antimicrobial action of powdered sprouted oats extracts containing phenolic acids and avenanthramides [90].

Phenolic acids are among the most abundant phenolic components in oats, and the most common among them are cinnamic acid and benzoic acid derivatives [77]. Phenolic acids are mostly bound to proteins, sugars, or fatty acids in oat grains [91]. Chen et al. [92] demonstrated that cellulase treatment increased the availability of the majority of the phenolic compounds from oat bran, while heating-only treatment had no significant influence. On the other hand, heat treatment can cause incorporation of phenolic acids and other polyphenolic compounds into high molecular weight polymers called melanoidins, which are formed as a result of Maillard reaction [93]. In our study, despite the fact that the content of phenolic acids in traditionally prepared extracts was significantly lower than in the hydrolyzed ones, their antimicrobial effect was almost comparable for the majority of the studied bacterial strains. This indicated that some other compounds, in addition to phenolic acids, contributed to the demonstrated effects. Previous studies suggested that Maillard reaction products such as melanoidins also possess antimicrobial effects [94,95]. Kukuminato et al. [94] showed that the antibacterial effect of some melanoidins is comparable to that of nisin solution in a concentration 350 IU/mL. Rurian-Henares and Morales [96] pointed out that water-soluble melanoidins eliminated pathogenic bacteria strains (*E. coli*) by generating irreversible changes in both the inner and outer membranes.

The antimicrobial activity of phenolic acids is presented in a number of studies and is connected with their ability to interact with cell membranes or to cross them. Campos et al. [97] showed that phenolics could damage cell membranes causing the efflux of cellular components. Furthermore, some studies reported that phenolic acids can cross cell membranes of *E. coli* and exert their antimicrobial activity inside the cytoplasm [98,99].

The best antimicrobial activity of yellow OH extract, obtained by hydrolysis, can be explained by the high share of phenolic acids, among all the analyzed oat samples, for which strong antimicrobial activity has been proven in numerous earlier studies. Among them, *p*-coumaric acid, ferulic acid, caffeic acid, and vanillic acid are the most prevalent (Table 4). The high contennt of ferulic, *p*-coumaric and vanillic acids was also reflected in the good antimicrobial activity of black and brown OHs, prepared according to the traditional recipe.

Many studies have proven the high antimicrobial activity of phenolic acids concerning bacteria that cause food spoilage and gastrointestinal problems. Borges et al. [100] found that ferulic acid had antimicrobial activity against *Escherichia coli*, *Listeria monocytogenes*, *Staphylococcus aureus*, and *Pseudomonas aeruginosa*. The results suggest that ferulic acid compromises the integrity of the cytoplasmic membrane and leads to irreversible changes in membrane properties (charge, intra and extracellular permeability, and physicochemical properties) through hydrophobicity changes, a decrease in the negative surface charge, and occurrence of local rupture or pore formation in the cell membranes, with consequent leakage of essential intracellular constituents. It is possible to observe that the percentage of cells with membrane damage considerably increased with the phenolic acid concentration. As ferulic acid, along with *p*-coumaric acid, is the most abundant in the most active OH samples in our tests, it can be concluded that its synergistic activity, along with other phenolic components, contributed to its good antimicrobial activity.

The *p*-coumaric acid, the most abundant in black and brown OHs, slightly less in the OHs of the yellow oat variety, has shown considerable antimicrobial activity through a dual mechanism of action. The results of Lou et al. [101] showed that *p*-coumaric acid effectively inhibited the growth of all test bacterial pathogens, and the MIC values ranged from 10 to 80 μg/mL. The *p*-coumaric acid was tested for antibacterial potential against Gram-positive (*S. pneumoniae* ATCC49619, *B. subtilis* 9372, *S. aureus* 6538) and Gram-negative bacteria (*S. dysenteriae* 51302, *E. coli* ATCC25922, *Salmonella typhimurium* 50013). These studies further showed that *p*-coumaric acid significantly increased the outer and plasma membrane permeability, resulting in the loss of the barrier function. The results demonstrate that *p*-coumaric acid has dual mechanisms of bactericidal activity: disrupting bacterial cell membranes and binding to bacterial genomic DNA to inhibit cellular functions, ultimately leading to cell death (apoptosis). Ojha and Patil [102] revealed that *p*-coumaric acid has the ability to inhibit the DNA repair mechanisms of *L. monocytogenes* by in vivo inhibition of its inducer—RecA expresion The results obtained in our study are in correlation with the research presented because the highest inhibitory activity on the growth of pathogenic bacteria was shown by OH samples in which *p*-coumaric acid, along with ferulic acid, is the most abundant.

Very important information was provided by the research, in which it was proven that the addition of *p*-coumaic acid to a probiotic may enhance the probiotic properties of bacteria such as *Lactobacillus acidophilus* LA-5 and *Lacticaseibacillus rhamnosus* GG [103]. It was also confirmed that this phenolic acid has a protective effect on colon pretumors induced by 1,2 dimethylhydrazine (DMH) in rats [104].

Moreover, many studies have proven the distinct antimicrobial activity of other phenolic acids that are slightly less abundant in our samples than the first two described, such as vanillic acid, caffeic acid, and 3,4-dihydrohybenzoic acid. Syafni et al. [105] have shown significant and dose-dependent inhibition activity of extracts and fractions of leaves, stems, and roots of *T. cinense* L. and their dominant compounds 3,4-dihydrohybenzoic acid and 3,4-dihydrohybenzaldehyde toward *E. coli* 25922, *Staphylococcus aureus* ATCC 25923, and *Vibrio cholera* Inaba. The experiments also revealed that the extracts and isolated 3,4-dihydrohybenzoic acid and 3,4-dihydrohybenzaldehyde are also active against *Salmonella thypimurium* ATCC 14028 NCTCC 12023, which caused not only diarrhea but also salmonellosis with fever and abdominal cramps as additional symptoms. These results support the traditional use of this plant and its two dominant components in treatment of diarrhea. Although 3,4-dihydrobenzoic acid was present only in traditionally prepared samples, the abovementioned results from previous studies support our outcomes in terms of the possible application of oat OHs in the treatment of gastrointestinal issues induced by, amongst others, bacterial infections.

Previous studies have shown that vanillic acid, the most abundant in black and brown OHs, and OHs of the yellow oat variety, has proved its efficacy as an antimicrobial compound against foodborne bacteria such as *Staphylococcus aureus*, *Escherichia coli*, and carbapenem-resistant *Enterobacter cloacae*, *Enterobacter homaechei*, *Sarcina* spp., and *Candida albicans* [106]. The existing literature indicates that vanillic acid can disrupt the biofilm formation of clinical pathogens. Additionally, disruption of cell membrane integrity by vanillic acid may result in the formation of pores and leakage of intracellular components into the extracellular space [107]. Moreover, Ingole et al. [108] indicated that vanillic acid could help patients with ulcerative colitis caused by dextran sulfate sodium (DSS), by reducing the severity of clinical symptoms, which justifies its use as a flavoring agent.

Prior investigations have demonstrated that the antimicrobial activity of the plant extracts is frequently based on the synergistic activity of both dominant and the less abundant compounds. Ijabadeniyi et al. [109] evaluated the antimicrobial effects of two phenolic acids (caffeic and ferulic) against foodborne pathogens, *Escherichia coli* O157:H7 ATCC 43888 and *Listeria monocytogenes* ATCC 7644, in ready-to-eat meats that are susceptible to pathogenic contamination during their production, distribution, and sale. These results indicate the ability of caffeic and ferulic acids, individually and in combination, to reduce pathogenic contamination and improve the safety of cold-cut meats. The combination of caffeic acid and ferulic acid in a 1:1 ratio demonstrated a synergistic effect. It was also reported that caffeic acid showed antimicrobial potential and/or synergistic effects with antibiotics against *S. aureus*, *S. epidermidis*, *K. pneumoniae*, *S. marcescens*, *P. mirabilis*, *E. coli*, *P. aeruginosa*, *B. cereus*, *M. luteus*, *L. monocytogenes*, and *C. albicans* strains [110]. When combined, phenolic acids could reflect considerably higher antimicrobial efficacy against foodborne pathogens compared to individual phenolic acids [111].

The beneficial effects of oats consumption on the conditions and diseases of the digestive tract have also been manifested by other mechanisms. Seether et al. [112] showed that oat OH alcohol extract has the potential to inhibit gastric acid secretion in pigs. Moreover, some physiological benefits of oatmeal were reported, such as reducing hypercholesterolemia, hyperglycemia, hyperinsulinemia, and hypertension. Those activities are often attributed to β-glucans and phenolics [9].

Based on our findings and all of the above facts, we can conclude that there is the potential for developing oat-based products that are beneficial to the digestive tract.

4. Conclusions

This study investigated the nutritional and bioactive properties of oat flour (OF) and ground oat husks (OHs) from three differently colored oat genotypes. Both OF and OHs were rich in essential nutrients, particularly protein, lipids, and dietary fiber. The phenolic profile analysis revealed a significant presence of phenolic acids, primarily ferulic acid, in both OF and OHs. OHs had 13 to 25 times higher contents of total insoluble-bound phenolic compounds. These bioactive compounds may contribute to the strong antioxidant and antimicrobial activities observed in the extracts. Notably, OH extracts exhibited superior antimicrobial properties against several gastrointestinal pathogens, *L. monocytogenes*, *E. coli*, and *S. haemolyticus*. These findings highlight the potential of OF and OHs as valuable ingredients in functional foods and nutraceuticals. The high nutritional content and the presence of bioactive compounds suggest that both the main constituent—oat grain as well as the byproduct—husk can offer numerous health benefits, such as improved gut health, reduced oxidative stress, and enhanced immunity. However, further research is needed to fully elucidate the mechanisms underlying these effects and to optimize the extraction procedures and formulation of oat-based products.

Author Contributions: Conceptualization, V.N. and S.Ž.; methodology, S.Ž. and K.Š.; investigation, M.S., B.S. and D.M.; data analysis, formal analysis, T.S., J.Ž., M.S., B.S. and V.N.; writing—original draft preparation, writing—review and editing, S.Ž. and K.Š.; supervision, S.Ž.; resources, V.K.R. All authors have read and agreed to the published version of the manuscript.

Funding: This study received no external funding.

Data Availability Statement: The original data presented in the study are openly available in RIK at https://rik.mrizp.rs/?locale-attribute=sr_RS accessed on 22 November 2024.

Acknowledgments: This study was supported by the Ministry of Science, Technological Development and Innovation of the Republic of Serbia (Grant No. 451-03-66/2024-03/200040 and 451-03-66/2024-03/200003).

Conflicts of Interest: The authors declare that they have no competing interests.

References

1. Batîr Rusu, D.C.; Murariu, D.; Gheorghita, R.; Graur, M. Some nutritional value aspects of barley and oat and their impact in human nutrition and healthy life. *Plants* **2024**, *13*, 2764. [CrossRef] [PubMed]
2. Venn, B.J.; Mann, J.I. Cereal grains, legumes and diabetes. *Eur. J. Clin. Nutr.* **2004**, *58*, 1443–1461. [CrossRef] [PubMed]
3. Anderson, J.W. Whole grains protect against atherosclerotic cardiovascular disease. *Proc. Nutr. Soc.* **2003**, *62*, 135–142. [CrossRef] [PubMed]
4. Slavin, J.L. Dietary fiber and body weight. *Nutrition* **2005**, *21*, 411–418. [CrossRef] [PubMed]
5. Slavin, J. Why whole grains are protective: Biological mechanisms. *Proc. Nutr. Soc.* **2003**, *62*, 129–134. [CrossRef]
6. Singh, R.; De, S.; Belkheir, A. *Avena sativa* (Oat), A potential neutraceutical and therapeutic agent: An overview. *Crit. Rev. Food Sci. Nutr.* **2012**, *53*, 126–144. [CrossRef]
7. Huang, Z.; Feng, W.; Zhang, T.; Miao, M. Structure and functional characteristics of starch from different hulled oats cultivated in China. *Carbohydr. Polym.* **2024**, *330*, 121791. [CrossRef]
8. Jágr, M.; Hofinger-Horvath, A.; Ergang, P.; Čepková, P.H.; Schönlechner, R.; Pichler, E.C.; D'Amico, S.; Grausgruber, H.; Vagnerová, K.; Dvořáček, V. Comprehensive study of the effect of oat grain germination on the content of avenanthramides. *Food Chem.* **2024**, *437*, 137807. [CrossRef]
9. Quiñones-Muñoz, T.A.; Villanueva-Rodríguez, S.J.; Torruco-Uco, J.G. Nutraceutical properties of *Medicago sativa* L., *Agave* spp., *Zea mays* L. and *Avena sativa* L. A review of metabolites and mechanisms. *Metabolites* **2022**, *12*, 806. [CrossRef]
10. Blumenthal, M.; Goldberg, A.; Brinckmann, J. *Herbal Medicine: Expanded Commission E Monographs*; Integrative Medicine Communications: Boston, MA, USA, 2000; p. xiii+519.
11. Abascal, K.; Yarnell, E. Nervine herbs for treating anxiety. *Altern. Complement. Ther.* **2004**, *10*, 309–315. [CrossRef]
12. Bi, Y.; Gao, F.; Guo, J.; Yao, X.; Wang, A.; Liu, H.; Sun, Y.; Yao, R.; Li, M. An ethnobotanical survey on the medicinal and edible plants used by the Daur people in China. *J. Ethnobiol. Ethnomed.* **2024**, *20*, 55. [CrossRef] [PubMed]
13. Sargin, S.A. Plants used against obesity in Turkish folk medicine: A review. *J. Ethnopharmacol.* **2021**, *270*, 113841. [CrossRef] [PubMed]
14. Mazzei, R.; De Marco, E.V.; Gallo, O.; Tagarelli, G. Italian folk plant-based remedies to heal headache (XIX–XX century). *J. Ethnopharmacol.* **2018**, *210*, 417–433. [CrossRef] [PubMed]
15. Tucakov, J. Ovas—*Avena sativa*. In *Lečenje Biljem, New, Amended and Supplemented Edition in Serbian;* Vulkan Izdavaštvo: Belgrade, Serbia, 2014; p. 439.
16. Rodríguez, J.M.; Estévez, V.; Bascuñán, K.; Ayala, J.; Araya, M. Commercial oats in gluten-free diet: A persistent risk for celiac patients. *Front. Nutr.* **2022**, *9*, 986282. [CrossRef]
17. Rostamabadi, H.; Karaca, A.C.; Deng, L.; Colussi, R.; Narita, I.M.P.; Kaur, K.; Aaliya, B.; Sunooj, K.V.; Falsafi, S.R. Oat starch—How physical and chemical modifications affect the physicochemical attributes and digestibility? *Carbohydr. Polym.* **2022**, *296*, 119931. [CrossRef]
18. Sunilkumar, B.A.; Leonova, S.; Oste, R.; Olsson, O. Identification and characterization of high protein oat lines from a mutagenized oat population. *J. Cereal Sci.* **2017**, *75*, 100–107. [CrossRef]
19. Bai, J.; Zhang, C.; Cui, J.; Fan, M.; Li, Y.; Huang, L.; Wang, L. Analysis of the effect of oat β-glucan on gut microbiota and clarification of their interaction relationship. *J. Cereal Sci.* **2024**, *117*, 103921. [CrossRef]

20. Guo, H.; Li, H.-Y.; Liu, L.; Wu, C.-Y.; Liu, H.; Zhao, L.; Zhang, Q.; Liu, Y.-T.; Li, S.-Q.; Qin, W.; et al. Effects of sulfated modification on the physicochemical properties and biological activities of β-glucans from *Qingke* (Tibetan hulless barley). *Int. J. Biol. Macromol.* **2019**, *141*, 41–50. [CrossRef]
21. Žilić, S.; Šukalović, V.H.T.; Dodig, D.; Maksimović, V.; Maksimović, M.; Basić, Z. Antioxidant activity of small grain cereals caused by phenolics and lipid soluble antioxidants. *J. Cereal Sci.* **2011**, *54*, 417–424. [CrossRef]
22. Schmitz, E.; Karlsson, E.N.; Adlercreutz, P. Ultrasound assisted alkaline pretreatment efficiently solubilises hemicellulose from oat hulls. *Waste Biomass Valorization* **2021**, *12*, 5371–5381. [CrossRef]
23. Zhou, Y.; Tian, Y.; Beltrame, G.; Laaksonen, O.; Yang, B. Ultrasonication-assisted enzymatic bioprocessing as a green method for valorizing oat hulls. *Food Chem.* **2023**, *426*, 136658. [CrossRef] [PubMed]
24. Savas, B.S.; Akan, E. Oat bran fortified raspberry probiotic dairy drinks: Physicochemical, textural, microbiologic properties, in vitro bioaccessibility of antioxidants and polyphenols. *Food Biosci.* **2021**, *43*, 101223. [CrossRef]
25. Perez-Ternero, C.; de Sotomayor, M.A.; Herrera, M.D. Contribution of ferulic acid, γ-oryzanol and tocotrienols to the cardiometabolic protective effects of rice bran. *J. Funct. Foods* **2017**, *32*, 58–71. [CrossRef]
26. Nemes, S.A.; Fărcas, A.C.; Ranga, F.; Teleky, B.E.; Călinoiu, L.F.; Dulf, F.V.; Vodnar, D.C. Enhancing phenolic and lipid compound production in oat bran via acid pretreatment and solid-state fermentation with *Aspergillus niger*. *New Biotechnol.* **2024**, *83*, 91–100. [CrossRef]
27. Tian, W.; Hu, R.; Chen, G.; Zhang, Y.; Wang, W.; Li, Y. Potential bioaccessibility of phenolic acids in whole wheat products during in vitro gastrointestinal digestion and probiotic fermentation. *Food Chem.* **2021**, *362*, 130135. [CrossRef]
28. Spinola, V.; Llorent-Martinez, E.J.; Castilho, P.C. Polyphenols of *Myrica faya* inhibit key enzymes linked to type II diabetes and obesity and formation of advanced glycation end-products (in vitro): Potential role in the prevention of diabetic complications. *Food Res. Int.* **2019**, *116*, 1229–1238. [CrossRef]
29. Ozdal, T.; Sela, D.A.; Xiao, J.; Boyacioglu, D.; Chen, F.; Capanoglu, E. The reciprocal interactions between polyphenols and gut microbiota and effects on bioaccessibility. *Nutrients* **2016**, *8*, 78. [CrossRef]
30. Dong, L.; Qin, C.; Li, Y.; Wu, Z.; Liu, L. Oat phenolic compounds regulate metabolic syndrome in high fat diet-fed mice via gut microbiota. *Food Biosci.* **2022**, *50*, 101946. [CrossRef]
31. Moorthy, M.; Chaiyakunapruk, N.; Jacob, S.A.; Palanisamy, U.D. Prebiotic potential of polyphenols, its effect on gut microbiota and anthropometric/clinical markers: A systematic review of randomised controlled trials. *Trends Food Sci. Technol.* **2020**, *99*, 634–649. [CrossRef]
32. Zhang, Y.; Li, Y.; Ren, X.; Zhang, X.; Wu, Z.; Liu, L. The positive correlation of antioxidant activity and prebiotic effect about oat phenolic compounds. *Food Chem.* **2023**, *402*, 134231. [CrossRef]
33. *ISO 10520*; Determination of Starch Content-Ewers Polarimetric Method. ISO—International Organization for Standardization: Geneva, Switzerland, 1997.
34. Horwitz, W.; Latimer, G.W. *Official Methods of Analysis of AOAC International*, 18th ed.; AOAC International: Gaithersburg, MD, USA, 2005.
35. Horwitz, W. *Official Methods of Analysis of AOAC International*, 17th ed.; AOAC International: Gaithersburg, MD, USA, 2000.
36. Lookhart, G.; Bean, S. Separation and characterization of wheat protein fractions by high-performance capillary electrophoresis. *Cereal Chem.* **1995**, *72*, 527–532.
37. Žilić, S.; Milašinović, M.; Terzić, D.; Barać, M.; Ignjatović Micić, D. Grain characteristics and composition of maize specialty hybrids. *Span. J. Agric. Res.* **2011**, *9*, 230–241. [CrossRef]
38. Megazyme, 2023. β-Glucan (Mixed Linkage) Assay Protocol K-BGLU. Available online: https://d1kkimny8vk5e2.cloudfront.net/documents/Assay_Protocol/K-BGLU_DATA.pdf (accessed on 1 November 2024).
39. Megazyme, 2023a. Resistant Starch (Rapid) Assay Protocol K-RAPRS. Available online: https://d1kkimny8vk5e2.cloudfront.net/documents/Assay_Protocol/K-RAPRS_DATA.pdf (accessed on 1 November 2024).
40. Žilić, S.; Akıllıoğlu, H.G.; Serpen, A.; Perić, V.; Gökmen, V. Comparisons of phenolic compounds, isoflavones, antioxidant capacity and oxidative enzymes in yellow and black soybeans seed coat and dehulled bean. *Eur. Food Res. Technol.* **2013**, *237*, 409–418. [CrossRef]
41. Singleton, V.L.; Orthofer, R.; Lamuela-Raventós, R.M. Analysis of total phenols and other oxidation substrates and antioxidants by means of Folin-Ciocalteu reagent. *Methods Enzymol.* **1999**, *299*, 152–178. [CrossRef]
42. Žilić, S.; Serpen, A.; Akillioglu, G.; Gökmen, V.; Vančetović, J. Phenolic compounds, carotenoids, anthocyanins, and antioxidant capacity of colored maize (*Zea mays* L.) kernels. *J. Agric. Food Chem.* **2012**, *60*, 1224–1231. [CrossRef]
43. Serpen, A.; Gökmen, V.; Pellegrini, N.; Fogliano, V. Direct measurement of the total antioxidant capacity of cereal products. *J. Cereal Sci.* **2008**, *48*, 816–820. [CrossRef]
44. Cornejo, F.; Rosell, C.M. Physicochemical properties of long rice grain varieties in relation to gluten free bread quality. *LWT-Food Sci. Technol.* **2015**, *62*, 1203–1210. [CrossRef]
45. Haynes, L.C.; Bettge, A.D.; Slade, L. Soft wheat and flour products methods review: Solvent retention capacity equation correction. *Cereal Foods World* **2009**, *54*, 174–175. [CrossRef]
46. Papillo, V.A.; Vitaglione, P.; Graziani, G.; Gokmen, V.; Fogliano, V. Release of antioxidant capacity from five plant foods during a multistep enzymatic digestion protocol. *J. Agric. Food Chem.* **2014**, *62*, 4119–4126. [CrossRef]

47. Hamzalıoğlu, A.; Gökmen, V. Formation and elimination reactions of 5-hydroxymethylfurfural during in vitro digestion of 635 biscuits. *Food Res. Int.* **2017**, *99*, 308–314. [CrossRef]
48. Mao, H.; Xu, M.; Ji, J.; Zhou, M.; Li, H.; Wen, Y.; Wang, J.; Sun, B. The utilization of oat for the production of wholegrain foods: Processing technology and products. *Food Front.* **2022**, *3*, 28–45. [CrossRef]
49. Kaur, P.; Kaur, K.; Basha, S.J.; Kennedy, J.F. Current trends in the preparation, characterization and applications of oat starch—A review. *Int. J. Biol. Macromol.* **2022**, *212*, 172–181. [CrossRef] [PubMed]
50. Li, N.; Huang, Y.; Zhao, Y.; Yang, Z.; Jia, Q.; Feng, B.; David, C.; Taylor, D.C.; Du, C.; Zhang, M. Lipidomics studies reveal dynamic changes in polar lipids of developing endosperm of oat and wheat varieties with differing oil contents. *Food Chem.* **2024**, *444*, 138597. [CrossRef] [PubMed]
51. Kouřimská, L.; Pokhrel, K.; Božik, M.; Tilami, S.K.; Horčička, P. Fat content and fatty acid profiles of recently registered varieties of naked and hulled oats with and without husks. *J. Cereal Sci.* **2021**, *99*, 103216. [CrossRef]
52. Ibrahim, M.S.; Ahmad, A.; Sohail, A.; Asad, M.J. Nutritional and functional characterization of different oat (*Avena sativa* L.) cultivars. *Int. J. Food Prop.* **2020**, *23*, 1373–1385. [CrossRef]
53. Usman, S.; Nazir, S.; Ali, S.; Nasreen, Z.; Najim, A. Determination of biochemical composition of *Avena sativa* (oat) and to estimate the effect of high fiber diet on hypercholesterolemic rats. *Bangladesh Res. Publ. J.* **2010**, *4*, 312–319.
54. Neitzel, N.; Eder, M.; Hosseinpourpia, R.; Walther, T.; Adamopoulos, S. Chemical composition, particle geometry, and micromechanical strength of barley husks, oat husks, and wheat bran as alternative raw materials for particleboards. *Mater. Today Commun.* **2023**, *36*, 106602. [CrossRef]
55. Zhu, Y.; Dong, L.; Huang, L.; Shi, Z.; Dong, J.; Yao, Y.; Shen, R. Effects of oat β-glucan, oat resistant starch, and the whole oat flour on insulin resistance, inflammation, and gut microbiota in high-fat-diet-induced type 2 diabetic rats. *J. Funct. Foods* **2020**, *69*, 103939. [CrossRef]
56. Ingvordsen, C.H.; Gislum, R.; Jørgensen, J.R.; Mikkelsen, T.N.; Stockmarr, A.; Jørgensen, R.B. Grain protein concentration and harvestable protein under future climate conditions. A Study of 108 Spring Barley Accessions. *J. Exp. Bot.* **2016**, *67*, 2151–2158. [CrossRef]
57. Miller, S.S.; Fulcher, R.G. *Microstructure and chemistry of the oat kernel, In Oats: Chemistry and Technology*, 2nd ed.; Webster, F.H., Wood, P.J., Eds.; American Association of Cereal Chemists: St. Paul, MN, USA, 2011; pp. 77–94. [CrossRef]
58. Mäkinen, O.E.; Ercili-Cura, D.; Poutanen, K.; Holopainen-Mantila, U.; Lehtinen, P.; Sozer, N. Protein from oat: Structure, processes, functionality, and nutrition. In *Sustainable Protein Sources*; Nadathur, S., Wanasundara, J.P.D., Scanlin, L., Eds.; Academic Press: Cambridge, MA, USA; Elsevier: Amsterdam, The Netherlands, 2024; pp. 121–141. [CrossRef]
59. Shewry, P.R. Avenins: The prolamins of oats. In *Seed Proteins*; Shewry, P.R., Casey, R., Eds.; Springer: Dordrecht, The Netherland, 1999; pp. 79–92. [CrossRef]
60. Wieser, H.; Koehler, P.; Konitzer, K. Gluten–the precipitating factor. In *Celiac Disease and Gluten: Multidisciplinary Challenges and Opportunities*, 1st ed.; Wieser, H., Koehler, P., Konitzer, K., Eds.; Elsevier: Amsterdam, The Netherland, 2014; pp. 97–148. [CrossRef]
61. Kosová, K.; Leišová-Svobodová, L.; Dvořáček, V. Oats as a safe alternative to *Triticeae* cereals for people suffering from celiac disease? A review. *Plant Foods Hum. Nutr.* **2020**, *75*, 131–141. [CrossRef]
62. Ahola, H.G.; Sontag-Strohm, T.S.; Schulman, A.H.; Tanhuanpää, P.; Viitala, S.; Huang, X. Immunochemical analysis of oat avenins in an oat cultivar and landrace collection. *J. Cereal Sci.* **2020**, *95*, 103053. [CrossRef]
63. Sha, X.; Sun, H.; Li, Y.; Wu, J.; Zhang, H.; Yang, R. The prolamins, from structure, property, to the function in encapsulation and delivery of bioactive compounds. *Food Hydrocoll.* **2023**, *149*, 109508. [CrossRef]
64. Zhu, F. Anthocyanin in cereals: Composition and health effects. *Food Res. Int.* **2018**, *109*, 232–249. [CrossRef] [PubMed]
65. Quirós-Sauceda, A.E.; Palafox-Carlos, H.; Sáyago-Ayerdi, S.G.; Ayala-Zavala, J.F.; Bello-Perez, L.A.; Alvarez-Parrilla, E.; de la Rosad, L.A.; González-Córdova, A.F.; González-Aguilar, G.A. Dietary fiber and phenolic compounds as functional ingredients: Interaction and possible effect after ingestion. *Food Funct.* **2014**, *5*, 1063–1072. [CrossRef]
66. de Oliveira, J.P.; Bruni, G.P.; Lima, K.O.; El Halal, S.L.M.; da Rosa, G.S.; Dias, A.R.G.; da Rosa Zavareze, E. Cellulose fibers extracted from rice and oat husks and their application in hydrogel. *Food Chem.* **2017**, *221*, 153–160. [CrossRef]
67. Žilić, S.; Dodig, D.; Milašinović-Šeremešić, M.; Kandić, V.; Kostadinović, M.; Prodanović, S.; Savić, Đ. Small grain cereals compared for dietary fibre and protein contents. *Genetika* **2011**, *43*, 381–395. [CrossRef]
68. Nikolić, V.; Žilić, S.; Radosavljević, M.; Vančetović, J.; Božinović, S. Properties of different silage maize hybrids. *Food Feed Res.* **2020**, *47*, 139–147. [CrossRef]
69. Pichler, E.C.; Schönlechner, R.; Różyło, R.; Dziki, D.; Świeca, M. Addition of micronized oat husk fiber to gluten-free bread-Effects on chemical, physical and physiological properties. *J. Cereal Sci.* **2024**, *118*, 103981. [CrossRef]
70. Majumdar, A.; Gil-González, A.B.; Grau, A.B.; Sardari, R.R.; Larsson, O.; Thyagarajan, A.; Hansson, A.; Hernández-Hernández, O.; Olsson, O.; Zambrano, J.A. Macromolecular characterization of high β-glucan oat lines. *Heliyon* **2024**, *10*, e24552. [CrossRef]
71. Zhang, K.; Dong, R.; Hu, X.; Ren, C.; Li, Y. Oat-based foods: Chemical constituents, glycemic index, and the effect of processing. *Foods* **2021**, *10*, 1304. [CrossRef]

72. EFSA Panel on Dietetic Products, Nutrition and Allergies (NDA). Scientific Opinion on the substantiation of health claims related to beta-glucans from oats and barley and maintenance of normal blood LDL-cholesterol concentrations (ID 1236, 1299), increase in satiety leading to a reduction in energy intake (ID 851, 852), reduction of post-prandial glycaemic responses (ID 821, 824), and "digestive function" (ID 850) pursuant to Article 13 (1) of Regulation (EC) No 1924/2006. *EFSA J.* **2011**, *9*, 2207.
73. Englyst, H.N.; Kingman, S.M.; Cummings, J. Classification and measurement of nutritionally important starch fractions. *Eur. J. Clin. Nutr.* **1992**, *46*, S33–S50. [PubMed]
74. Xia, J.; Zhang, Y.; Huang, K.; Cao, H.; Sun, Q.; Wang, M.; Zhang, S.; Sun, Z.; Guan, X. Different multi-scale structural features of oat resistant starch prepared by ultrasound combined enzymatic hydrolysis affect its digestive properties. *Ultrason. Sonochem.* **2023**, *96*, 106419. [CrossRef] [PubMed]
75. Alemayehu, G.F.; Forsido, S.F.; Tola, Y.B.; Amare, E. Nutritional and phytochemical composition and associated health benefits of oat (*Avena sativa*) grains and oat-based fermented food products. *Sci. World J.* **2023**, *1*, 2730175. [CrossRef]
76. Varga, M.; Jójárt, R.; Fónad, P.; Mihály, R.; Palágyi, A. Phenolic composition and antioxidant activity of colored oats. *Food Chem.* **2018**, *268*, 153–161. [CrossRef]
77. Emmons, C.L.; Peterson, D.M. Antioxidant activity and phenolic contents of oat groats and hulls. *Cereal Chem.* **1999**, *76*, 902–906. [CrossRef]
78. Hodzic, Z.; Pasalic, H.; Memisevic, A.; Srabovic, M.; Saletovic, M.; Poljakovic, M. The influence of total phenols content on antioxidant capacity in the whole grain extracts. *Eur. J. Sci. Res.* **2009**, *28*, 471–477.
79. Emmons, C.L.; Peterson, D.M. Antioxidant activity and phenolic content of oat as affected by cultivar and location. *Crop Sci.* **2001**, *41*, 1676–1681. [CrossRef]
80. Xue, B.; Zhao, B.; Luo, S.; Wu, G.; Hui, X. Inducing apoptosis in human hepatocellular carcinoma cell lines via Nrf2/HO-1 signalling pathway of blueberry and blackcurrant powder manipulated oat bran paste extracts. *J. Funct. Foods* **2022**, *89*, 104967. [CrossRef]
81. Majzoobi, M.; Raiss Jalali, A.; Farahnaky, A. Impact of whole oat flour on dough properties and quality of fresh and stored part baked bread. *J. Food Qual.* **2016**, *39*, 620–626. [CrossRef]
82. Smuda, S.S.; Mohsen, S.M.; Olsen, K.; Aly, M.H. Bioactive compounds and antioxidant activities of some cereal milling by-products. *J. Food Sci. Technol.* **2018**, *55*, 1134–1142. [CrossRef] [PubMed]
83. Nikolić, V.; Simić, M.; Kandić, V.; Dodevska, M.; Titan, P.; Dodig, D.; Žilić, S. Pasting properties and the baking functionality of whole-grain wheat flour with different amylose and dietary fibers content. *J. Food Process. Preserv.* **2022**, *46*, e15805. [CrossRef]
84. Hammed, A.M.; Ozsisli, B.; Ohm, J.; Simsek, S. Relationship between solvent retention capacity and protein molecular weight distribution, quality characteristics, and breadmaking functionality of hard red spring wheat flour. *Cereal Chem.* **2016**, *92*, 466–474. [CrossRef]
85. Pasha, I.; Anjum, F.M.; Butt, M.S. Genotypic variation of spring wheats for solvent retention capacities in relation to end-use quality. *LWT-Food Sci. Technol.* **2009**, *42*, 418–423. [CrossRef]
86. Schefer, S.; Oest, M.; Rohn, S. Interactions between phenolic acids, proteins, and carbohydrates-influence on dough and bread properties. *Foods* **2021**, *10*, 2798. [CrossRef]
87. Zhang, X.; Zhang, M.; Dong, L.; Jia, X.; Liu, L.; Ma, Y.; Huang, F.; Zhang, R. Phytochemical profile, bioactivity, and prebiotic potential of bound phenolics released from rice bran dietary fiber during in vitro gastrointestinal digestion and colonic fermentation. *J. Agric. Food Chem.* **2019**, *67*, 12796–12805. [CrossRef]
88. Leszczyńska, D.; Wirkijowska, A.; Gasiński, A.; Średnicka-Tober, D.; Trafiałek, J.; Kazimierczak, R. Oat and oat processed products—Technology, composition, nutritional value, and health. *Appl. Sci.* **2023**, *13*, 11267. [CrossRef]
89. Alrahmany, R.; Avis, T.J.; Tsopmo, A. Treatment of oat bran with carbohydrases increases soluble phenolic acid content and influences antioxidant and antimicrobial activities. *Food Res. Int.* **2013**, *52*, 568–574. [CrossRef]
90. Lim, S.; Yun, C.-I.; Song, D.; Kang, S.-S.; Kim, Y.-J. Validation of avenanthramide and other phenolic compounds in oats and sprouted oats and their antimicrobial properties against *Escherichia coli* O157:H7. *Food Sci. Biotechnol.* **2022**, *31*, 1145–1155. [CrossRef]
91. Verardo, V.; Serea, C.; Segal, R.; Caboni, M.F. Free and bound minor polar compounds in oats: Different extraction methods and analytical determinations. *J. Cereal Sci.* **2011**, *54*, 211–217. [CrossRef]
92. Chen, D.; Shi, J.; Hu, X. Enhancement of polyphenol content and antioxidant capacity of oat (*Avena nuda* L.) bran by cellulase treatment. *Appl. Biol. Chem.* **2016**, *59*, 397–403. [CrossRef]
93. Brudzynski, K.; Miotto, D. Honey melanoidins: Analysis of the compositions of the high molecular weight melanoidins exhibiting radical-scavenging activity. *Food Chem.* **2011**, *127*, 1023–1030. [CrossRef] [PubMed]
94. Kukuminato, S.; Koyama, K.; Koseki, S. Antibacterial properties of melanoidins produced from various combinations of Maillard reaction against pathogenic bacteria. *Microbiol. Spectr.* **2021**, *9*, e01142-21. [CrossRef] [PubMed]
95. Nooshkam, M.; Varidi, M.; Verma, D.K. Functional and biological properties of Maillard conjugates and their potential application in medical and food: A review. *Food Res. Int.* **2020**, *131*, 109003. [CrossRef]
96. Rurian-Henares, J.A.; Morales, F.J. Antimicrobial activity of melanoidins against *Escherichia coli* is mediated by a membrane-damage mechanism. *J. Agric. Food Chem.* **2008**, *56*, 2357–2362. [CrossRef]
97. Campos, F.M.; Couto, J.A.; Figueiredo, A.R.; Toth, I.V.; Rangel, A.O.S.S.; Hogg, T.A. Cell membrane damage induced by phenolic acids on wine lactic acid bacteria. *Int. J. Food Microbiol.* **2009**, *135*, 144–151. [CrossRef]

98. Ikigai, H.; Nakae, T.; Hara, Y.; Sshmamura, T. Bactericidal catechins damage the lipid bilayer. *Biochim. Biophys. Acta* **1993**, *1147*, 132–136. [CrossRef]
99. Cueva, C.; Victoria Moreno-Arribas, M.; Martin-Alvarez, P.J.; Bills, G.; Francisca Vicente, M.; Basilio, A.; López Rivas, C.; Requena, T.; Rodríguez, J.M.; Bartolomé, B. Antimicrobial activity of phenolic acids against commensal, probiotic and pathogenic bacteria. *Res. J. Microbiol.* **2010**, *161*, 372–382. [CrossRef]
100. Borges, A.; Ferreira, C.; Saavedra, M.; Simoes, M. Antibacterial activity and mode of action of ferulic and gallic acids against pathogenic bacteria. *Microb. Drug Resist.* **2013**, *19*, 256–265. [CrossRef]
101. Lou, Z.; Wang, H.; Rao, S.; Sun, J.; Ma, C.; Li, J. *p*-Coumaric acid kills bacteria through dual damage mechanisms. *Food Control* **2012**, *25*, 550–554. [CrossRef]
102. Ojha, D.; Patil, K.N. *p*-Coumaric acid inhibits the *Listeria monocytogenes* RecA protein functions and SOS response: An antimicrobial target. *Biochem. Biophys. Res. Commun.* **2019**, *517*, 655–661. [CrossRef] [PubMed]
103. Derebasi, B.; Bulut, S.; Erden, B.; Sadeghian, N.; Taslimi, P.; Celebioglu, H. Effects of *p*-coumaric acid on probiotic properties of *Lactobacillus acidophilus* LA-5 and *Lacticaseibacillus rhamnosus* GG. *Arch. Microbiol* **2024**, *206*, 223. [CrossRef] [PubMed]
104. Sharma, S.; Chellappan, D.; Chinnaswamy, P.; Nagarajan, S. Protective effect of *p*-coumaric acid against 1, 2 dimethylhydrazine induced colonic preneoplastic lesions in experimental rats. *Biomed. Pharmacother.* **2017**, *94*, 577–588. [CrossRef] [PubMed]
105. Syfani, N.; Prima Putra, D.; Arbain, D. 3,4-dihydroxybenzaldehyde from the fern *Trichomanes chinense* L.; isolation antimicrobial and antioxidant properties. *Indones. J. Chem.* **2012**, *12*, 273–278. [CrossRef]
106. Matejczyk, M.; Ofman, P.; Juszczuk-Kubiak, E.; Świsłocka, R.; Shing, W.L.; Kesari, K.K.; Prakash, B.; Lewandowski, W. Biological effects of vanillic acid, iso-vanillic acid, and orto-vanillic acid as environmental pollutants. *Ecotoxicol. Environ. Saf.* **2024**, *277*, 116383. [CrossRef]
107. Duskaev, D.; Kurilkina, M.; Zavyalov, O. Growth-stimulating and antioxidant effects of vanillic acid on healthy broiler chickens. *Vet. World* **2023**, *16*, 518–525. [CrossRef]
108. Ingole, A.; Kadam, M.P.; Dalu, A.P.; Kute, S.M.; Mange, P.R.; Theng, V.D.; Patil, P.A. A review of the pharmacological characteristics of vanillic acid. *J. Drug Deliv. Ther.* **2021**, *11*, 200–204. [CrossRef]
109. Ijabadeniyi, O.; Govender, A.; Olagunju, O.; Oyedeji, A. The antimicrobial activity of two phenolic acids against foodborne *Escherichia coli* and *Listeria monocytogenes* and their effectiveness in a meat system. *Ital. J. Food Sci.* **2021**, *33*, 39–45. [CrossRef]
110. Kepa, M.; Miklasińska-Majdanik, M.; Wojtyczka, R.; Idzik, D.; Korzeniowski, K.; Smoleń-Dzirba, J.; Wąsik, T. Antimicrobial potential of caffeic acid against *Staphylococcus aureus* clinical strains. *Biomed. Res. Int.* **2018**, *2018*, 7413504. [CrossRef]
111. Vaquero, M.; Aredes, P.; Nadra, M. Effect of phenolic compound mixtures on the viability of *Listeria monocytogenes* in meat model. *Food Technol. Biotechnol.* **2011**, *49*, 83–88.
112. Seether, K.A.; Miya, T.S.; Perry, T.W.; Boehm, P.N. Extraction of an ulcer-preventing principle from oat hulls. *J. Anim. Sci.* **1971**, *32*, 1160–1163. [CrossRef]

foods

MDPI

Article

Bioactive Composition of Tropical Flowers and Their Antioxidant and Antimicrobial Properties

Elena Coyago-Cruz [1,*], Alejandro Barrigas [2], Aida Guachamin [1], Jorge Heredia-Moya [3], Johana Zuñiga-Miranda [3] and Edwin Vera [4]

1. Carrera de Ingeniería en Biotecnología de los Recursos Naturales, Universidad Politécnica Salesiana, Sede Quito, Campus El Girón, Av. 12 de Octubre N2422 y Wilson, Quito 170109, Ecuador
2. Maestría en Productos Farmacéuticos Naturales, Universidad Politécnica Salesiana, Sede Quito, Campus El Girón, Av. 12 de Octubre N2422 y Wilson, Quito 170109, Ecuador
3. Centro de Investigación Biomédica (CENBIO), Facultad de Ciencias de la Salud Eugenio Espejo, Universidad UTE, Quito 170527, Ecuador
4. Departamento de Ciencia de los Alimentos y Biotecnología, Facultad de Ingeniería Química, Escuela Politécnica Nacional, Quito 170524, Ecuador

* Correspondence: ecoyagoc@ups.edu.ec; Tel.: +593-995129321

Abstract: This study evaluated tropical flower petals' bioactive compounds and antioxidant and antimicrobial properties. The physicochemical characteristics, carotenoids, phenolics, anthocyanins, organic acids, and antioxidant activity of 67 flowers were analyzed. In addition, the antimicrobial activity against *Escherichia coli*, *Staphylococcus aureus*, *Pseudomonas aeruginosa*, *Streptococcus mutans*, *Candida albicans*, and *Candida tropicalis* of 35 species was determined. A 2 $\times$ 3 experimental design was used for the extraction of carotenoids and phenolics, including solvents and ultrasonic agitation times. The mixture of methanol–acetone–dichloromethane (1:1:2) and acetone–methanol (2:1) resulted in the highest concentration of carotenoids, while acidified 80% methanol favoured phenolic extraction. *Renealmia alpinia* was extremely rich in carotenoids (292.5 mg β-carotene/g DW), *Pleroma heteromallum* in anthocyanins (7.35 mg C-3-gl/g DW), while a high content of citric acid was found in *Hibiscus rosa-sinensis* (17,819 mg/100 g DW). On the other hand, *Thibaudia floribunda* showed the highest antioxidant activity (7.8 mmol Trolox equivalent/g DW). The main phenolics were *m*-coumaric acid in *Acalypha poiretii* (12,044 mg/100 g DW), 4-hydroxybenzoic acid in *Brugmansia arborea* (10,729 mg/100 g DW), and kaempferol in *Dahlia pinnata* (8236 mg/100 g DW). The extract of *Acalypha poiretii*, *Brownea macrophylla*, and *Cavendishia nobilis* showed antibacterial activity, while the extract of *Pleroma heteromallum* was the only one active against *Candida albicans*. These findings highlight the potential health benefits from certain tropical flowers.

Keywords: carotenoids; phenolics; organic acids; anthocyanins; micro-extraction; experimental design; PCA

1. Introduction

Tropical forests are characterized by average temperatures of 20–25 °C, abundant sunlight, and regular rainfall. These ecosystems cover 6% of the Earth's land surface and are home to various species, including plants that provide food and medicine for local communities [1–3].

Within the vast diversity of plants, the flowers of several tropical species stand out for their content of bioactive compounds such as carotenoids, phenolic compounds, monophenols, and terpenes, which provide functional benefits beyond basic nutrition [4,5]. These compounds act as antioxidants and contribute to well-being through anti-inflammatory, immunomodulatory, anticarcinogenic, and cardiovascular protective activities. In addition, some of the carotenoids found in these plants can be converted into vitamin A, essential for eye health and the immune system, reinforcing the importance of these natural sources in the human diet [5–9]. Recent research has shown that the synergistic effect of bioactive compounds is more effective than the consumption of individual molecules. For example, eating foods high in carotenoids and phenolics is beneficial for cardiovascular disease, neurodegenerative disease, and certain types of cancer [4,10–13].

Families such as Asteraceae, Lamiaceae, Fabaceae, Apiaceae, Araliaceae, Ericaceae, Zingiberaceae, and others have been identified as valuable sources for the treatment of liver disease due to the presence of bioactive compounds in their floral structures. For example, the Asteraceae family, which contains more than 2500 flowering species, is important in traditional medicine and dietary practises. Similarly, the Lamiaceae family contains more than 7000 species, many of which are used for their essential oil content and medicinal applications [7,12,14].

On the other hand, extracting bioactive compounds for food and medicinal purposes is a significant challenge. Green chemistry methods using ultrasound, microwave-assisted extraction, and enzymatic extraction have significantly reduced solvent consumption and processing time. In particular, carotenoids require organic solvents such as hexane, methanol, acetone, petroleum ether, etc. For phenolic compounds, ethanol, methanol, water, and acetone are generally used, depending on the polarity of the target phenolics [15,16]. Another parameter to consider in this context is the extraction time, which is also critical, as a more extended time may result in a complete extraction but must be balanced with the efficiency of the process [17].

As the concentration of carotenoids and phenolics varies according to species, maturity, and growth conditions, it is essential to adapt extraction methods to maximize the yield of each compound [9]. In this context, this study focused on the evaluation of bioactive compounds, antioxidants, and antimicrobials in tropical flowers, using different assays to identify species with a high content of bioactive compounds that can be used in the food and pharmaceutical industries and contribute to the development of functional foods.

2. Materials and Methods

2.1. Reagents and Standards

The chemicals used in this investigation included acetone (CAS 67-64-1) and fluconazole (86386-73-4), which were reagent grade. At the same time, HPLC grade chemicals included acetonitrile (CAS 75-05-8), ethanol (CAS 64-17-5), and methanol (CAS 67-56-1), which were purchased from Fisher Chemical (Fischer Scientific Inc., Madrid, Spain). ABTS (2,2′-azino-bis-(3-ethylbenzothiazoline-6-sulfonic acid) (CAS 30931-67-0), DPPH (2,2-Diphenyl-1-picrylhydrazyl) (CAS 1898-66-4), formic acid (CAS 64-18-6), Folin–Ciocalteu (CAS 7732-18-5), metaphosphoric acid (CAS 37267-86-0), potassium chloride (CAS 7447-40.7), potassium persulphate (CAS 7727-21-1), sodium acetate try hydrate (CAS 6131-90.4), sodium carbonate (CAS 497-19-8), and sodium hydroxide (CAS 1310-73-2) were also used; all analytical grades were purchased from Sigma (Merck, Darmstadt, Germany). Hydrochloric acid (CAS 7647-01-0) was also obtained as an analytical grade from Labscan (RCI Labscan group, Dublin, Republic of Ireland). Brain Heart Infusion (BHI), Mueller Hinton agar (MHA), and Sabouraud dextrose agar (SDA) were purchased from BD DifcoTM (Fisher Scientific Inc., Madrid, Spain). Yeast Peptone Dextrose Broth (YPDB) from SRL (Sisco Research Laboratories Pvt. Ltd., Mumbai, India) and streptomycin sulphate sulphate (CAS 3810-74-0) from Phytotech (PhytoTechnology Laboratories ®, Lenexa, KS, USA). Water was purified using a NANOpureDiamondTM system (Barnsted Inc., Dubuque, IO, USA).

Standards include β-carotene 93.0% (CAS 7235-40-7), caffeic acid 98.0% (CAS 331-39-5), chlorogenic acid 95.0% (CAS 327-97-9), chrysin 97.0% (CAS 480-40-0), cyanidin-e-glucoside chloride 97.0% (CAS 7084-24-4), ferulic acid 100.0% (CAS 1135-24-6), gallic acid 100.0% (CAS 149-91-7), 2,5-dihydroxybenzoic acid 98.0% (CAS 490-79-9), 3-hydroxybenzoic acid 99.0% (CAS 99-06-3), kaempferol 97.0% (CAS 520-18-3), luteolin 98% (CAS 491-70-3), *m*-coumaric acid 99.0% (CAS 588-30-7), naringin 95.0% (CAS 10236-47-2), *o*-coumaric acid 97.0% (CAS 614-60-8), *p*-coumaric acid 98.0% (CAS 501-98-4), *p*-hydroxybenzoic acid 99,0% (CAS 99-06-3), quercetin 95.0% (CAS 849061-97-8), rutin 94.0% (CAS 153-18-

4), shikimic acid 99.0% (CAS 138-59-0), syringic acid 95,0% (CAS 530-57-4), vanillic acid 97.0% (CAS 121-34-6), and Trolox 98% (CAS 53188-07-1), which were purchased from Sigma (Merck, Darmstadt, Germany). *Candida albicans* ATCC 1031, *Candida tropicalis* ATCC 13803, *Escherichia coli* ATCC 8739, *Pseudomonas aeruginosa* ATCC 9027, *Staphylococcus aureus* ATCC 6538P and, *Streptococcus mutans* ATCC 25175 were purchased from ATTC (ATTC, Manassas, VA, USA).

2.2. Physicochemical Quantification

For this study, 67 different flower species from tropical areas of Ecuador were selected (Table 1 and Figure 1) and collected throughout the year 2020. Thirty flowers were randomly sampled from each species and considered for physicochemical analysis. These were placed in containers provided with floral foam to prevent deterioration. At the same time, for other studies, approximately 200 g of petals were selected and carefully stored in Falcon tubes and subjected to lyophilisation using the Christ Alpha 1-4 LDplus apparatus (Martin Gefriertrocknungsanlagen GmbH, Osterode am Harz, Germany). In addition, fresh samples of flowers, leaves, stems, and fruits were collected, pressed, and dried for botanical identification in QUPS-Ecuador Herbarium. The physicochemical analysis included the determination of weight, equatorial and longitudinal diameters, soluble solids, pH, total titratable acid, moisture., and ash following the protocol of Coyago et al. [9].

Table 1. Floral species are under study.

N°	Family	Species	Altitude (masl)
1	Acanthaceae	*Megaskepasma erythrochlamys* Lindau	1020
2	Acanthaceae	*Odontonema cuspidatum* (Nees) Kuntze	562
3	Acanthaceae	*Pachystachys lutea* Nees	981
4	Acanthaceae	*Sanchezia oblonga* Ruiz & Pav.	1382
5	Acanthaceae	*Thunbergia grandiflora* Roxb.	1240
6	Amaranthaceae	*Celosia argentea* L.	136
7	Apiaceae	*Eryngium aquaticum* L.	1260
8	Apocynaceae	*Allamanda cathartica* L.	40
9	Araceae	*Philodendron fragrantissimum* (Hook.) G.Don	562
10	Araceae	*Philodendron strictum* G.S. Bunting	562
11	Araceae	*Xanthosoma robustum* Schott	620
12	Asteraceae	*Acmella papposa* (Hemsl.) R.K. Jansen	1486
13	Asteraceae	*Dahlia pinnata* Cav.	2222
14	Asteraceae	*Dendrophorbium balsapampae* (Cuatrec.) B. Nord.	2447
15	Asteraceae	*Pseudogynoxys chenopodioides* (Kunth) Cabrera	1360
16	Balsaminaceae	*Impatiens hancockii* C.H.Wright.	2593
17	Balsaminaceae	*Impatiens hawkeri* W. Bull	630
18	Balsaminaceae	*Impatiens sodenii* Engl. & Warb. ex Engl.	2814
19	Balsaminaceae	*Impatiens walleriana* Hook.f.	630
20	Bromeliaceae	*Aechmea corymbosa* (Mart. Ex Schult. & Schult.f.) Mez	1506
21	Bromeliaceae	*Guzmania nicaraguensis* Mez & Baker	1580
22	Bromeliaceae	*Ronnbergia veitchii* (Baker) Aguirre-Santoro	1337
23	Campanulaceae	*Centropogon cornutus* (L.) Druce	562
24	Caricaceae	*Carica papaya* L.	562
25	Convolvulaceae	*Ipomoea triloba* L.	16
26	Costaceae	*Costus spiralis* (Jacq.) Roscoe	1320
27	Clusiaceae	*Clusia nitida* Bittrich & F.N. Cabral	1477
28	Ericaceae	*Cavendishia bracteata* (Ruiz & Pav. Ex J.St.-Hil.) Hoerold	2100
29	Ericaceae	*Cavendishia cuatrecasasii* A.C.Sm.	1910
30	Ericaceae	*Cavendishia nobilis* Lindl.	2060
31	Ericaceae	*Thibaudia floribunda* Kunth	1641
32	Euphorbiaceae	*Acalypha poiretii* Spreng.	620
33	Fabaceae	*Brownea macrophylla* Linden	940
34	Fabaceae	*Calliandra angustifolia* Spruce ex Benth.	1598
35	Fabaceae	*Erythrina americana* Mill.	562

Table 1. *Cont.*

N°	Family	Species	Altitude (masl)
36	Gentianaceae	*Macrocarpaea sodiroana* Gilg	1490
37	Gesneriacea	*Glossoloma ichthyoderma* (Hanst.) J.L. Clark	1495
38	Heliconiaceae	*Heliconia episcopalis* Vell.	620
39	Heliconiaceae	*Heliconia collinsiana* Griggs	1641
40	Heliconiaceae	*Heliconia latispatha* Benth.	1494
41	Heliconiaceae	*Heliconia rostrata Ruiz & Pav.*	562
42	Heliconiaceae	*Heliconia wagneriana* Petersen	562
43	Lamiaceae	*Prunella vulgaris* L.	1850
44	Malvaceae	*Hibiscus rosa-sinensis* L.	500
45	Marantaceae	*Stromanthe stromanthoides* (J.F. Macbr.) L.Andersson	530
46	Melastomataceae	*Pleroma heteromallum* (D.Don) D. Don	1260
47	Melastomataceae	*Andesanthus lepidotus* (Humb. & Bonpl.) P.J.F.Guim. & Michelang.	1742
48	Melastomataceae	*Chaetogastra mollis* (Bonpl.) DC.	1850
49	Melastomataceae	*Plerona urvilleanum (DC.) PJ:F:Gui. & Michelang.*	1234
50	Musaceae	*Musa velutina* H. Wendl. & Drude	562
51	Nyctaginaceae	*Bougainvillea spectabilis* Willd.	14
52	Orchidaceae	*Elleanthus aurantiacus* (Lindl.) Rchb.f.	2530
53	Orchidaceae	*Oncidium sp.*	306
54	Orchidaceae	*Sobralia liliastrum* Lindl.	1260
55	Rubiaceae	*Mussaenda erythrophylla Schumach. & Thonn.*	510
56	Rubiaceae	*Mussaenda philippica* A. Rich.	940
57	Rubiaceae	*Palicourea angustifolia* Kunth	510
58	Solanaceae	*Brugmansia arborea* (L.) Sweet	1641
59	Solanaceae	*Iochroma gesnerioides* (Kunth) Miers	2054
60	Solanaceae	*Browallia americana* L.	1477
61	Solanaceae	*Brugmansia versicolor* Lagerh.	2833
62	Solanaceae	*Solanum dulcamara L.*	1910
63	Verbenaceae	*Lantana rugulosa* Kunth	2798
64	Zingiberaceae	*Alpinia purpurata* (Vieill.) K. Schum.	14
65	Zingiberaceae	*Hedychium coronarium* J. Koening	940
66	Zingiberaceae	*Hedychium coccineum* Buch.-Ham.exSm.	1240
67	Zingiberaceae	*Renealmia alpinia* (Rottb.) Maas	940

Note: masl, metres above sea level.

2.3. Optimization of Extraction Parameters and Quantification of Carotenoids

Micro-extractions were carried out in the dark and in triplicate. A 2 × 3 experimental design was implemented with solvent type and ultrasonic extraction time. Different solvent combinations, including acetone–methanol (2:1), *n*-hexane–acetone (1:1) and methanol–acetone–dichloromethane (1:1:2), were considered. In addition, the ultrasonic extraction time varied between 1, 2, and 3 min. Only the petals of *Taraxacum officinale*, *Pyrostegia venusta*, and *Buddleja globosa*, species known for their yellow colour and carotenoid content [18,19], were selected to optimize the ultrasonic extraction time and extraction solvent for the 67 samples. For the extraction of carotenoids according to the specific combinations of the experimental design, a mixture of 20 mg of lyophilised powder with one mL of the chosen solvent. This solution was then vortexed using a VM-300 vortex (Interbiolab Inc., Orlando, FL, USA) and shaken in a Fisher Scientific FS60 ultrasonicator (Fisher Scientific, Waltham, MA, USA) for the time specified in the experimental design. Subsequently, the mixture was centrifuged for 3 min at 1400 rpm and 4 °C using a MiniSpin microcentrifuge (Eppendorf, Bochum, Germany). The coloured phase was collected, and the extraction process was repeated until the solid residue became colourless. The coloured phase evaporated to dryness (below 30 °C) using a Buchi TM R-100 (Fisher Scientific, Waltham, MA, USA).

Figure 1. Geographical distribution of the flowers under study. Note: The numbers correspond to the number of blossoms examined (Table 1).

To quantify carotenoids by spectrophotometry, the dried extract was dissolved in 2 mL of ethanol and transferred to a 10 mm light path quartz cell. The absorbance of the solution was measured at 450 nm using a ThermoSpectromic Genesys 10 UV-Vis spectrophotometer (ThermoFisher Scientific, Waltham, MA, USA). A calibration curve was prepared using a 5 mg β-carotene standard dissolved in 25 mL of ethanol to ensure accurate measurements. The concentration of total carotenoids in the samples was expressed as micrograms of β-carotene per gram of freeze-dried petal weight (DW) (μg β-carotene/g DW) [20]. Finally, following the above procedure, the remaining samples were extracted using the solvent and for a time that maximized extraction.

2.4. Optimization of Extraction Parameters Andquantification of Phenolic Compounds

Micro-extraction was performed in triplicate. A 3 × 2 experimental design was implemented with solvent type and ultrasonic extraction time. Different aqueous solvents, including ethanol (75%), methanol (75%), and methanol (80%) acidified with hydrochloric acid (0.1%), were evaluated to determine the most efficient mixture for phenol extraction. In addition, the ultrasonic extraction time varied between 1, 3, and 5 min. Only the petals of *Dahlia pinnata, Dianthus caryophyllus, Plerona urvilleanum, Agapanthus africanus,* and *Lupinus mutabilis*, known for their red to purple colours and phenolic content [18,19] were selected to optimize the ultrasonic extraction time and extraction solvent for the 67 samples. To extract phenolics according to the specific combinations of the experimental design, 40 mg of lyophilised powder was mixed with 1 mL of the chosen solvent. This solution was then vortexed using a VM-300 vortex mixer (Interbiolab Inc., Orlando, FL, USA) and subjected to ultrasound in a Fisher Scientific FS60 (Fisher Scientific, Waltham, MA, USA) for the time specified in the experimental design. The supernatant was recovered by centrifugation at 1400 rpm for 5 min at 4 °C. The extraction was repeated twice using 500 μL of the methanolic solution. To remove impurities, the resulting methanolic extract was filtered through a 0.45 μm PVDF filter [5,21].

To quantify total phenolics, 20 μL of the methanolic extract was mixed with 100 μL of a 1:4 Folin–Cioacalteu solution and 75 μL of a sodium carbonate solution (100 g/L) in a 96-well plate. The mixture was homogenized and allowed to stand for two hours at room temperature. Absorbance was measured at 750 nm. The calibration curve was established with gallic acid at 10 and 200 mg/L concentrations. Total phenolics were expressed as mg gallic acid equivalent per 100 g of freeze-dried petal weight (mg GAE/100 g DW) [22]. Finally, using the solvent and time that maximized the extraction of phenolics, the rest of the samples under investigation were extracted according to the above process.

To identify individual phenolics, 20 μL of the methanolic extract was applied to an Agilent 1200 series RRLC with a DAD-UV-Vis detector operating between 220 and 500 nm and a Zorbax Eclipse Plus C18 column (4.6 × 150 mm, 5 μm) (Agilent Technologies, Santa Clara, CA, USA) at 30 °C [5]. The mobile phase was a linear gradient at a flow rate of 1 mL/min of an aqueous solution of formic acid (0.01%) (solvent A) and acetonitrile (solvent B). During the run, the mobile phase was pumped as follows: 100% A at 0 min; 95% A + 5% B at 5 min; 50% A + 50% B at 20 min; washing and re-equilibration of the column at 30 min. Spectra stored in the Open Lab ChemStation software (version 2.15.26) were used to identify the individual phenolics. In contrast, calibration curves were used for quantification at concentrations ranging from 0. 33 to 1 mg/mL of shikimic acid, benzoic acids (*p*-hydroxybenzoic acid, 3-hydroxybenzoic acid, 2-methoxybenzoic acid, 3-methoxybenzoic acid, 2,5-dihydroxybenzoic acid, gallic acid, vanillic acid, syringic acid, *p*-coumaric acid, *o*-coumaric acid, *m*-coumaric acid), hydroxycinnamic acids (caffeic acid, chlorogenic acid, ferulic acid), flavonols (kaempferol, quercetin, quercetin glycoside, rutin), and flavones (chrysin, luteolin, flavanones (naringin)). Each phenolic compound was expressed as milligrammes per hundred grammes of freeze-dry petal weight (mg/100 g DW). Total phenolics were calculated as the sum of the individual compounds.

2.5. Quantification of Total Anthocyanins

To extract anthocyanins from the freeze-dried flower powder, 20 mg of freeze-dried powder was weighed and mixed with 2 mL of absolute ethanol. The mixture was shaken for 3 min in an ultrasonic bath. Residual solids were removed by centrifugation at 1400 rpm for 5 min at 4 °C [23–25]. Quantification of anthocyanins was performed using the pH differential method, which is widely used for this type of analysis. For the procedure, 50 μL of the extract was diluted with 200 μL of a 0.025 M potassium chloride buffer at pH 1, while 50 μL of the extract was mixed with 200 μL of a 0.4 M sodium acetate buffer at pH 4.5 [26,27]. The absorbances of each dilution were measured at 520 nm and 700 nm in a spectrophotometer with a microplate reader. The calibration curve was constructed over a 0.05 to 0.2 mg/mL concentration range. The total anthocyanin concentration was expressed as mg cyanidin-3-glucoside chloride per gram dry weight (mg C-3-gl/g DW).

2.6. Quantification of Organic Acids

Organic acids were extracted from 40 mg of lyophilised powder with 1.5 mL of 0.02 N sulphuric acid containing metaphosphoric acid (0.05%) and *DL*-homocysteine (0.02%). The mixture was homogenized and vortexed under ultrasound for 3 min, and 500 µL of deionised water was added. The supernatant was collected by centrifugation at 1400 rpm for 5 min at 4 °C and filtered through a 0.45 µm PVDF filter. Separation of organic acids was performed using 10 µL of the extract applied to an Agilent 1200 series RRLC, DAD-UV-Vis detector at 210 nm, and a YMC-Triart C18 column (150 × 4.6 mm, 3, 12 nm, 400 bar) (YMC Europe GmbH, Dinslaken, Germany) at 30 °C. The mobile phase was a 0.027% sulphuric acid solution pumped at 1 mL/min for 30 min. Organic acids were analyzed using spectra stored in Open Lab ChemStation software and retention time. Calibration curves with concentrations between 33.3 and 100 mg/mL of citric acid, malic acid, and L-(+)-tartaric acid standards were used for quantification [18]. Each organic acid was expressed in grammes per 100 g of freeze-dried petal weight (g/100 g DW).

2.7. Antioxidant Activity (ABTS and DPPH)

To extract the ABTS antioxidant activity assay sample, 20 mg of lyophilised powder was mixed with 400 µL of methanol and 400 µL of distilled water separately. The mixture was homogenized by vortexing and shaken in an ultrasonic bath for 3 min. Afterwards, the solution was separated by microcentrifugation at 14000 rpm for 5 min at 4 °C. The resulting solid was mixed separately with 560 µL of acetone and 240 µL of distilled water. The process was repeated to obtain the supernatant, which was then combined with the previous supernatant. The final mixture was refrigerated until quantification.

To prepare the $ABTS^{•+}$ radical, a 1:1 solution of 7 mM ABTS was mixed with 2.45 mM potassium persulfate and left to stand in the dark for 16 h. The resulting $ABTS^{•+}$ radical solution was diluted approximately 1/10 with absolute ethanol or until an absorbance of 0.7 at 754 nm was achieved [20]. For the calibration curve, a stock solution of 0.25 mM Trolox was prepared and diluted to 0.01 to 0.07 mM concentrations. To quantify the samples, 20 µL of the final supernatant solution was added to a 96-well VWR tissue culture plate (Novachen, Connersville, IN, USA). Subsequently, 280 µL of the $ABTS^{•+}$ radical solution was added to each well following the procedure described by Chan et al. [28]. The absorbance was measured at 754 nm using a spectrophotometer, specifically a Thermo Scientific Multiskan GO microplate reader (Agilent Scientific Instruments, Santa Clara, CA, USA). Finally, the antioxidant activity was expressed as mmol Trolox equivalents per gram of freeze-dried petal weight (mmol TE/g DW).

To extract the DPPH antioxidant activity assay sample, 20 mg de lyophilised powder was mixed with 2 mL of methanol. The mixture was homogenized by vortexing and shaken in an ultrasonic bath for 3 min. Afterwards, the solution was separated by microcentrifugation at 14,000 rpm for 5 min at 4 °C. For the calibration curve, a stock solution of 0.25 mM Trolox was prepared and diluted to 0.01 to 0.07 mM concentrations. To quantify the samples, 20 µL of the final supernatant solution was added to a 96-well VWR tissue culture plate. Subsequently, 280 µL of the $DPPH^{•}$ (10 mg of DPPH in 50 mL of methanol) was added. The absorbance was measured at 560 nm after 30 min [29]. Antioxidant activity was expressed as mmol Trolox equivalents per gram of freeze-dried petal weight (mmol TE/g DW).

2.8. Antimicrobial Activity

The extract was prepared by weighing 2 g of the freeze-dried sample of the 35 species with the largest sample size and mixing it with 10 mL of 50% ethanol. The mixture was homogenized and vortexed in an FS60 ultrasonic bath (Fisher Scientific, USA) for 6 min. The resulting supernatant was collected by centrifugation at 14,000 rpm for 3 min using a microcentrifuge (Eppendorf, Germany). This extraction procedure was performed twice, with 10 mL of the ethanol solution for each extraction. The final supernatant was filtered through PDVF filters with a pore size of 0.45 µm and a diameter of 25 mm. The extract was dried using a Christ Alpha 1-4 LDplus freeze dryer (GmbH, Germany). Finally, the dried extract was reconstituted in 1 mL of sterile distilled water to evaluate antimicrobial activity using the well diffusion method according to Clinical and Laboratory Standards Institute (CLSI) guidelines with some modifications [30–32].

The antibacterial properties of the flower extracts were evaluated against Gram-positive bacteria such as *Staphylococcus aureus* ATCC 6538P and Gram-negative bacteria such as *Escherichia coli* ATCC 8739, *Pseudomonas aeruginosa* ATCC 9027, and *Streptococcus mutans* ATCC 25175. In addition, antifungal activity was evaluated against two pathogenic fungal species, *Candida albicans* ATCC 1031 and *Candida tropicalis* CC 13803. All bacterial strains used in this study were obtained from the American Type Culture Collection (ATCC, Manassas, VA, USA) and maintained at −80 °C in a 25% (v/v) glycerol solution. Both Gram-positive and Gram-negative bacteria were pre-cultured in Brain Heart Infusion (BHI) overnight at 37 °C on a rotary shaker. Each strain was then calibrated to a concentration of 0.5 MacFarland standard (10^8 cells/mL). The fungal inoculum was obtained from a 24 h culture of the isolated fungal strains in Yeast Peptone Dextrose Broth (YPDB). Each strain was then adjusted to the 0.5 MacFarland standard (resulting in a final concentration of 10^6 cells/mL).

The agar well diffusion method was used to evaluate the antimicrobial activity of the flower extracts. The suspension of active microorganisms was spread evenly on solidified Müller-Hinton agar (MHA) for bacterial strains and on Sabouraud dextrose agar (SDA) for fungal strains using a sterile swab. 5 mm diameter agar wells were prepared using a sterile punch in each plate. Then, 80 µL of flower extracts were added to the wells, and the Petri dishes were incubated at 37 °C/18 h for bacteria and 35 °C/48 h for fungi. The zone of inhibition obtained was measured in millimetres. Streptomycin and fluconazole were used as controls for growth inhibition at recommended working concentrations for the bacterial and fungal strains, respectively. In addition, distilled water was used as a negative control. These assays were performed in, at least, triplicate.

2.9. Statistical Analysis

Statistical analysis was conducted using Statgraphics Centurion XVII, Rstudio, and Sigmaplot 14.0 software. Results are given as the mean ± standard deviation. A simple ANOVA was employed to identify significant differences, with a significance level set at $p < 0.05$. Furthermore, correlation and principal component analyses explored potential relationships among the study parameters—this analysis aimed to uncover any associations between the variables under investigation.

3. Results

3.1. Physicochemical Quantification

Table 2 shows the results of the physicochemical analyses on the petals studied. This study evaluated weight, size, pH, soluble solids, titratable acidity, moisture, and ash.

Flower weights ranged from 0.01 g in *P. vulgaris* and *L. rugulosa* to 72.79 g in *X. ro-bustum*. Concerning flower size, longitudinal and equatorial diameters displayed notable variations, with values ranging from 1.07 cm in *L. rugulosa* to 25.30 cm in *I. hancockii* and from 0.36 cm in *O. cuspidatum* and *P. vulgaris* to 62.85 cm in *A. purpurata*. The pH levels ranged from 2.00 in *A. poiretii*, *B. macrophylla*, and *I. walleriana* to 11.40 in *P. urvilleana*. Soluble solids ranged from 0.21 °Brix in *P. fragrantissimum* to 15.80 °Brix in *P. lutea*. The total titratable acidity spanned from 0.05% in *X. robustum* and *H. episcopalis* to 3.81% in *A. corymbosa*. Moisture content varied from 14.26% in *M. sodiroana* to 96.05% in *I. walleriana*. The ash content ranged from 0.03% in *I. hancockii* to 8.02% in *S. oblonga*.

Table 2. Average values of the physicochemical parameters of the floral species analyzed.

N°	Scientific Name	Weight (g)	DL (cm)	DE (cm)	pH	SS (°Brix)	TA (%)	Humidity (%)	Ash (%)
1	*Megaskepasma erythrochlamys*	0.17 ± 0.03	4.21 ± 0.80	1.36 ± 0.36	9.15 ± 0.24	1.74 ± 0.22	1.44 ± 0.08	85.06 ± 3.32	0.18 ± 0.00
2	*Odontonema cuspidatum*	0.03 ± 0.00	1.69 ± 0.23	0.36 ± 0.08	4.60 ± 0.46	5.04 ± 0.05	0.25 ± 0.00	85.59 ± 0.18	3.53 ± 0.20
3	*Pachystachys lutea*	0.05 ± 0.02	1.77 ± 0.58	1.54 ± 0.82	7.00 ± 0.00	15.80 ± 0.42	1.44 ± 0.19	77.72 ± 4.23	3.46 ± 0.00
4	*Sanchezia oblonga*	4.16 ± 0.37	4.23 ± 1.01	4.16 ± 0.37	7.00 ± 0.00	4.00 ± 0.00	0.48 ± 0.05	81.8 ± 1.22	8.02 ± 0.29
5	*Thunbergia grandiflora*	1.44 ± 0.15	3.99 ± 0.25	7.93 ± 0.74	8.50 ± 0.00	7.00 ± 0.00	0.28 ± 0.02	56.98 ± 4.08	1.85 ± 0.08
6	*Celosia argentea*	21.33 ± 0.92	7.68 ± 0.33	9.52 ± 0.90	6.50 ± 0.00	3.94 ± 0.10	0.12 ± 0.00	80.53 ± 0.25	1.44 ± 0.21
7	*Eryngium aquaticum*	0.10 ± 0.02	1.32 ± 0.22	3.75 ± 1.01	5.50 ± 0.00	10.60 ± 0.97	0.59 ± 0.11	73.77 ± 1.40	2.21 ± 0.13
8	*Allamanda cathartica*	2.29 ± 0.42	9.22 ± 0.24	11.30 ± 0.44	6.00 ± 0.00	3.00 ± 0.00	0.94 ± 0.05	89.55 ± 0.27	0.51 ± 0.04
9	*Philodendron fragrantissimum*	5.72 ± 1.33	8.82 ± 1.25	1.06 ± 0.17	6.00 ± 0.00	0.21 ± 0.07	0.11 ± 0.01	84.35 ± 0.97	3.54 ± 0.29
10	*Philodendron strictum*	6.42 ± 1.92	8.74 ± 2.47	2.54 ± 0.48	4.00 ± 0.00	2.71 ± 0.22	0.19 ± 0.05	91.52 ± 0.34	1.28 ± 0.04
11	*Xanthosoma robustum*	72.79 ± 15.16	23.68 ± 3.11	2.24 ± 0.22	4.30 ± 0.48	3.00 ± 0.00	0.05 ± 0.01	91.77 ± 0.11	1.32 ± 0.01
12	*Acmella papposa*	0.06 ± 0.02	1.28 ± 0.08	1.62 ± 0.24	6.00 ± 0.00	5.14 ± 0.10	1.01 ± 0.15	76.63 ± 0.23	2.39 ± 0.00
13	*Dahlia pinnata*	18.70 ± 3.95	9.96 ± 1.19	9.93 ± 0.93	4.40 ± 0.52	5.40 ± 0.52	0.17 ± 0.06	92.63 ± 0.09	0.67 ± 0.00
14	*Dendrophorbium balsapampae*	0.32 ± 0.03	1.39 ± 0.11	0.89 ± 0.39	6.00 ± 0.00	3.85 ± 0.24	0.34 ± 0.04	76.71 ± 0.05	1.72 ± 0.08
15	*Pseudogynoxys chenopodioides*	0.46 ± 0.17	1.19 ± 0.24	1.48 ± 0.52	5.7 ± 0.48	2.40 ± 0.52	0.32 ± 0.08	73.16 ± 1.17	2.58 ± 0.22
16	*Impatiens hancockii*	1.02 ± 0.00	25.33 ± 3.02	25.33 ± 3.02	4.30 ± 0.48	3.92 ± 0.10	0.75 ± 0.02	93.85 ± 0.36	0.03 ± 0.00
17	*Impatiens hawkeri*	0.22 ± 0.03	5.7 ± 0.45	2.73 ± 0.58	2.40 ± 0.84	3.00 ± 0.00	0.83 ± 0.53	95.74 ± 0.32	0.53 ± 0.01
18	*Impatiens sodenii*	0.91 ± 0.21	3.51 ± 0.19	5.15 ± 0.33	4.00 ± 0.00	3.55 ± 0.44	0.58 ± 0.07	94.02 ± 1.33	0.34 ± 0.08
19	*Impatiens walleriana*	0.22 ± 0.07	1.51 ± 0.63	3.27 ± 1.07	2.00 ± 0.00	3.00 ± 0.00	0.29 ± 0.03	96.05 ± 0.19	0.60 ± 0.02
20	*Aechmea corymbosa*	0.27 ± 0.01	2.62 ± 0.26	0.65 ± 0.06	5.80 ± 0.42	2.00 ± 0.00	3.81 ± 0.45	90.10 ± 0.29	0.25 ± 0.02
21	*Guzmania nicaraguensis*	13.84 ± 2.41	10.30 ± 0.98	2.42 ± 0.26	7.00 ± 0.00	2.85 ± 0.24	0.27 ± 0.07	86.67 ± 0.08	0.98 ± 0.10
22	*Ronnbergia veitchii*	2.18 ± 0.63	5.31 ± 0.53	1.46 ± 0.17	6.60 ± 0.52	2.00 ± 0.00	0.62 ± 0.18	78.51 ± 2.15	0.81 ± 0.05
23	*Centropogon cornutus*	0.47 ± 0.08	4.80 ± 1.01	0.78 ± 0.12	5.00 ± 0.00	8.88 ± 0.27	0.29 ± 0.03	83.40 ± 0.14	0.15 ± 0.01
24	*Carica papaya*	0.33 ± 0.53	1.76 ± 0.81	0.38 ± 0.17	6.00 ± 0.00	1.00 ± 0.00	0.15 ± 0.01	87.73 ± 0.07	0.70 ± 0.37
25	*Ipomoea triloba*	0.17 ± 0.03	4.37 ± 0.48	1.83 ± 0.28	5.45 ± 0.44	10.06 ± 0.10	2.01 ± 0.31	90.12 ± 1.44	0.10 ± 0.01
26	*Costus spiralis*	13.37 ± 4.69	6.60 ± 1.30	2.45 ± 0.40	3.00 ± 0.00	3.70 ± 0.48	0.14 ± 0.01	66.10 ± 5.02	1.30 ± 0.09
27	*Clusia nitida*	7.23 ± 0.93	2.20 ± 0.34	3.45 ± 0.28	4.87 ± 0.02	4.08 ± 0.10	0.49 ± 0.02	90.52 ± 0.85	0.21 ± 0.00
28	*Cavendishia bracteata*	1.60 ± 0.15	4.46 ± 0.20	0.99 ± 0.11	2.73 ± 0.41	4.75 ± 0.18	0.27 ± 0.02	92.30 ± 0.31	0.83 ± 0.02
29	*Cavendishia cuatrecasasii*	0.61 ± 0.07	3.34 ± 0.16	0.52 ± 0.10	3.00 ± 0.00	4.56 ± 1.32	0.99 ± 0.09	92.28 ± 0.12	0.54 ± 0.01
30	*Cavendishia nobilis*	0.73 ± 0.14	3.27 ± 0.36	0.84 ± 0.12	4.00 ± 0.00	4.75 ± 0.41	0.20 ± 0.02	87.04 ± 0.06	0.81 ± 0.05
31	*Thibaudia floribunda*	0.16 ± 0.04	1.67 ± 0.19	0.47 ± 0.11	4.00 ± 0.00	2.53 ± 0.21	1.53 ± 0.18	91.10 ± 0.21	0.47 ± 0.00
32	*Acalypha poiretii*	2.08 ± 1.07	16.70 ± 7.39	0.70 ± 0.17	2.00 ± 0.00	6.10 ± 0.88	1.19 ± 0.09	75.26 ± 0.70	2.19 ± 0.14
33	*Brownea macrophylla*	2.38 ± 0.50	13.45 ± 2.88	32.48 ± 8.65	2.00 ± 0.94	12.00 ± 1.33	0.23 ± 0.04	84.17 ± 0.12	0.70 ± 0.07
34	*Calliandra angustifolia*	0.69 ± 0.12	3.32 ± 0.52	1.74 ± 0.33	3.45 ± 0.44	6.08 ± 0.29	1.15 ± 0.45	80.87 ± 1.64	0.75 ± 0.06
35	*Erythrina americana*	1.07 ± 0.17	8.10 ± 0.33	0.98 ± 0.16	6.00 ± 0.00	2.00 ± 0.00	0.36 ± 0.01	78.69 ± 0.89	0.87 ± 0.05
36	*Macrocarpaea sodiroana*	0.07 ± 0.02	4.00 ± 0.00	1.28 ± 0.55	2.18 ± 0.52	6.00 ± 0.00	2.00 ± 0.00	14.26 ± 1.48	0.46 ± 0.12
37	*Glossoloma ichthyoderma*	0.72 ± 0.28	3.38 ± 0.51	0.97 ± 0.19	5.00 ± 0.00	0.48 ± 0.23	0.25 ± 0.15	81.34 ± 0.54	1.81 ± 0.03
38	*Heliconia episcopalis*	2.18 ± 0.63	5.31 ± 0.53	1.46 ± 0.17	5.60 ± 0.39	3.68 ± 0.54	0.05 ± 0.01	84.74 ± 2.13	2.89 ± 0.15

Table 2. *Cont.*

N°	Scientific Name	Weight (g)	DL (cm)	DE (cm)	pH	SS (°Brix)	TA (%)	Humidity (%)	Ash (%)
39	*Heliconia collinsiana*	19.93 ± 2.42	13.43 ± 0.21	3.61 ± 0.30	4.45 ± 0.44	2.86 ± 0.05	0.53 ± 0.09	70.34 ± 0.74	0.66 ± 0.03
40	*Heliconia latispatha*	3.25 ± 0.62	11.20 ± 1.95	1.99 ± 0.12	8.00 ± 0.00	3.00 ± 0.00	0.37 ± 0.08	82.65 ± 0.99	3.11 ± 0.22
41	*Heliconia rostrata*	11.36 ± 2.46	8.51 ± 0.26	3.44 ± 0.18	6.00 ± 0.00	1.00 ± 0.00	0.21 ± 0.03	70.52 ± 0.56	3.98 ± 0.27
42	*Heliconia wagneriana*	29.04 ± 1.37	12.38 ± 0.69	7.78 ± 0.36	7.00 ± 0.00	1.04 ± 0.15	0.22 ± 0.04	85.64 ± 2.89	0.86 ± 0.03
43	*Prunella vulgaris*	0.01 ± 0.00	1.42 ± 0.11	0.36 ± 0.07	7.80 ± 0.26	6.70 ± 0.48	0.51 ± 0.01	84.43 ± 1.28	2.31 ± 0.01
44	*Hibiscus rosa-sinensis*	1.01 ± 0.00	24.43 ± 1.21	5.45 ± 0.91	5.00 ± 0.82	3.20 ± 0.26	0.28 ± 0.02	91.66 ± 0.52	0.35 ± 0.00
45	*Stromanthe stromanthoides*	0.19 ± 0.03	1.61 ± 0.21	0.45 ± 0.08	5.00 ± 0.00	4.60 ± 0.52	0.21 ± 0.01	76.92 ± 3.80	2.09 ± 0.06
46	*Pleroma heteromallum*	0.17 ± 0.23	1.65 ± 0.31	0.47 ± 0.06	6.00 ± 0.00	6.00 ± 0.00	2.73 ± 0.50	86.16 ± 0.50	1.02 ± 0.01
47	*Andesanthus lepidotus*	0.45 ± 0.09	5.12 ± 0.56	5.32 ± 0.69	2.45 ± 0.44	4.01 ± 0.09	1.10 ± 0.04	93.96 ± 0.00	0.33 ± 0.01
48	*Chaetogastra molis*	0.07 ± 0.01	2.08 ± 0.20	1.60 ± 0.41	3.00 ± 0.00	4.80 ± 0.42	2.8 ± 0.75	86.82 ± 0.05	2.53 ± 0.03
49	*Plerona urvilleanum*	0.46 ± 0.05	2.03 ± 0.32	5.18 ± 0.59	11.4 ± 0.52	9.20 ± 1.03	0.30 ± 0.08	88.21 ± 1.03	1.20 ± 0.04
50	*Musa velutina*	39.73 ± 6.53	9.89 ± 0.22	4.31 ± 0.41	4.30 ± 0.48	2.46 ± 0.05	1.03 ± 0.10	91.66 ± 0.52	1.13 ± 0.12
51	*Bougainvillea spectabilis*	0.36 ± 0.08	3.61 ± 0.44	3.06 ± 0.42	4.00 ± 0.00	3.00 ± 0.00	0.41 ± 0.02	83.57 ± 0.47	2.11 ± 0.24
52	*Elleanthus aurantiacus*	0.10 ± 0.03	1.37 ± 0.12	0.58 ± 0.10	5.50 ± 0.75	6.10 ± 0.32	0.43 ± 0.04	90.5 ± 1.69	1.60 ± 0.25
53	*Oncidium sp.*	1.29 ± 0.40	2.82 ± 0.53	5.65 ± 0.38	2.70 ± 0.48	11.20 ± 1.32	0.39 ± 0.02	89.80 ± 0.56	0.39 ± 0.07
54	*Sobralia liliastrum*	6.62 ± 2.23	12.09 ± 1.77	2.09 ± 0.34	5.50 ± 0.41	4.58 ± 1.46	0.14 ± 0.02	92.70 ± 0.14	0.56 ± 0.02
55	*Mussaenda erythrophylla*	0.61 ± 0.16	5.64 ± 0.43	7.89 ± 1.32	5.60 ± 0.52	5.00 ± 1.05	0.29 ± 0.02	77.96 ± 1.92	3.35 ± 0.19
56	*Mussaenda philippica*	0.39 ± 0.18	4.25 ± 1.21	2.93 ± 0.53	6.40 ± 0.52	11.00 ± 0.67	0.61 ± 0.06	78.56 ± 0.69	1.36 ± 0.09
57	*Palicourea angustifolia*	0.24 ± 0.04	2.52 ± 0.27	0.52 ± 0.05	5.60 ± 0.21	4.00 ± 0.27	0.24 ± 0.03	87.85 ± 0.35	1.20 ± 0.01
58	*Brugmansia arborea*	11.54 ± 2.67	22.93 ± 3.84	4.09 ± 0.36	5.23 ± 0.56	6.62 ± 1.27	0.60 ± 0.04	87.34 ± 5.40	0.49 ± 0.05
59	*Iochroma gesnerioides*	0.30 ± 0.05	3.14 ± 0.30	0.45 ± 0.04	5.00 ± 0.00	5.41 ± 0.41	2.68 ± 0.12	86.14 ± 0.15	0.30 ± 0.03
60	*Browallia americana*	0.02 ± 0.00	1.67 ± 0.07	1.35 ± 0.26	6.45 ± 0.44	4.00 ± 0.00	0.31 ± 0.01	87.39 ± 0.69	0.95 ± 0.06
61	*Brugmansia versicolor*	10.91 ± 2.52	17.66 ± 2.99	8.76 ± 0.96	6.40 ± 0.52	7.00 ± 0.00	0.60 ± 0.04	88.96 ± 0.45	0.85 ± 0.02
62	*Solanum dulcamara*	0.10 ± 0.02	2.20 ± 0.36	1.78 ± 0.51	10.10 ± 0.77	3.68 ± 0.99	1.24 ± 0.25	91.00 ± 0.32	0.98 ± 0.03
63	*Lantana rugulosa*	0.01 ± 0.00	1.07 ± 0.11	1.02 ± 0.09	7.00 ± 0.00	5.46 ± 0.47	0.20 ± 0.01	70.49 ± 0.05	3.35 ± 0.25
64	*Alpinia purpurata*	28.97 ± 4.45	23.75 ± 2.06	62.85 ± 14.6	3.00 ± 0.00	1.40 ± 0.52	0.15 ± 0.03	87.88 ± 2.76	4.02 ± 0.14
65	*Hedychium coronarium*	2.31 ± 0.66	10.92 ± 1.78	13.08 ± 5.38	6.60 ± 0.21	2.64 ± 0.31	0.47 ± 0.04	94.70 ± 0.11	1.09 ± 0.01
66	*Hedychium coccineum*	0.23 ± 0.03	4.29 ± 0.61	1.73 ± 0.32	7.00 ± 0.00	5.00 ± 0.00	1.06 ± 0.09	42.04 ± 4.81	6.12 ± 1.32
67	*Renealmia alpinia*	0.24 ± 0.04	2.52 ± 0.27	0.52 ± 0.05	5.60 ± 0.21	4.00 ± 0.27	0.24 ± 0.03	53.70 ± 1.70	1.20 ± 0.01

Note: DL, longitudinal diameter; DE, equatorial diameter; SS, soluble solid; TA, total titratable acidity.

3.2. Optimization of Extraction Parameters and Quantification of Carotenoids

Figure 2A,B shows the experimental design results for the carotenoids' micro-extraction. The solvents selected for optimization in this study were derived from a literature review of 119 species and previous studies [9,33].It is highlighted that both the plant species and the solvent used for extraction have a significant influence, with extraction time being a minor effect compared to the other variables. It is important to note that the maximum carotenoid extraction was obtained using the solvent combination of methanol–acetone–dichloromethane (1:1:2) and acetone–methanol (2:1), with an extraction time of one minute.

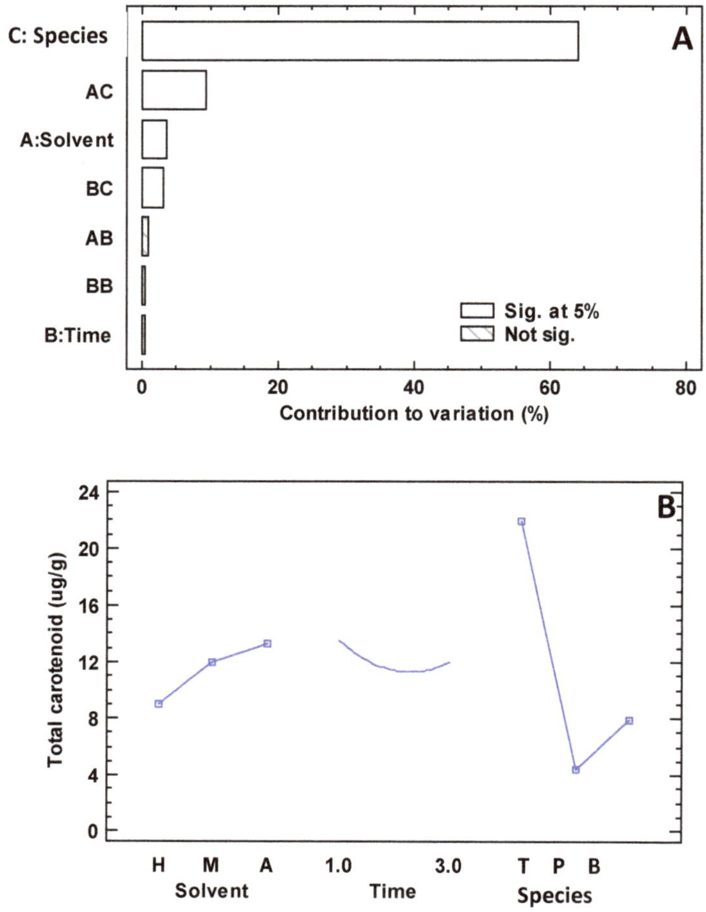

Figure 2. Results of the experimental design for the extraction of carotenoids. Note: In (**A**), the capital letters on the y-axis represent: A, solvent; B, time; C, species. In (**B**), the capital letters on the x-axis represent: H, *n*-hexane–acetone (1:1); M, methanol–acetone–dichloromethane (1:1:2); A, acetone–methanol (2:1); time in minutes; T, *Taraxacum officinale*; P, *Pyrostegia venusta*; B, *Buddleja globose*.

Table 3 shows the analysis results to quantify the total carotenoid content in various flower species. The results show a significant variation in the levels of total carotenoids among the multiple species examined. The recorded values ranged from 0.25 mg β-carotene/g DW in *C. cuatrecasasii*, which exhibited a pink visual colour, to 292.5 mg β-carotene/g DW in *R. alpinia*, characterized by a visible orange colour. These results highlight the diverse and substantial differences in carotenoid accumulation within the studied flower species, emphasizing the influence of pigmentation on carotenoid content.

Table 3. Average values of carotenoids, phenolics, and antioxidant activity concentration of the floral species analyzed.

N°	Scientific Name	Total Carotenoids (mg β-Carotene/g DW)	Total Phenolics (mg GAE/g DW)	Total Anthocyanins (mg C-3-gl/g DW)	Citric Acid (mg/100 g DW)	Malic Acid (mg/100 g DW)	Tartaric Acid (mg/100 g DW)	Antioxidant Activity (mmol TE/g) ABTS	Antioxidant Activity (mmol TE/g) DPPH
1	*M. erythrochlamys*	1.53 ± 0.02	76.09 ± 1.01	0.28 ± 0.03	116.4 ± 12.3	341.6 ± 5.2	213.4 ± 8.0	6.04 ± 0.70	31.89 ± 2.79
2	*O. cuspidatum*	3.04 ± 0.00	223.24 ± 4.66	0.20 ± 0.04	320.9 ± 7.5	141.7 ± 3.1	59.2 ± 3.4	5.96 ± 1.04	20.76 ± 0.57
3	*P. lutea*	83.03 ± 8.81	104.41 ± 0.75	0.13 ± 0.02	142.6 ± 1.0	112.8 ± 3.8	54.8 ± 1.2	2.50 ± 0.40	23.42 ± 3.49
4	*S. oblonga*	2.95 ± 0.02	194.55 ± 6.28	0.16 ± 0.01	42.9 ± 3.5	322.8 ± 10.9	344.4 ± 5.9	4.63 ± 0.02	31.50 ± 2.89
5	*T. grandiflora*	0.82 ± 0.02	110.25 ± 0.54	0.16 ± 0.01	20.2 ± 1.8	74.0 ± 1.3	106.6 ± 6.6	5.22 ± 0.04	48.42 ± 3.75
6	*C. argentea*	1.33 ± 0.01	140.63 ± 0.85	0.10 ± 0.02	51.6 ± 3.3	122.3 ± 10.6	63.6 ± 1.2	2.47 ± 0.57	28.00 ± 0.67
7	*E. aquaticum*	15.04 ± 0.56	85.81 ± 2.03	0.36 ± 0.08	84.8 ± 1.5	4605.0 ± 308.3	103.0 ± 3.9	4.96 ± 0.07	32.26 ± 3.90
8	*A. cathartica*	32.53 ± 0.09	167.12 ± 3.72	0.16 ± 0.01	613.6 ± 19.9	261.3 ± 21.3	106.7 ± 0.0	3.70 ± 0.85	25.73 ± 2.07
9	*P. fragrantissimum*	1.80 ± 0.02	291.75 ± 10.84	0.59 ± 0.08	44.1 ± 0.9	249.1 ± 1.3	22.0 ± 0.3	6.09 ± 0.92	33.41 ± 1.14
10	*P. strictum*	1.81 ± 0.05	353.57 ± 17.01	0.72 ± 0.15	923.3 ± 64.9	241.3 ± 7.1	13.6 ± 0.3	4.84 ± 0.60	46.05 ± 2.11
11	*X. robustum*	0.59 ± 0.01	94.45 ± 3.11	0.50 ± 0.16	391.9 ± 3.5	37.8 ± 0.9	34.3 ± 1.0	5.05 ± 0.03	47.75 ± 1.88
12	*A. papposa*	211.00 ± 0.57	92.39 ± 1.21	0.31 ± 0.06	112.7 ± 6.6	262.6 ± 43.8	104.8 ± 0.3	3.96 ± 0.02	36.53 ± 4.46
13	*D. pinnata*	1.11 ± 0.03	199.65 ± 2.50	1.84 ± 0.19	106.8 ± 23.7	1735.6 ± 275.3	92.7 ± 15.2	4.46 ± 0.20	65.79 ± 7.14
14	*D. balsapampae*	129.43 ± 12.75	133.07 ± 2.89	2.04 ± 0.12	168.5 ± 18.9	156.9 ± 10.2	226.9 ± 2.5	5.86 ± 0.08	64.46 ± 9.82
15	*P. chenopodioides*	1.87 ± 0.01	114.09 ± 4.62	0.31 ± 0.03	206.5 ± 2.8	431.4 ± 10.8	316.2 ± 5.8	4.55 ± 0.04	21.02 ± 0.87
16	*I. hancockii*	2.07 ± 0.11	165.15 ± 18.36	0.40 ± 0.01	39.1 ± 3.5	30.3 ± 0.0	35.7 ± 1.2	7.33 ± 0.10	32.73 ± 5.26
17	*I. hawkeri*	18.82 ± 0.65	215.17 ± 2.00	0.69 ± 0.09	105.9 ± 4.6	360.5 ± 20.3	222.9 ± 17.2	5.04 ± 0.03	77.77 ± 5.82
18	*I. sodenii*	23.86 ± 2.34	253.43 ± 11.18	0.30 ± 0.02	125.8 ± 9.2	95.5 ± 0.7	116.8 ± 2.3	4.07 ± 0.15	54.92 ± 6.94
19	*I. walleriana*	19.83 ± 1.83	288.17 ± 4.94	0.77 ± 0.03	76.7 ± 1.7	229.0 ± 2.6	129.5 ± 1.6	4.22 ± 1.00	73.23 ± 6.44
20	*A. corymbosa*	12.95 ± 0.14	302.67 ± 14.21	0.50 ± 0.05	62.4 ± 4.7	298.4 ± 1.8	31.3 ± 1.2	4.59 ± 0.04	67.63 ± 2.15
21	*G. nicaraguensis*	3.97 ± 0.06	87.75 ± 1.11	0.20 ± 0.02	254.3 ± 2.6	301.3 ± 28.6	48.7 ± 0.0	2.02 ± 0.08	32.61 ± 5.38
22	*R. veitchii*	1.35 ± 0.01	72.62 ± 0.49	0.04 ± 0.00	272.1 ± 3.1	979.4 ± 16.4	836.6 ± 16.4	4.00 ± 0.06	16.29 ± 1.00
23	*C. cornutus*	1.11 ± 0.01	262.11 ± 15.33	0.23 ± 0.04	956.7 ± 34.6	11655.9 ± 251.3	75.6 ± 1.6	2.86 ± 0.46	34.61 ± 2.44
24	*C. papaya*	2.43 ± 0.02	223.18 ± 7.29	0.23 ± 0.03	435.0 ± 10.6	490.9 ± 12.3	313.7 ± 13.2	5.02 ± 0.26	27.81 ± 2.79
25	*I. triloba*	0.62 ± 0.08	157.49 ± 14.03	0.37 ± 0.09	1333.2 ± 89.2	651.9 ± 2.1	30.5 ± 0.1	5.74 ± 0.05	23.96 ± 4.85
26	*C. spiralis*	0.69 ± 0.04	121.13 ± 0.44	0.76 ± 0.06	137.7 ± 10.7	109.8 ± 5.2	4.8 ± 0.3	5.26 ± 0.01	18.48 ± 1.44
27	*C. nitida*	1.18 ± 0.01	292.34 ± 8.24	0.91 ± 0.05	713.8 ± 76.6	225.5 ± 11.7	170.7 ± 23.4	3.88 ± 0.90	33.25 ± 3.28
28	*C. bracteata*	1.00 ± 0.09	232.83 ± 6.52	0.08 ± 0.00	443.7 ± 1.5	186.7 ± 3.4	143.4 ± 9.5	5.03 ± 0.02	25.84 ± 4.21
29	*C. cuatrecasasii*	0.25 ± 0.02	108.28 ± 6.40	0.16 ± 0.00	48.7 ± 7.8	1002.6 ± 5.8	188.6 ± 20.3	5.02 ± 0.03	40.99 ± 3.91
30	*C. nobilis*	0.73 ± 0.04	151.12 ± 4.14	0.03 ± 0.00	260.8 ± 18.6	278.3 ± 12.0	157.7 ± 29.7	5.01 ± 0.07	17.89 ± 2.61
31	*T. floribunda*	1.39 ± 0.03	259.37 ± 3.35	0.29 ± 0.06	3516.1 ± 154.1	209.4 ± 18.9	87.8 ± 10.2	7.80 ± 0.06	52.46 ± 3.96
32	*A. poiretii*	2.54 ± 0.06	269.57 ± 2.06	0.73 ± 0.06	205.4 ± 19.6	15613.6 ± 1238.4	55.2 ± 4.6	2.95 ± 0.47	44.11 ± 5.50
33	*B. macrophylla*	1.58 ± 0.05	299.60 ± 4.87	2.44 ± 0.38	114.3 ± 0.8	85.9 ± 1.3	51.8 ± 0.3	5.06 ± 0.02	56.74 ± 7.54
34	*C. angustifolia*	2.18 ± 0.05	337.21 ± 3.78	0.17 ± 0.01	158.8 ± 10.3	123.5 ± 12.8	196.0 ± 6.2	4.68 ± 0.03	49.59 ± 1.38

Table 3. *Cont.*

N°	Scientific Name	Total Carotenoids (mg β-Carotene/g DW)	Total Phenolics (mg GAE/g DW)	Total Anthocyanins (mg C-3-gl/g DW)	Citric Acid (mg/100 g DW)	Malic Acid (mg/100 g DW)	Tartaric Acid (mg/100 g DW)	Antioxidant Activity (mmol TE/g) ABTS	Antioxidant Activity (mmol TE/g) DPPH
35	*E. americana*	0.84 ± 0.04	258.43 ± 14.93	0.35 ± 0.05	2874.9 ± 101.0	656.1 ± 13.8	439.3 ± 13.6	4.31 ± 1.00	39.36 ± 3.48
36	*M. sodiroana*	1.41 ± 0.16	175.38 ± 36.04	0.88 ± 0.13	142.3 ± 0.5	60.7 ± 1.2	6.9 ± 0.2	3.31 ± 0.48	16.74 ± 1.50
37	*G. ichthyoderma*	0.66 ± 0.06	251.00 ± 13.57	0.37 ± 0.04	28.0 ± 0.4	186.5 ± 1.9	334.7 ± 2.4	3.86 ± 0.89	47.84 ± 2.16
38	*H. episcopalis*	1.55 ± 0.13	56.86 ± 0.50	0.09 ± 0.00	346.9 ± 6.0	761.3 ± 12.6	17.4 ± 0.3	3.02 ± 0.03	29.28 ± 3.04
39	*H. collinsiana*	1.78 ± 0.01	300.94 ± 13.94	0.15 ± 0.04	92.1 ± 7.3	1553.1 ± 8.4	104.1 ± 0.8	4.57 ± 0.02	24.92 ± 3.65
40	*H. latispatha*	0.43 ± 0.01	86.40 ± 1.09	0.11 ± 0.06	45.0 ± 10.0	114.6 ± 0.5	5488.2 ± 70.6	4.65 ± 0.02	15.98 ± 0.70
41	*H. rostrata*	2.49 ± 0.08	155.21 ± 9.89	0.57 ± 0.01	164.9 ± 14.1	637.7 ± 4.1	357.0 ± 75.7	2.56 ± 0.26	22.53 ± 1.77
42	*H. wagneriana*	2.43 ± 0.00	288.17 ± 4.94	0.18 ± 0.04	3800.5 ± 104.4	1327.4 ± 40.0	61.0 ± 1.1	4.22 ± 1.00	46.45 ± 3.57
43	*P. vulgaris*	1.92 ± 0.05	110.69 ± 10.96	0.34 ± 0.06	90.5 ± 0.2	141.8 ± 14.7	25.0 ± 3.1	4.99 ± 0.04	59.43 ± 6.32
44	*H. rosa-sinensis*	3.64 ± 0.13	290.30 ± 9.67	1.93 ± 0.45	17818.6 ± 475.5	1693.8 ± 48.2	46.1 ± 7.2	7.67 ± 0.12	62.15 ± 1.08
45	*S. stromanthoides*	2.64 ± 0.00	50.60 ± 1.55	0.18 ± 0.03	39.6 ± 5.6	55.5 ± 6.8	5.6 ± 1.9	2.63 ± 0.08	24.73 ± 3.20
46	*P. heteromallum*	0.93 ± 0.05	344.22 ± 2.63	7.35 ± 0.43	88.5 ± 7.9	2254.1 ± 33.6	51.3 ± 0.9	4.18 ± 0.95	56.30 ± 2.91
47	*A. lepidotus*	1.42 ± 0.04	246.19 ± 16.01	0.29 ± 0.05	56.1 ± 0.7	5591.2 ± 64.0	137.4 ± 9.8	7.43 ± 0.13	31.76 ± 4.02
48	*C. mollis*	0.86 ± 0.07	354.20 ± 1.20	1.67 ± 0.08	51.1 ± 3.3	161.5 ± 6.8	708.2 ± 32.6	5.01 ± 0.07	55.69 ± 5.02
49	*P. urvilleana*	1.00 ± 0.04	510.53 ± 34.59	1.01 ± 0.01	55.2 ± 13.4	3540.8 ± 173.8	255.7 ± 25.7	4.66 ± 0.03	53.66 ± 4.49
50	*M. velutina*	0.92 ± 0.04	219.08 ± 8.51	0.42 ± 0.01	105.3 ± 10.6	846.2 ± 62.5	39.6 ± 8.1	6.58 ± 0.08	26.27 ± 3.13
51	*B. spectabilis*	2.65 ± 0.02	327.01 ± 8.42	0.06 ± 0.00	175.2 ± 12.5	394.0 ± 4.4	3197.1 ± 401.1	5.65 ± 0.07	29.85 ± 2.64
52	*E. aurantiacus*	1.12 ± 0.08	74.77 ± 2.06	0.13 ± 0.02	59.6 ± 7.2	293.1 ± 2.8	154.2 ± 11.4	4.98 ± 0.04	28.42 ± 3.65
53	*Oncidium sp.*	13.76 ± 0.12	99.11 ± 4.48	0.75 ± 0.06	201.0 ± 9.2	216.0 ± 1.8	171.3 ± 12.0	5.01 ± 0.08	26.56 ± 1.05
54	*S. liliastrum*	0.62 ± 0.01	97.89 ± 3.01	0.14 ± 0.03	365.5 ± 8.0	735.3 ± 56.3	30.6 ± 0.1	3.37 ± 0.11	38.61 ± 5.56
55	*M. erythrophylla*	0.92 ± 0.11	153.47 ± 3.39	0.74 ± 0.06	171.8 ± 5.7	740.0 ± 16.4	350.4 ± 14.9	5.11 ± 0.03	55.21 ± 9.54
56	*M. philippica*	1.99 ± 0.05	202.06 ± 7.07	0.17 ± 0.03	24.9 ± 1.6	87.3 ± 7.6	7.5 ± 0.58	5.03 ± 0.03	25.33 ± 5.74
57	*P. angustifolia*	0.84 ± 0.09	79.42 ± 1.43	0.21 ± 0.05	126.2 ± 9.6	3121.3 ± 260.7	92.5 ± 7.6	2.54 ± 0.04	18.64 ± 1.46
58	*B. arborea*	0.89 ± 0.04	153.37 ± 16.54	0.16 ± 0.05	878.1 ± 25.9	13.2 ± 0.5	197.1 ± 6.1	4.29 ± 0.03	27.70 ± 3.26
59	*I. gesnerioides*	1.15 ± 0.03	273.93 ± 5.84	0.47 ± 0.05	2662.8 ± 1581	666.3 ± 8.0	276.8 ± 9.7	5.74 ± 0.02	30.59 ± 6.11
60	*B. americana*	2.01 ± 0.02	165.91 ± 19.98	0.16 ± 0.01	782.6 ± 54.2	483.3 ± 22.8	1519.5 ± 33.5	4.40 ± 0.02	75.87 ± 4.75
61	*B. versicolor*	3.10 ± 0.16	120.74 ± 9.98	0.13 ± 0.04	1366.5 ± 67.7	764.5 ± 29.9	116.3 ± 0.1	4.70 ± 0.37	52.89 ± 4.57
62	*S. dulcamara*	3.65 ± 0.02	110.94 ± 7.18	0.40 ± 0.01	498.5 ± 34.6	182.1 ± 7.9	67.3 ± 0.2	4.96 ± 0.03	19.35 ± 4.53
63	*L. rugulosa*	0.89 ± 0.05	268.80 ± 13.73	0.31 ± 0.01	195.4 ± 15.4	1152.5 ± 13.5	34.0 ± 0.09	4.78 ± 1.13	40.97 ± 2.21
64	*A. purpurata*	1.33 ± 0.04	344.68 ± 5.80	1.32 ± 0.08	58.1 ± 0.4	189.5 ± 9.6	38.6 ± 1.0	5.72 ± 0.06	37.98 ± 3.63
65	*H. coronarium*	1.19 ± 0.00	75.57 ± 1.11	0.52 ± 0.07	214.9 ± 13.8	89.1 ± 2.9	16.0 ± 1.2	3.78 ± 0.05	30.00 ± 4.07
66	*H. coccineum*	1.71 ± 0.10	86.60 ± 0.81	0.23 ± 0.04	245.2 ± 1.51	100.7 ± 4.9	14.9 ± 0.6	4.90 ± 0.20	24.57 ± 1.36
67	*R. alpinia*	292.50 ± 0.89	34.50 ± 0.77	0.29 ± 0.00	142.1 ± 1.10	98.3 ± 3.2	9.2 ± 0.6	4.85 ± 0.20	42.11 ± 2.78

3.3. Optimization of Extraction Parameters and Quantification of Phenolic Compounds

Figure 3A,B show the experimental design results applied to the phenolic compound micro-extraction. The solvents selected for optimization in this study were derived from a literature review of 119 species and previous studies [9,33]. It is highlighted that both the solvent and the plant species used for extraction have a significant influence, with extraction time being a minor effect compared to the other variables. Notably, the maximum phenolics extraction was obtained using 80% methanol acidified with 0.1% hydrochloric acid, with an extraction time of five minutes. The concentration of total phenolics determined by the spectrophotometric method ranged from 50.60 mg GAE/g DW in *S. stromanthoides*, which had an orange visual colour, to 510.53 mg GAE/g DW in *P. urvilleana*, characterized by a yellow-orange visual colour (Table 3).

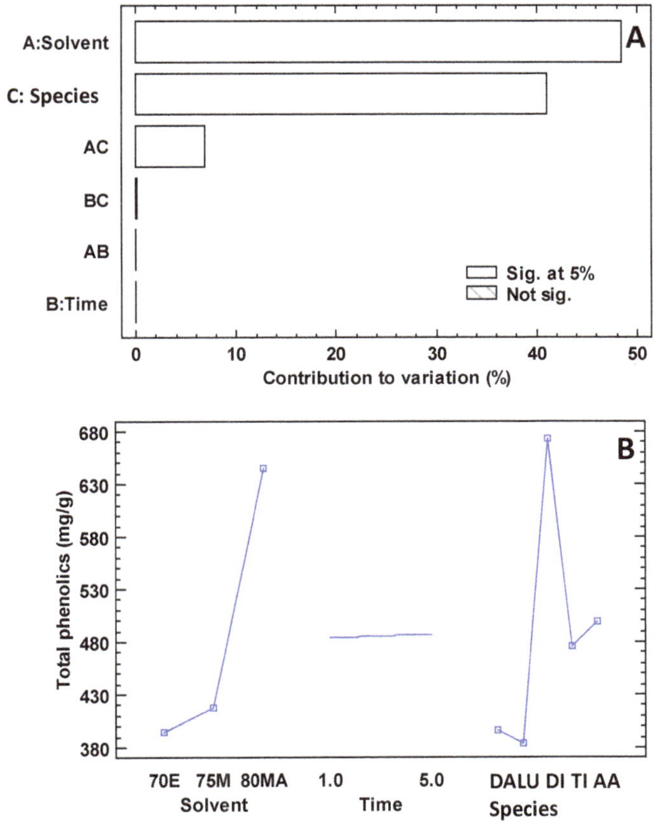

Figure 3. Results of the experimental design for the extraction of phenolics. Note: In (**A**), the capital letters on the y-axis represent: A, solvent; B, time; C, species. In (**B**), the capital letters on the x-axis represent: 70E, 75% ethanol; 75M, 75% methanol; 80MA, 80% methanol acidified with 0.1% hydrochloric acid; time in minutes; DA, *Dahlia pinnata*; LU, *V*; DI, *Dianthus caryophyllus*; TI, *Plerona urvilleanum*; AA, *Agapanthus africanus*.

Table 4 shows the concentration of gallic acid (ranging from 3.6 mg/100 g DW in *G. ichtyoderma* to 869.4 mg/100 g DW in *C. cornutus*), protocatechuic acid (ranging from 4.6 mg/100 g DW in *H. coronarium* to 231.0 mg/100 g DW in *P. angustifolia*.), *p*-coumaric acid (ranging from 30.2 mg/100 g DW in *H. wagneriana* to 2929.0 mg/100 g DW in *C. bracteata*), *m*-coumaric acid (ranged from 30.2 mg/100 g DW in *H. wagneriana* to 2929.0 mg/100 g DW in *C. bracteata*), *m*-coumaric acid (ranged from 23.5 mg/100 g DW in

I. hancockii to 12,044.0 mg/100 g DW in *A. poiretii*), syringic acid (ranged from 11.0 mg/100 g DW in *S. stromanthoides* to 3225.0 mg/100 g DW in *A. poiretii*), chlorogenic acid (ranged from 37.1 mg/100 g DW in *M. erythrophylla* to 3435.0 mg/100 g DW in *S. dulcamara*), 4-hydroxybenzoic acid (range from 17.6 mg/100 g DW in *S. oblonga* to 10,729 mg/100 g DW in *E. arborea*), caffeic acid (range from 108. 6 mg/100 g DW in *I. hancockii* to 5893.0 mg/100 g DW in *A. corymbosa*), ferulic acid (ranged from 148.2 mg/100 g DW in *I. hancockii* to 5849.0 mg/100 g DW in *T. grandiflora*), rutin (ranged from 16. 9 mg/100 g DW in *H. episcopalis* to 3814.0 mg/100 g DW in *C. angustifolia*), quercetin (ranged from 28.1 mg/100 g DW in *H. latispatha* to 5119.0 mg/100 g DW in *D. pinnata*), quercetin glucoside (ranged from 32.8 mg/100 g DW in *I. triloba* to 2950.0 mg/100 g DW in *L. rugulosa*), and kamferol (ranged from 39.4 mg/100 g DW in *H. coccineum* to 8236.0 mg/100 g DW in *D. pinnata*). The total concentration as the sum of the individual phenolics ranged from 6.3 mg/100 g DW in *R. alpinia* to 110364 mg/100 g DW in *M. erythrophylla*. These results show the diversity in the accumulation of phenolic compounds within the flower species studied.

3.4. Quantification of Total Anthocyanins

Anthocyanins are a subclass of flavonoids [9]. Table shows the results of the quantified analysis of the content of total anthocyanins in different species of flowers. Thus, the concentrations ranged from 0.03 mg C-3-gl/g DW (*C. nobilis*) to 7.35 mg C-3-gl/g DW (*P. heteromallum*). In addition, species such as *B. macrophylla* (2.44 mg C-3-gl/g DW), *D. balsapampae* (2.04 mg C-3-gl/g DW), *H. rosa-sinensis* (1.93 mg C-3-gl/g DW), and *D. pinnata* (1.84 mg C-3-gl/g DW) also showed high anthocyanin concentrations.

3.5. Quantification of Organic Acids

Table 3 shows the results of the quantified analysis of the content of organic acids in different species of flowers. The data show a significant variation in the content of these compounds among the species studied. Citric acid content ranged from 20.2 mg/100 g DW in *T. grandiflora* to 17,819 mg/100 g DW in *H. rosa-sinensis*, the latter being particularly high. For malic acid, values ranged from 13.2 mg/100 g DW in *B. arborea* to 16,614 mg/100 g DW in *A. piretii*. Finally, tartaric acid showed values ranging from 4.8 mg/100 g DW in *C. spiralis* to 5488 mg/100 g DW in *H. latispatha*. These results show the remarkable diversity in the composition of organic acids among the flower species analyzed, highlighting their potential for specific applications based on their biochemical profile.

3.6. Antioxidant Activity (ABTS and DPPH)

Table 3 presents the results of the percentage inhibition of the ABTS radical and the corresponding antioxidant activity using aqueous methanol and acetone extracts of different flower species. The antioxidant activity by the ABTS method, expressed in concentration, ranged from 2.02 mmol ET/g DW in *G. nicaraguensis* to 7.80 mmol TE/g DW in *T. floribunda*. In addition, *H. rosa-sinensis* (7.67 mmol TE/g DW), *A. lepidotus* (7.43 mmol TE/g DW), and *I. hancockii* (7.33 mmol TE/g DW) also showed high antioxidant activity values. In turn, the antioxidant activity of the DPPH method ranged from 14.98 mmol TE/g DW (*H. latispatha*) to 77.77 mmol TE/g DW (*I. hawkeri*). In addition, species such as *B. americana* (75.87 mmol TE/g DW), *I. walleriana* (73.23 mmol TE/g DW), *A. corymbosa* (67.63 mmol TE/g DW), and *D. pinnata* (65.79 mmol TE/g DW) showed a high antioxidant concentration.

Table 4. Average values of individual phenolics (mg/100 g DW) of the floral species analyzed.

N°	Scientific Name	Gallic Acid	Protocatechuic Acid	*p*-Coumaric Acid	*m*-Coumaric Acid	Syringic Acid	Chlorogenic Acid	4-Hydroxibenzoic Acid	Caffeic Acid	Ferulic Acid	Rutin	Quercetin	Quercetin Glucoside	Kamferol	Total
1	*M. erythrochlamys*	41.1 ± 1.3				778.8 ± 7.6			725.3 ± 53.8	2196.2 ± 39.8	135.0 ± 6.1	450.0 ± 3.1	118.4 ± 9.6		4444.8 ± 89.8
2	*O. cuspidatum*	24.0 ± 1.0		349.1 ± 8.4					2536.1 ± 11.7	1957.5 ± 34.9			712.1 ± 16.5		5578.8 ± 72.4
3	*P. lutea*	92.3 ± 4.1		117.8 ± 1.3		46.8 ± 4.5		666.3 ± 41.2			225.5 ± 3.4		235.8 ± 4.4		1384.6 ± 39.6
4	*S. oblonga*	17.6 ± 1.2		159.4 ± 1.3			60.6 ± 1.5	17.6 ± 1.1	820.8 ± 60.5	2605.1 ± 48.5	33.6 ± 0.4				3714.7 ± 108.9
5	*T. grandiflora*	212.6 ± 2.1			302.2 ± 6.0			106.5 ± 0.2	509.9 ± 35.9	5848.8 ± 8.6	814.8 ± 4.5				7795.2 ± 36.1
6	*C. argentea*	81.8 ± 2.7				41.5 ± 1.2		164.7 ± 3.7	242 ± 30.1		67.2 ± 0.5		431.2 ± 11.7	89.6 ± 3.2	1118.0 ± 45.2
7	*E. aquaticum*	13.2 ± 0.6	43.3 ± 1.7				621 ± 4.9		227.5 ± 1.0				602.6 ± 22.9		1507.6 ± 31.1
8	*A. cathartica*	21.1 ± 2.0		503.9 ± 6.1	219.9 ± 13.9		66.3 ± 5.0	133.6 ± 12.7	307.7 ± 31.5	637 ± 32.9	140.4 ± 23.2		138.9 ± 13.5		2168.6 ± 128.7
9*	*P. fragrantissimum*	15.5 ± 0.6				22.5 ± 3.4		148.7 ± 11.8							655.0 ± 7.1
10	*P. strictum*	78.4 ± 4.1									427.3 ± 0.4	793.8 ± 1.1			1299.5 ± 4.8
11	*X. robustum*	46.2 ± 1.3	36.3 ± 0.2				224.7 ± 0.8	141.6 ± 22.2			380.6 ± 12.9		152.1 ± 4.2		981.6 ± 30.1
12*	*A. papposa*	24.3 ± 0.1		164.2 ± 1.3	165.0 ± 6.5			128.1 ± 2.7	1514.7 ± 56.8	204.8 ± 22.3					2208.3 ± 86.9
13	*D. pinnata*	114.1 ± 0.1				122.7 ± 2.7					601.3 ± 23	5118.6 ± 23.8	1047.4 ± 161.9	8235.5 ± 214.8	15239.6 ± 8.9
14*	*D. balsapampae*						314.3 ± 6.6	308.8 ± 4.8			657.5 ± 1.3				2712.3 ± 66
15	*P. chenopodioides*	37.2 ± 2.2		104.0 ± 2.2				118.7 ± 0.8			207.1 ± 4.8		159.1 ± 10.3		626.0 ± 6.3
16*	*I. hancockii*	14.5 ± 0.4	28.6 ± 0.5		23.5 ± 0.5			133.5 ± 6.8	108.6 ± 15.0	148.2 ± 7.5	152 ± 1.3		117.1 ± 0.9	117.8 ± 0.3	851.1 ± 26.7
17	*I. hawkeri*	42.5 ± 1.3	42.7 ± 0.4	632.3 ± 0.8			104.4 ± 1.5	185.9 ± 7.8	2202.9 ± 32.7					100.0 ± 4.6	3310.7 ± 5.5
18	*I. sodenii*	32.7 ± 2.9	55 ± 4.2								378.5 ± 17.8	450.0 ± 3.1	2405 ± 115.7	577.5 ± 25.2	3448.7 ± 165.2
19	*I. walleriana*	53.6 ± 5.3	59.6 ± 0.6	613.0 ± 14.4			126.4 ± 2.2	183.9 ± 6.1	2838.4 ± 42.5					164.6 ± 2.0	4039.4 ± 17.2
20	*A. corymbosa*	52.3 ± 2.0				25.6 ± 0.3		138.3 ± 4.2	5892.5 ± 172.2	757 ± 2	123.7 ± 13.1				6989.3 ± 159.1
21	*G. nicaraguensis*	30.5 ± 1.1		186.1 ± 4.6				736.5 ± 46.9	1443.6 ± 7.0	557.8 ± 56.2					2954.5 ± 113.6
22*	*R. veitchii*	89.6 ± 1.5		100.9 ± 8.5			477.2 ± 0.7	333.2 ± 8.8			258.2 ± 0.3		176.0 ± 10.5		1460.9 ± 12.7
23	*C. cornutus*	869.4 ± 2.0		1267.6 ± 96.0					329 ± 4.4		168.4 ± 0.3		2013.0 ± 2.8		4647.4 ± 99.3
24	*C. papaya*	80.6 ± 3.1							4005 ± 113.3		810.2 ± 1.5				4895.8 ± 108.7
25	*I. triloba*	117.3 ± 3.4						304.6 ± 3.0	947.5 ± 99.3				32.8 ± 3.0		1402.2 ± 102.7
26*	*C. spiralis*	32.9 ± 1.1				314.3 ± 1.8		147.1 ± 13.2					59.8 ± 3.2		583.4 ± 16.3
27*	*C. nitida*	27.4 ± 0.0	38.7 ± 2.1				356 ± 24.3				55.3 ± 0.0		59.1 ± 3.4		1449.2 ± 60.7
28	*C. bracteata*			2928.5 ± 97.2			240.2 ± 9.2		291.2 ± 10.6						3459.9 ± 117
29*	*C. cuatrecasasii*		27.2 ± 1.1	1170.4 ± 34.0			77.5 ± 2.1				91.4 ± 4.5		192.3 ± 6.5	46.7 ± 6.0	1640.4 ± 12.3
30	*C. nobilis*			2149.5 ± 14.8			158.8 ± 2.1		204 ± 8.8						2512.2 ± 35.8
31*	*T. floribunda*	10.4 ± 0.2		252.4 ± 18.5					1133 ± 63.5	552 ± 30.6					1960.3 ± 112.5
32	*A. poiretii*	81.5 ± 0.7			12,044.3 ± 58.7	3224.7 ± 9.5									15,350.2 ± 35.6
33	*B. macrophylla*	4.3 ± 0.1	133.2 ± 1.0	345.8 ± 9.7		439.3 ± 4.4		65.7 ± 1.2			59.1 ± 1.1		106.1 ± 1.7	41.5 ± 1.4	1194.9 ± 18.7
34	*C. angustifolia*	51.4 ± 3.4			1960.5 ± 1.2		709.4 ± 12.3	113.2 ± 0.9			3814.4 ± 10.1		175.6 ± 1.6		6824.5 ± 29.5
35	*E. americana*	92.4 ± 6.2				402.9 ± 41.9			4811.2 ± 24.3		150.1 ± 2.2		908.5 ± 21.3	306.0 ± 10.8	6671 ± 58.2
36	*M. sodiroana*	42.6 ± 0.1	196.7 ± 10.4	463.9 ± 18.8	5390.4 ± 28.8						73.5 ± 0.2	177.4 ± 3.0	101.5 ± 1.4	281.0 ± 13.3	6727.1 ± 328.6
37*	*G. ichthyoderma*	3.6 ± 0.1							3561.9 ± 79.8						6952.4 ± 153.6
38*	*H. episcopalis*	65.6 ± 0.9			59.3 ± 1.9	11.5 ± 0.9		185.5 ± 8.1			16.9 ± 1.0				338.9 ± 7.2
39	*H. collinsiana*	30.8 ± 1.4				52.6 ± 0.3					45.1 ± 1.7	73.0 ± 1.6	180.4 ± 0.6		431.0 ± 1.2
40	*H. latispatha*	36.1 ± 0.2			89.2 ± 7.2			125.8 ± 8.3		234.1 ± 2.6	18.5 ± 0.1	28.1 ± 0.6			531.8 ± 12.6
41	*H. rostrata*	44.7 ± 4.6		102.5 ± 6.4	387.5 ± 3.3						272.0 ± 17.3		403.6 ± 24.6	469.0 ± 6.2	1679.3 ± 49.6
42	*H. wagneriana*	359 ± 15.7		30.2 ± 1.3					432.3 ± 6.5		33.8 ± 1.6		140.6 ± 4.4	69.7 ± 1.6	1065.5 ± 6.0
43	*P. vulgaris*	18.3 ± 0.6					77.9 ± 1.2		771.8 ± 36.2	3784.4 ± 268.3	57.3 ± 0.6		80.9 ± 2.4		4790.5 ± 308.2
44*	*H. rosa-sinensis*	20.6 ± 0.3			7022.9 ± 330.0						263.8 ± 1.8		81.8 ± 1.0		8428.7 ± 334.9
45	*S. stromanthoides*	20.4 ± 1.4				11.0 ± 0.5	45.3 ± 2.4	214.9 ± 3.1	130.2 ± 7.2	636.7 ± 12.9	54.0 ± 0.1	66.6 ± 2.3	68.5 ± 1.7		1247.4 ± 9.9
46	*P. heteromallum*	54 ± 0.4		1850.7 ± 56.8		68.5 ± 2.3									1973.2 ± 54.1
47	*A. lepidotus*	32 ± 1.1	24.3 ± 0.8		313.4 ± 7.2			2799.6 ± 67.3							3169.3 ± 60.4
48*	*C. mollis*	172.8 ± 9.3		822 ± 13.4							724.4 ± 2.2		472.1 ± 5.7		2907.3 ± 19.2
49	*P. urvilleana*	97.9 ± 1.9		175.1 ± 7.5		480.8 ± 2.9					207.9 ± 10.3		229.2 ± 19.9		1190.8 ± 6.9
50	*M. velutina*	48.4 ± 0.3				20.0 ± 0.6		177.2 ± 11.0			90.5 ± 1.6				336.1 ± 12.9
51	*B. spectabilis*	429.7 ± 14.9			755.9 ± 10.9	92.6 ± 1.9			435.2 ± 19.5	2850.1 ± 66					4563.5 ± 40.6
52*	*A. comata*	57.4 ± 0.4						653.7 ± 20.4							5824 ± 68.2
53	*Oncidium sp.*	47.31 ± 1.2				262.2 ± 3.1		207.2 ± 17.8			110.7 ± 2.4		1071.4 ± 6.6	236.5 ± 6.3	1935.3 ± 37.4
54	*S. liliastrum*	68.5 ± 2.3		191.1 ± 0.2		345.9 ± 12.7	280.3 ± 3.8	163.1 ± 12.6			102.9 ± 4.1		354.3 ± 9.0		1506.1 ± 6.3
55	*M. erythrophylla*	5.0 ± 0.2	16.1 ± 0.3			226.7 ± 29.2	37.1 ± 3.3	702.4 ± 54.6	2959.5 ± 144.2	4814.2 ± 137.3	176.9 ± 17.3		383.1 ± 86.5	587.8 ± 68.5	110,363.7 ± 431.8
56	*M. philippica*						547.6 ± 28.2				1349.8 ± 32.6		940.8 ± 26.3	3552.1 ± 23.8	6390.3 ± 110.8
57	*P. angustifolia*	11.5 ± 1.6	231.4 ± 11.9			425.7 ± 7.3	173.3 ± 1.3		430.3 ± 3.0						1272.3 ± 19
58	*B. arborea*	316.3 ± 1.9						10,729.0 ± 167.0			91.1 ± 2.0		78.3 ± 2.8		11,214.7 ± 173.7
59	*I. gesnerioides*						245.8 ± 5.7		2738.4 ± 190.6		83.1 ± 2.7		55.1 ± 2.4		3122.4 ± 191.2

Table 4. *Cont.*

N°	Scientific Name	Gallic Acid	Protocatechuic Acid	*p*-Coumaric Acid	*m*-Coumaric Acid	Syringic Acid	Chlorogenic Acid	4-Hydroxibenzoic Acid	Caffeic Acid	Ferulic Acid	Rutin	Quercetin	Quercetin Glucoside	Kamferol	Total
60	*B. americana*	95.5 ± 3.5						1279.4 ± 114.5			596.3 ± 0.0		825.9 ± 24.6		2797.1 ± 142.6
61	*B. versicolor*	106.2 ± 0.9			1231.2 ± 57.3			809.2 ± 30.9	2556.1 ± 293.8	775.9 ± 45.2	545.6 ± 42		203.3 ± 11.9		6227.5 ± 758.4
62	*S. dulcamara*			898.7 ± 2.4			3435 ± 65.4								4333.7 ± 67.8
63*	*L. rugulosa*	32.4 ± 0.8							2326.9 ± 1.6				2949.5 ± 27.9	998.0 ± 14.7	6368.6 ± 9.6
64*	*A. purpurata*	15.5 ± 0.6				558.3 ± 43.2		1410.9 ± 65.0			498.9 ± 19.0		235.6 ± 13.3	282.8 ± 11.8	3147.4 ± 172.4
65	*H. coronarium*	24.0 ± 0.6	4.6 ± 0.5				43.9 ± 0.8	112.9 ± 1.7	232.0 ± 2.9		56.9 ± 0.8		602.0 ± 1.4	88.6 ± 2.8	1164.8 ± 2.0
66	*H. coccineum*	13.3 ± 0.4				17.5 ± 0.4	43.1 ± 1.2	146.5 ± 17.0					41.8 ± 3.1	39.4 ± 0.7	
67	*R. alpinia*	6.3 ± 0.1													6.3 ± 0.2

Additional phenolics detected in species with *: 9*, catechin (468.3 ± 7.5 mg/100g); 12*, vanillic acid (7.3 ± 0.0 mg/100 g); 14*, synapinic acid (1431.8 ± 62.9 mg/100 g); 16*, vanillic acid (7.3 ± 0.4 mg/100 g); 22*, vanillic acid (25.8 ± 0.7 mg/100 g); 26*, vanillic acid (6.3 ± 0.7 mg/100 g) and catechin (23.0 ± 0.1 mg/100 g); 27*, vanillic acid (912.7 ± 30.9 mg/100 g); 29* catechin (34.9 ± 1.5 mg/100 g); 31*, vanillic acid (12.5 ± 0.0 mg/100 g). 37*, *o*-coumaric acid (3387.0 ± 73.9 mg/100 g); 38*, 49.0 ± 0.3 mg/100 g); 44*, naringenin (1039.7 ± 3.7 mg/100 g); 48*, myricetin (716.0 ± 6.9 mg/100 g); 52*, catechin (130.2 ± 9.1 mg/100 g) and naringenin (4982.7 ± 39.0 mg/100 g); 63*, vanillic acid (61.8 ± 1.2 mg/100 g); 64*, catechin (145.3 ± 20.7 mg/100 g).

3.7. Antimicrobial Activity

Antimicrobial susceptibility testing is critical for the effective management of pathogenic microorganisms. The well diffusion method demonstrated the magnitude of susceptibility of the pathogenic microorganisms (Figure 4). Thus, Table 5 presents the inhibition zone measurements for the antimicrobial activity of various flower extracts against bacterial and fungal strains. The inhibition zones ranged from 0 mm, indicating no activity in several species, to a maximum of 22.0 mm in *B. spectabilis*, which exhibited significant antibacterial efficacy. This variability highlights the broad spectrum of antimicrobial activity across the flower species tested. Notably, the extracts exhibited no antifungal activity against the tested fungi, except *S. heteromallum*, which displayed selective inhibition against *Candida albicans*. These findings suggest that while several floral species demonstrate potential as antibacterial agents, their antifungal properties are more limited and species-specific.

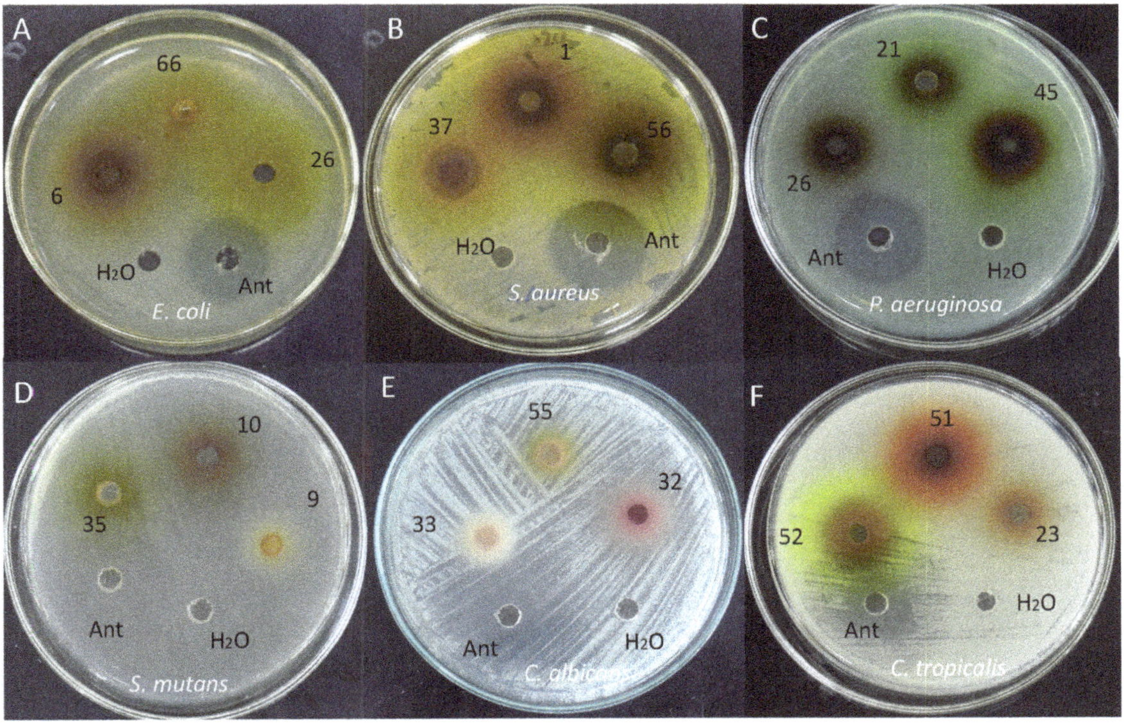

Figure 4. Antimicrobial activity of flower extracts against (**A**) Escherichia coli; (**B**) Staphylococcus aureus; (**C**) Pseudomonas aeruginosa; (**D**) Streptococcus mutans; (**E**) Candida albicans; (**F**) Candida tropicalis. Note: 1, *M. erythrochlamys*; 6, *C. argentea*; 9, *P. fragrantissimum*; 10, *P. strictum*; 21, *G. nicaraguensis*; 22, *R. veitchii*; 23, *C. cornutus*, 26, *C. spiralis*, 27, *C. spiralis*; 32, *A. poiretii*; 33, *B. macrophylla*; 35, *E. americana*; 37, *G. ichthyoderma*; 45, *S. stromanthoides*; 51, *B. spectabilis*; 52, *E. aurantiacus*; 55, *M. erythrophylla*; 56, *M. philippica*, 66, *H. coccineum*.

Table 5. Average values of the zone of inhibition of the floral species were analyzed.

N°	Flower Extracts	Extract Concentration (mg/mL)	Zone of Inhibition (mm) Bacterial strain				Fungal strain	
			E. coli ATCC 8739	*S. aureus* ATCC 6538P	*P. aeruginosa* ATCC 9027	*S. mutans* ATCC 25175	*C. albicans* ATCC 1031	*C. tropicalis* ATCC 13803
1	*M. erythrochlamys*	162.4	-	12.5 ± 2.1	-	-	-	-
2	*O. cuspidatum*	127.5	-	-	-	-	-	-
4	*S. oblonga*	157.8	-	11.5 ± 0.7	-		-	-
5	*T. grandiflora*	130.2	-	-	-	10.0 ± 0.0	-	-
6	*C. argentea*	104.4	-	11.5 ± 0.7	-	-	-	-
8	*A. cathartica*	66.7	-	7.5 ± 0.7	-	-	-	-
9	*P. fragrantissimum*	97.0	-	14.0 ± 0.0	10.5 ± 0.7	10.5 ± 0.7	-	-
10	*P. strictum*	127.9	10.5 ± 0.7	12.5 ± 0.7	-	10.0 ± 2.8	-	-
11	*X. robustum*	127.2	-	10.5 ± 0.7	10.0 ± 0.0	-	-	-
13	*D. pinnata*	132.7	-	8.5 ± 2.1		-	-	-
20	*A. corymbosa*	75.0	-	11.5 ± 0.7	-	-	-	-
21	*G. nicaraguensis*	128.6	-	-	-	-	-	-
22	*R. veitchii*	81.0	-	7.5 ± 0.7	-	-	-	-
26	*C. spiralis*	334.9	13.0 ± 0.0	11.0 ± 1.4	11.0 ± 0.0	-	-	-
30	*C. nobilis*	251.2	15.5 ± 0.7	20.5 ± 0.7	13.0 ± 2.8	16.0 ± 0.0	-	-
32	*Acalypha poiretii*	300.0	10.5 ± 0.7	17.0 ± 1.4	12.0 ± 4.2	13.0 ± 0.0	-	-
33	*B. macrophylla*	550.5	10.5 ± 0.7	16.0 ± 1.4	10.5 ± 0.7	17.0 ± 0.0	-	-
35	*E. americana*	223.1	10.5 ± 0.7	10.0 ± 0.0	-	15.5 ± 0.7	-	-
37	*G. ichthyoderma*	174.8	-	14.5 ± 0.7	-	-	-	-
38	*H. episcopalis*	255.6	-	11.5 ± 0.7	-	-	-	-
41	*H. rostrata*	281.2	-	13.5 ± 2.1	-	10.0 ± 0.0	-	-
42	*H. wagneriana*	71.9	-	7.0 ± 0.0	-	-	-	-
47	*S. stromanthoides*	336.9	-	10.0 ± 1.4	-	-	-	-
46	*P. heteromallum*	255.6	10.5 ± 0.7	18.5 ± 0.8	15.0 ± 1.4	21.0 ± 0.0	13.0 ± 0.0	-
50	*M. velutina*	159.2	-	12.0 ± 1.4	-	12.0 ± 0.0	-	-
51	*B. spectabilis*	143.3	-	13.0 ± 0.0	-	22.0 ± 0.0	-	-
52	*E. aurantiacus*	131.2	-	11.5 ± 0.7		-	-	-
53	*Oncidium sp*	110.1	-	-			-	-
55	*M. erythrophylla*	127.2	15.0 ± 0.0	14.5 ± 0.7			-	-
56	*M. philippica*	302.9	-	12.0 ± 1.4	-	-	-	-
57	*P. angustifolia*	125.6	-	8.5 ± 0.7	-	-	-	-
61	*B. versicolor*	132.5	9.0 ± 1.4	12.0 ± 0.0	-	-	-	-
62	*S. dulcamara*	111.3	-	-	-	-	-	-
65	*H. coronarium*	127.8	8.5 ± 0.7	15.0 ± 1.4	-	-	-	-
66	*H. coccineum*	146.2	-	8.0 ± 1.4	-	-	-	-
	Control *		21.0 ± 2.1	24.8 ± 1.8	24.0 ± 1.2	30.0 ± 3.9	12.9 ± 1.4	18.9 ± 3.0

Note: -, non-active at the tested concentration; *, Streptomycin for bacteria and fluconazole for fungi.

3.8. Statistical Analysis

Figure 5A shows the principal component analysis (PCA) considering the visual colour of the flowers analyzed, while Figure 5B shows the PCA considering the total set of variables in the study. Figure 5C shows a correlation plot including all variables considered in the study, and Figure 5D shows the specific correlation plot for these variables.

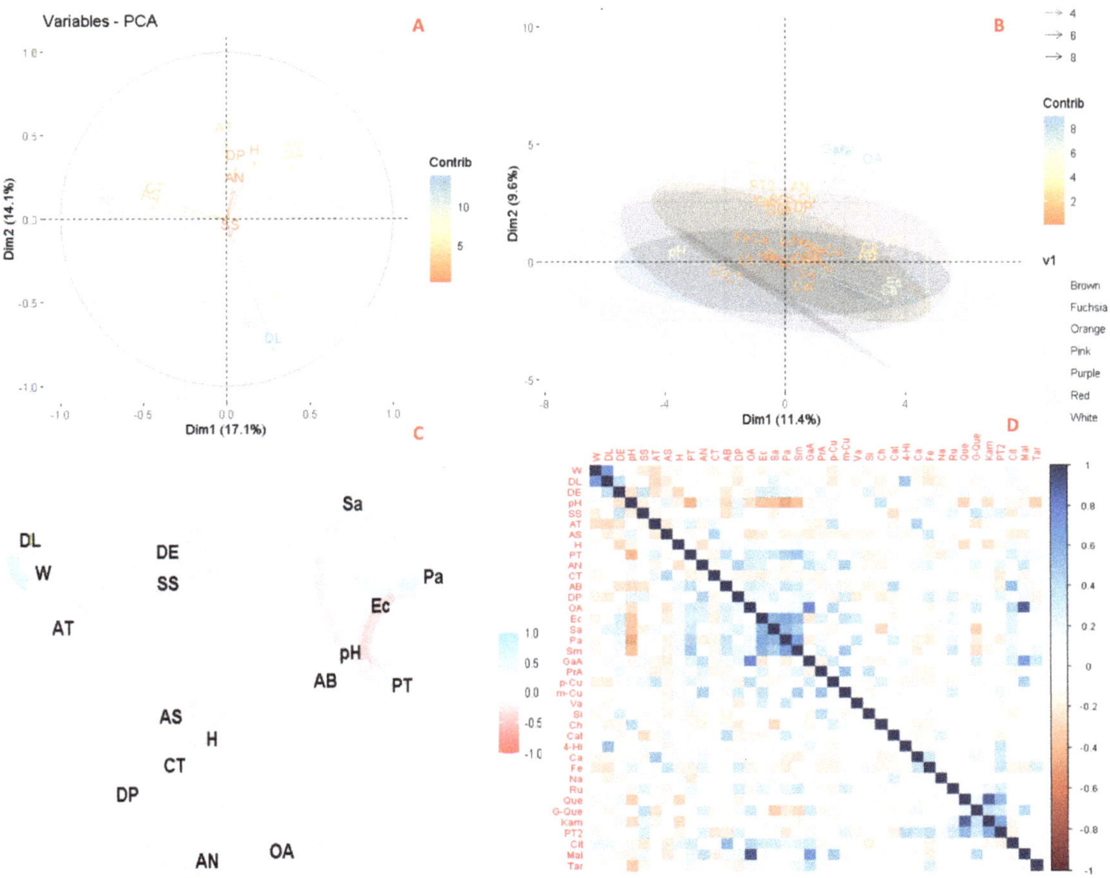

Figure 5. Exploratory multivariate analysis using principal components (**A**,**B**), and correlation analysis (**C**,**D**) of physicochemical parameters, bioactive compounds, and antioxidant activity of the floral species analyzed. Note: W, weight; DL, longitudinal diameter; DE, equatorial diameter; SS, soluble solids; AT, titratable acidity; H, humidity; AS; ash; CT, total carotenoid; PT, total phenolics; AB, antioxidant activity of ABTS, DP, antioxidant activity of DPPH, OA, organic acids; Ga, Gallic acid; Pr, protocatechuic acid; p-Cu, *p*-coumaric acid; o-Cu, *o*-coumaric acid; Va; vanillic acid; Si, syringic acid; Ch, chlorogenic acid; Cat, caffeic acid; 4-Hi, 4-hydroxy-benzoic acid; Ca, caffeic acid; Fe, ferulic acid; Na, naringin; Ru; rutin; Que, quercetin; G-Que, quercetin glucoside; Kam, kaempferol; PT2, total phenolics; Cit, citric acid; Mal, malic acid; Tar, tartaric acid; Ec, *Escherichia coli*; Sa, *Staphylococcus aureus*; Pa, *Pseudomonas aeruginosa*; Sm, *Streptococcus mutans*; Ca, *Candida albicans*; Ct, *Candida tropicalis*.

4. Discussion

4.1. Physicochemical Quantification

The flower in this study varied widely in shape, size, and weight, which are characteristics of each species. For example, *X. robustum*, which had the highest weight (72.79 g), is a species with very fleshy and spongy flowers [34], which gives it a high weight. In addition, these morphological characteristics are related to the degree of maturity, which is why in this study we considered fully developed (open) flowers, as they have the highest weight and size, as suggested by others [35]. In turn, the characteristics are also influenced by plant size, age, tissue water content, inherent species variability, and environmental influences [36,37]. For instance, the literature data suggest that *H. rosa-sinensis*, typically found in tropical and subtropical regions, exhibits a red variety with lengths of up to 15 cm [38–40]; however, the present study recorded a value of 24.43 cm. Another study focusing on the genus *Heliconia* reveals that tropical species such as *H. latispatha*, *H. wagneriana*, and *H. rostrata* have bract sizes ranging from 12 to 14 cm, 11 to 13 cm, and 6.5 to 7.5 cm, respectively [41]. This study's corresponding values were 11.20 cm, 12.38 cm, and 8.51 cm, respectively. Conversely, *O. cuspidatum*, commonly found in sub-deciduous and deciduous tropical forests, typically exhibits a red cluster-shaped inflorescence measuring 2.1 to 3.5 cm in length, surpassing the values found in this study (1.69 cm) [42]. A wide range of pH values was observed, from very acidic to slightly alkaline samples, reflecting the influence of various chemical and biological processes within these species, including nutrient absorption, metal availability, and enzymatic activity [35]. The variation in pH can be attributed to the specific characteristics of the soil in which the plants are grown and the adaptive strategies employed by each species. For instance, species like *C. nobilis* and *T. floribunda* from the Ericaceae family thrive in slightly acidic soils abundant in organic matter, typically found near the Antisana volcano in Ecuador [43]. On the other hand, *M. erythrochlamys*, belonging to the Acanthaceae family, grows in clay soil with a pH close to neutral [44], resulting in flowers with a pH of 4.00 and 9.15.

The species *P. lutea* (15.8 °Brix), *B. macrophylla* (12.00 °Brix), and *Oncidium* sp. (11.20 °Brix) showed a notably high concentration of soluble solids, even exceeding the soluble solids content of fruits such as watermelon (*Citrullus lanatus*), where values of 10.43 to 13.56 °Brix have been reported [45]. In this sense, soluble solids include the ratio between sugars, organic acids, and other soluble compounds, which play an essential role in plant metabolism as a source of energy and are also responsible for the taste and quality of plant species. Variations in their content may reflect plants' different chemical composition and degree of maturity [46,47].

In terms of total titratable acidity, *A. corymbose* (3.81%), *C. mollis* (2.8%), and *P. heteromallum* (2.73%) showed the highest concentration. Total titratable acidity is a quantitative measure of the organic acids present in a plant species, such as citric acid, malic acid, tartaric acid, and other acidic compounds. These compounds can significantly affect plant species' flavour, shelf life, and stability. Variations in titratable acidity values can be attributed to plants' unique metabolic and physiological characteristics, environmental factors, and the stage of flower development [46,47]. However, it is essential to note that this study specifically focused on fully developed flowers, as titratable acid content tends to be higher, as demonstrated in a study examining different stages of feijoa (*Acca sellowiana*) flower development [47].

Species such as *I. walleriana* (96.05%), *I. hawkeri* (95.70%), and *H. coronarium* (94.70%) showed a high moisture content. This parameter is critical to plant growth and development by influencing water availability and transpiration. Optimal moisture levels are crucial for the efficient functioning of physiological processes in plants [48]. In turn, literature sources showed that *H. rosa-sinensis* had an average moisture content of 90%, which is consistent with the results of this study, where a moisture content of 91% was reported. In addition, species such as *S. oblonga* (8.02%), *H. coccineum* (6.12%), and *A. purpurata* (4.02%) showed high ash concentrations. This parameter indicates the percentage of inorganic residues remaining after combustion and provides information on the mineral composition of the plants, which may vary between cultivars [48].

4.2. Optimization of Extraction Parameters and Quantification of Carotenoids

Carotenoids are plant pigments responsible for photosynthesis. Carotenoids are essential for plant health because they act as antioxidants and increase resistance to adverse conditions such as ultraviolet radiation and oxidative stress. These secondary metabolites protect against oxidative stress and benefit human health. e.g., β-carotene is a precursor to vitamin A and has antioxidant properties [9]. Thus, the efficacy of carotenoid extraction was significantly influenced by the type of solvent used and the plant species. The highest extraction was obtained using the mixtures of methanol–acetone–dichloromethane (1:1:2) and acetone–methanol 2:1, with an extraction time of 1 min, especially for the species *T. officinale*. These results align with previous research by other scientists, who have also highlighted the importance of the choice of solvent in the extraction of carotenoids to ensure accurate quantification of these compounds [33].

Species such as *R. alpinia* (orange visual colour), *A. papposa* (yellow visual colour), *D. balsapampae* (yellow visual colour), and *H. coccineum* (orange visual colour) showed remarkably elevated levels of carotenoids, with values of 1925.0 mg β-carotene/g DW, 211.00 mg β-carotene/g DW, 129.43 mg β-carotene/g DW, and 126.26 mg β-carotene/g DW, respectively. These variations can be attributed to species genetics, growth conditions, exposure to sunlight, and environmental influences [5].

Species such as *X. robustum* (white visual colour), *H. latispatha* (red visual colour), and *C. cuatrecasasii* (pink visual colour) showed low concentrations of total carotenoids, with values of 0.59 mg β-carotene/g DW, 0.43 mg β-carotene/g DW, and 0.25 mg β-carotene/g DW, respectively. It is worth noting that the literature reports a total carotenoid content of 162.00 μg/g fresh weight for *H. rosa-sinensis* cultivated at the Cairo Faculty of Agriculture [49]. In contrast, the present study reports a higher concentration (3.64 mg β-carotene/g DW). The difference in concentration may be due to differences in cultivation conditions, as the study in question used average environmental conditions of 35 °C and 40% relative humidity. In contrast, this study collected samples from Ecuador's species' natural habitat, with an average temperature of 25 °C and relative humidity between 85 and 90%. Furthermore, an investigation into *A. cathartica* demonstrated varying carotenoid concentrations across different plant parts, with 9.01 mg/g in flowers, 12.41 mg/g in leaves, 4.53 mg/g in roots, and 2.91 mg/g in shoots [50]. This shows the significant influence of the analyzed plant part on the total carotenoid content. In this study, the concentration in the petals of *A. cathartica* was found to be 32.53 mg β-carotene/g DW.

In turn, *R. alpinia* presented the highest concentrations of total carotenoids compared to the other floral species studied [5]. These results are of great importance for various industries, as they may lead to the creation of new products, as evidenced by the literature, which mentions the use of this species in the ancestral treatment of fevers caused by snake bites and to relieve the pain of bruises [33]. It is also important to note that the Zingiberaceae family, to which *R. alpinia* belongs, is characterized by leaves with antifungal, cytotoxic, anti-inflammatory, antipyretic, antioxidant, insecticidal, hepatoprotective, and immunomodulatory properties [51].

4.3. Optimization of Extraction Parameters and Quantification of Total Phenolic Compounds

The efficiency of the phenolic extraction was significantly influenced by the type of solvent and the plant species used. It is evident that the highest extraction was obtained using the 80% methanol solution acidified with 0.1% hydrochloric acid, with an extraction time of 3 min, particularly in the case of the species *D. caryophyllus*. These results align with previous studies by other researchers, who have also highlighted the importance of the choice of solvent in the phenolic extraction process to ensure accurate quantification of these compounds [19]. Thus, a previous study reported total phenolic concentrations in *H. rosa-sinensis* grown in Cairo using different solvents. The concentrations ranged from 186.17 mg GAE/100 g FW, 235.77 mg GAE/100 g FW, and 281.23 mg GAE/100 g FW using absolute ethanol, water, and 80% ethanol, respectively [49]. In contrast, this study determined a concentration of 290.3 mg GAE/g DW using an acidified methanolic solution.

Some species are characterized by high levels of total phenolics; e.g., *T. sorensis*, *P. urvilleana*, and *C. mollis* exhibited remarkable concentrations of 581.48 mg GAE/g, 510.53 mg GAE/g, and 354.2 mg GAE/g, respectively. These species can be considered rich sources of phenolic compounds, as phenolics are widely known for their remarkable antioxidant properties and potential benefits for human health. These compounds function as potent defenders against oxidative stress and have been suggested to play a role in preventing chronic diseases like heart, cancer, and neurodegenerative disorders [33]. At the same time, other species showed lower levels of total phenolics; e.g., *S. stromanthoides*, *H. episcopalis*, and *R. veithchii* recorded levels of 50.6 mg GAE/g, 56.86 mg GAE/g, and 72.62 mg GAE/g, respectively. In this regard, the variation in total phenolic content observed among the different species may be due to genetic and environmental factors, soil type, nutrient availability, exposure to sunlight, and other environmental conditions, as suggested by other authors [5]. In this regard, a study on *H. rostrata* flowers identified the presence of phenolics through phytochemical screening, with a concentration of 155.21 mg GAE/100 g DW [9]. Similarly, in the case of *C. argentea*, a study reported a concentration of 47.00 mg GAE/100 mg using an ethanolic solution of flower extract [52], whereas in this study, a concentration of 140.63 mg GAE/g DW. Another investigation on *A. cathartica* reported a concentration of 2839 μM/g in flowers, 11,906 μM/g in leaves, 19344 μM/g in roots, and 3455 μM/g in shoots [50], whereas in this study a concentration of 167.12 mg GAE/g DW was observed.

Certain species in this study showed significant concentrations of individual phenolic compounds, e.g., in gallic acid, the species *C. cornutus* (869.4 mg/100 g DW), in protocatechuic acid the species *P. angustifolia* (231.4 mg/100 g DW), in *p*-coumaric acid the species *C. bracteate* (2929 mg/100 g DW), in *m*-coumaric acid and syringic acid the species *A. poiretii* with a concentration of 12,044 mg/100 g DW and 3225 mg/100 g, respectively, in chlorogenic acid the species *S. dulcamara* (3435 mg/100 g DW), in 4-hydroxybenzoic acid the species *B. arborea* (10,729 mg/100 g DW), in caffeic acid the species *A. corymbose* (5893 mg/100 g DW), in ferulic acid the species *T. grandiflora* (5848 mg/100 g DW), for quercetin glucoside the species *L. rugulosa* (2950 mg/100 g DW), and for kaempferol the species *D. pinnata* (8236 mg/100 g DW). In this context, information was presented on the widespread use of *C. cornutus* for lymphatic disorders, canker sores, and cold sores [6]. It was reported that various parts of the non-traditional shrub of *C. bracteate* had been traditionally used in nutrition and the roots to control diarrhea and diabetes [53]. *A. poiretii* has been shown to have a beneficial effect on the homeostatic mechanisms [54]. *p*-Coumaric, caffeic acid and ferulic acids are phenolic compounds with antimicrobial properties against various microorganisms when used in polymeric food packaging materials [55].

4.4. Quantification of Total Anthocyanins

Anthocyanins are a type of phenolic compound in the flavonoid group. They are responsible for the colours of many flowers, which contribute to their beauty, and they offer health benefits thanks to their antioxidant properties. In this regard, studies on flowers have shown that anthocyanins are predominant in dahlias, with variations in concentration between varieties (1.14 mg cyanidin-3-*O*-glucoside chloride/g in the variety 'Colorado Classic' and 1.25 mg cyanidin-3-*O*-glucoside chloride/g in the variety 'La Baron') [56]. These values were lower than those reported in this study for *D. pinnata* (1.84 mg C-3-gl/g dry weight). Furthermore, in *Hibiscus rosa-sinensis*, the concentration in the extract was 0.46 mg cyanidin-3-glucoside/g [57], which was lower than the value reported in this study (1.93 mg C-3-gl/g dry weight).

4.5. Quantification of Organic Acids

The results of the analysis of organic acids in flower species reveal considerable variation between species, highlighting the inherent biochemical diversity of each species and their contribution to the nutritional value and functional properties of flowers. These properties are increasingly recognized for their health benefits. A prominent example is *H. sabdariffa*, which contains citric, malic, and tartaric acids, all contributing to its diverse pharmacological effects [58]. The variability in the contents of these acids may be influenced by factors such as genetics, growing environment, flower maturity, and environmental conditions such as light and soil type [59,60].

In this study, several species showed high concentrations of organic acids. Citric acid, which plays a vital role in the tricarboxylic acid cycle in plants [61], showed a wide range of concentrations among the species tested. The highest concentrations were observed in

H. rosa-sinensis (17,818 mg/100 g DW), *H. wagneriana* (3800 mg/100 g DW), *T. floribunda* (3516 mg/100 g DW), *E. americana* (2874 mg/100 g DW), and *I. gesnerioides* (2662 mg/100 g DW). These results suggest that these species can accumulate high levels of organic acids, favouring their possible use in food and medicinal applications as natural acidifying agents.

For malic acid, known for its role in plant metabolism [61], significant variability was also observed. Species such as *A. piretii* (15,613 mg/100 g DW), *C. cornutus* (11,655 mg/100 g DW), *A. lepidotus* (5591 mg/100 g DW), *E. aquaticum* (4605 mg/100 g DW), *P. urvilleanum* (3540 mg/100 g DW), and *P. angustifolia* (3121 mg/100 g DW) stood out for their high content of this acid, positioning them as potential sources of malic acid in industrial and food applications.

Tartaric acid, generally found in lower concentrations in most plant species [61], also showed considerable variability among the flowers analyzed. The species with the highest concentrations of tartaric acid were *H. latispatha* (5488 mg/100 g DW), *B. spectabilis* (3197 mg/100 g DW), *B. americana* (1519 mg/100 g DW), *R. veitchii* (836 mg/100 g DW), and *C. mollis* (708 mg/100 g DW). Tartaric acid is known to have antioxidant properties, suggesting that these species may offer additional health benefits to consumers by providing a natural source of antioxidants [62].

4.6. Antioxidant Activity (ABTS and DPPH)

In this study, aqueous solutions of methanol and acetone were used to extract both water-soluble and liposoluble compounds associated with antioxidant activity. Since organic compounds have different polarities and solubilities, no single extraction method is optimal for all types of molecules, such as phenolic compounds and carotenoids [63–65]. This variability affects the extraction efficiency and the interaction of the antioxidant compounds with the ABTS and DPPH radical in the antioxidant activity assay. For example, solvents such as ethanol and methanol effectively extract phenolic compounds and flavonoids, whereas acetone and mixtures with water enhance the extraction of carotenoids and other lipophilic antioxidants. However, some of these compounds may show reduced reactivity in the ABTS assay due to variations in solubility and compatibility with the medium [66,67].

Antioxidant activity is a vital indicator of a substance's ability to prevent or reduce oxidative processes. However, these measures are influenced by factors such as the concentration of the sample, the method of quantification used, the specific effect measured, the culture conditions, and the specific part of the plant analyzed [68], as can be seen from the values obtained for the concentration of the antioxidant activity when using the ABTS and DPPH method.

In terms of antioxidant activity, remarkable values were observed in *T. floribunda*, *H. rosa-sinensis*, and *Tibouchina* sp., exhibiting values of 7.80 mmol TE/g DW, 7.67 mmol TE/g DW, and 7.43 mmol TE/g DW, respectively, by ABTS. Conversely, species such as *P. lutea*, *C. argentea*, and *G. nicaraguensis* displayed lower antioxidant activity, with values of 2.5 mmol TE/g DW, 2.47 mmol TE/g DW, and 2.02 mmol TE/g DW. However, *I. hawkeri* (77.77 mmol TE/g DW), *B. americana* (75.87 mmol TE/g DW), *I. walleriana* (73.23 mmol TE/g DW), *A. corymbosa* (67.63 mmol TE/g DW) and *D. pinnata* (65.79 mmol TE/g DW) showed the high antioxidant activity of DPPH. It is important to note that the antioxidant activity of a plant is not solely due to a single compound but results from the synergistic interaction of several bioactive compounds present in the plant [69,70]. These results are essential for the identification of plant candidates with potential applications in the pharmaceutical or cosmetic industry, where natural antioxidants are highly valued for their beneficial effects on human health, and also in the food industry, where phenolic compounds have been used for their antimicrobial and antioxidant activity in food packaging [55].

In this context, the antioxidant activity of *H. rosa-sinensis*, assessed through the DPPH assay, using water, 80% ethanol, and absolute ethanol as solvents at concentrations of 500, 1000, and 2000 mg/L, yielded a range between 2.78% and 80.78% [49], which is comparable to the value reported in this study of 23.33% using methanol. This difference may be due to variations in antioxidant activity caused by different extraction parameters and culture conditions. Similarly, a review on *C. argentea* showed that this species has a marked antioxidant activity [71], which is consistent with the results of this study (51.7%). In addition, a study on edible flowers of different species reported that *H. rosa-sinensis*, grown in areas with temperatures below 25 °C and in varieties of different colours, showed a range of antioxidant activity from 0.44 to 0.77 mmol TE/g DW, values lower than those observed in this work [20]. This suggests that environmental temperature may significantly influence the composition and concentration of antioxidants.

4.7. Antimicrobial Activity

Evaluating the flower extracts' antimicrobial properties included antibacterial and antifungal tests. The bacterial strains *Escherichia coli*, *Staphylococcus aureus*, *Pseudomonas aeruginosa*, and *Streptococcus mutans* and the pathogenic fungi *Candida albicans* and *Candida tropicalis* were tested. These microorganisms are responsible for several human infections, including those of the urinary, respiratory, skin, and oral systems. The ability to prevent their proliferation is of great interest to the healthcare sector, particularly given the increasing resistance to traditional therapeutic approaches [72].

The use of 50% ethanol for extraction and testing of antimicrobial activity provides an optimal balance between extraction efficiency of bioactive compounds and compatibility with the microbiological assay, giving reliable results. This is because a 50% ethanolic solution is sufficiently polar to extract a wide range of phenolic compounds and flavonoids known for their antimicrobial activity [73]. In turn, at high concentrations of ethanol, direct microbial inhibition has been observed, attributed to the solvent itself.

In this context, most of the flower extracts show remarkable antibacterial activity, especially against *S. aureus*, a Gram-positive bacterium widely known for its ability to develop resistance to several antibiotics [74]. Among the most effective extracts were *C. nobilis* (20.5 mm), *P. heteromallum* (18.5 mm), and *A. poiretii* (17.0 mm), which showed significant zones of inhibition. This antimicrobial activity could be attributed to bioactive compounds such as *p*-coumaric acid, caffeic acid, gallic acid, and chlorogenic acid, each of which has been shown to inhibit the growth of *S. aureus* [72,75]. In addition, a study of the methanolic extract of *A. poiretii* flowers confirmed its antibacterial activity against *S. aureus* [76], supporting the potential use of these extracts in developing natural treatments to combat resistant bacterial infections. In turn, a literature review of *C. argentea* suggests that this species has antibacterial activity due to the presence of 4-hydroxybenzoic acid and caffeic acid [71], two compounds present in the species in this study. Another study reported that *Bougainvillea* has activity against *S. aureus*, *E. coli*, *P. aeruginosa*, and other microorganisms [12].

For *E. coli* and *P. aeruginosa*, the results are more heterogeneous. Only a few extracts showed inhibitory activity against *E. coli*, in particular *C. nobilis* (15.5 mm), *M. erythrophylla* (15.0 mm), and *C. spiralis* (13.0 mm). Regarding *P. aeruginosa*, a bacterium notorious for its high antibiotic resistance, some extracts such as *P. heteromallum* (15.0 mm), *C. nobilis* (13.0 mm), and *A. poiretti* (12.0 mm) showed antimicrobial activity. However, most of the extracts did not show significant inhibition against these Gram-negative bacteria, suggesting that the bioactive compounds present in the flowers may have less efficacy against these bacteria's complex cell wall structure. In this context, previous studies have reported anthelmintic activity of *M. erythophylla* [77] and antimicrobial activity against *S. aureus* and *E. coli* [76], as well as inhibition of *E. coli* by *B. spectabilis* [78,79], possibly due to the presence of isophitol [80]; however, in the present study, the latter species did not show antimicrobial activity against *E. coli*.

S. mutans, a key pathogen in the development of dental caries [81], was significantly inhibited by several flower extracts. The most effective extract was *B. spectabilis*, with an inhibition halo of 22.0 mm, followed by *P. heteromallum* (21.0 mm) and *B. macrophylla*

(17.0 mm). These results suggest a remarkable antimicrobial potential of these extracts, which could make them suitable for developing products focused on oral health.

On the other hand, the antifungal activity of the extracts studied was limited. Only the extract of *P. heteromallum* showed significant activity against *C. albicans* (13.0 mm), while no extract was effective against *Candida tropicalis*. This suggests that the antifungal compounds present in the flowers are specific in their action and do not have a broad spectrum against different fungal species. In this regard, other studies on *C. argentea* showed antifungal activity against *C. albicans*, as well as antibacterial activity against *S. aureus*, *P. aeruginosa*, *E. coli*, and other microorganisms [71].

These showed superior inhibition zones compared to reference antibiotics, particularly against *S. mutans* (30.0 mm) and *S. aureus* (24.8 mm). Although some flower extracts showed promising results, the data suggest that they may require optimization or synergy with other bioactive compounds to match or exceed the efficacy of conventional antimicrobial treatments.

4.8. Statistical Analysis

Principal component analysis included weight (W), longitudinal (DL) and equatorial diameter (DE), pH, soluble solids (SS), total titratable acid (AT), moisture (H), ash (AS), total carotenoids (CT), total phenolics (PT), total anthocyanins (AN), organic acids (OA), antioxidant activity of ABTS (AB), antioxidant activity of DPPH (DP), and inhibition size for *Escherichia coli* (Ec), *Staphylococcus aureus* (Sa), *Pseudomonas aeruginosa* (Pa), *Streptococcus mutans* (Sm), *Candida albicans* (Ca), and *Candida tropicalis* (Ct). Principal component analysis (Figure 5A,B) showed that the highest variance was explained by Dim1 (11.4%), followed by Dim2 (9.6%) when all variables were considered, and when the total sum of bioactive compounds was considered, Dim1 explained 17.1% and Dim2 14.1%. A relationship between the antioxidant activity of ABTS, organic acids, and total phenolics and *E. coli* and *S. aureus* was observed. There was also a relationship between pH and total carotenoids and ash, titratable acidity, antioxidant activity of DPPH, total anthocyanins, and moisture. In addition, Figure 5A showed an inverse relationship between flower size and flower weight with titratable acidity. These relationships agree with the results reported by other authors [29]. On the other hand, the principal component analysis, considering all the study variables, showed that Fuchsia species showed an important contribution of organic compounds, especially malic acid and the contribution of gallic acid.

The correlation analysis included weight (W), longitudinal (DL) and equatorial diameter (DE), pH, soluble solids (SS), total titratable acid (AT), moisture (H), ash (AS), total carotenoids (CT), total phenolics (PT), total anthocyanins (AN), organic acids (OA), antioxidant activity of ABTS (AB), antioxidant activity of DPPH (DP), Gallic acid (Ga), protocatechuic acid (Pr), p-coumaric acid (p-Cu), o-coumaric acid (o-Cu), vanillic acid(Va), syringic acid (Si), chlorogenic acid (Ch), caffeic acid (Cat), 4-hydroxi-benzoic acid (4-Hi), caffeic acid (Ca), ferulic acid (Fe), naringin (Na), rutin (Ru), quercetin (Que), quercetin glucoside (G-Que), kamferol (Kam), PT2, total phenolics, citric acid (Cit), malic acid (Mal), tartaric acid (Tar), and inhibition size for *Escherichia coli* (Ec), *Staphylococcus aureus* (Sa), *Pseudomonas aeruginosa* (Pa), *Streptococcus mutans* (Sm), *Candida albicans* (Ca), and *Candida tropicalis* (Ct).

The correlation analysis in Figure 5C,D revealed several relationships between the studied variables. Thus, Figure 5C shows a positive correlation between weight and size, an association widely documented in plant species. Similarly, the antioxidant activity of the DPPH method was correlated with total anthocyanins, ash with total carotenoids, *S. aureus* with *E. coli* and *P. aeruginosa*. In contrast, soluble solids were inversely correlated with pH and titratable acidity, pH with *S. aureus* and *E. coli*, and total carotenoids with total anthocyanins. These results suggest that smaller fruits tend to have a higher concentration of soluble solids, which could inhibit glycolysis, reduce phenolic content, and increase pH [9].

Moreover, specific correlations were observed between the study variables (Figure 5D), the most representative of which are described below. A positive correlation was observed between organic acids with gallic acid, titratable acids with caffeic acid, antioxidant acids with malic acid, gallic acid with malic acid. A negative correlation was also found between pH and total phenolics and *E. coli*, *S. aureus*, *P. aeruginosa*, *S. mutans*, and *S. aureus* with chlorogenic acid and quercetin glycosides. These results highlight the complex relationship between antimicrobial activity and pH, as the antimicrobial agent variably influences pH. In the dental context, low pH conditions can promote the development of caries [82]. Likewise, pH plays a crucial role in skin health and wound healing, affecting protease activity, bacterial growth, and the efficacy of antibacterial agents, which has important implications for dressing design and diagnosis of pH-sensitive wounds [83].

5. Conclusions

Flowers are an important source of bioactive compounds. Certain species studied showed high values; *P. urvilleana* for pH, *P. lutea* for soluble solids, *A. corymbosa* for titratable acidity. Furthermore, extraction optimization showed that the concentration of carotenoids is strongly influenced by the plant matrix, whereas phenolic compounds are influenced by the type of solvent. In this respect, *R. alpinia* showed high concentrations of total carotenoids, *T. sorensis* of total phenolics, *P. heteromallum* of total anthocyanins, *H. rosa-sinensis* of citric acid, *A. piretti* of malic acid, and *C. spiralis* of tartaric acid. In addition, the highest concentrations of individual phenolic compounds were found in *A. poiretti* (*m*-coumaric acid), *B. arborea* (4-hydroxybenzoic acid), and *D. pinnata* (kaempferol). Regarding antioxidant activity, high values were observed in *T. floribunda* using the ABTS method and in *I. hawkeri* using the DPPH method. Finally, species such as *A. poiretti*, *B. macrophylla*, and *C. nobilis* showed inhibitory activity against *Escherichia coli*, *Staphylococcus aureus*, *Pseudomonas aeruginosa*, and *Streptococcus mutans*. In contrast, *P. heteromallum* showed inhibitory activity against *Candida albicans*. These results highlight the complex relationship between the different organic compounds in flower petals and indicate the need for further studies to unravel the potential health benefits.

Author Contributions: Conceptualization, E.C.-C.; methodology, E.C.-C.; software, E.V.; validation, E.C.-C.; formal analysis, A.B., A.G., J.H.-M. and J.Z.-M.; investigation, E.C.-C.; resources, E.C.-C.; data curation, E.C.-C.; writing—original draft preparation, A.G.; writing—review and editing, E.C.-C.; project administration, E.C.-C.; funding acquisition, E.C.-C. All authors have read and agreed to the published version of the manuscript.

Funding: This research was carried out under the framework contract MAE-DNB-CM-2017-0080-UTE. Project MAE-DNB-2019-0911-O was financed by the Ecuadorian Corporation for the Development of Research and the Academy (CEDIA) within the CEPRA-XII-2019-Flores Andinas Project "Physical-chemical characterization and bioactivity tests of Andean floral species with nutritional potential and preventive effect of certain human diseases.

Institutional Review Board Statement: Not applicable.

Informed Consent Statement: Not applicable.

Data Availability Statement: The original contributions presented in this study are included in the article. Further inquiries can be directed to the corresponding author.

Conflicts of Interest: The authors declare no conflicts of interest.

References

1. WWF. ¿Qué Son los Bosques Tropicales? Available online: https://www.wwf.org.co/en/?329990/Glosario-ambiental-Que-son-los-bosques-tropicales (accessed on 16 May 2023).
2. Paz, A. Bosques Tropicales: Cada Día Es Más Urgente Su Conservación. Available online: https://es.mongabay.com/2021/06/bosques-tropicales-urgente-su-conservacion/ (accessed on 16 May 2023).
3. Yusupova, Z.; Baratjon, S.; Rahimjon, Y. Chemical Composition of Medicinal Plants and Use in Medicine. *Pedago. Resp. Ilmiy Jurnali* **2023**, *5*, 30–36.

4. Nwozo, O.S.; Effiong, E.M.; Aja, P.M.; Awuchi, C.G. Antioxidant, Phytochemical, and Therapeutic Properties of Medicinal Plants: A Review. *Int. J. Food Prop.* **2023**, *26*, 359–388. [CrossRef]
5. Meléndez-Martínez, A.; Benítez, A.; Corell, M.; Hernanz, D.; Mapelli-Brahm, P.; Stinco, C.; Coyago-Cruz, E. Screening for Innovative Sources of Carotenoids and Phenolic Antioxidants among Flowers. *Foods* **2021**, *10*, 2625. [CrossRef] [PubMed]
6. Horackova, J.; Elena, M.; Zans, C.; Kokoska, L.; Sulaiman, N.; Mirella, Z.; Peralta, C. Ethnobotanical Inventory of Medicinal Plants Used by Cashinahua (Huni Kuin) Herbalists in Purus Province, Peruvian Amazon. *J. Ethnobiol. Ethnomed.* **2023**, *19*, 16. [CrossRef] [PubMed]
7. Foghis, M.; Bungau, S.G.; Bungau, A.F.; Vesa, C.M.; Purza, A.L.; Tarce, A.G.; Tit, D.M.; Pallag, A.; Behl, T.; ul Hassan, S.S.; et al. Plants-Based Medicine Implication in the Evolution of Chronic Liver Diseases. *Biomed. Pharmacother.* **2023**, *158*, 114207. [CrossRef]
8. Dincheva, I.; Badjakov, I.; Galunska, B. New Insights into the Research of Bioactive Compounds from Plant Origins with Nutraceutical and Pharmaceutical Potential. *Plants* **2023**, *12*, 258. [CrossRef]
9. Coyago-Cruz, E. Estudio Sobre el Contenido en Carotenoides y Compuestos Fenólicos de Tomates y Flores en el Contexto de la Alimentación Funcional. Doctoral Thesis, Departamento de Ciencias Agroforestales, Universidad de Sevilla, Seville, Spain, 2017.
10. Williamson, G. The Role of Polyphenols in Modern Nutrition. *Nutr. Bull.* **2017**, *42*, 226–235. [CrossRef]
11. Xu, D.P.; Li, Y.; Meng, X.; Zhou, T.; Zhou, Y.; Zheng, J.; Zhang, J.J.; Li, H. Bin Natural Antioxidants in Foods and Medicinal Plants: Extraction, Assessment and Resources. *Int. J. Mol. Sci.* **2017**, *18*, 96. [CrossRef]
12. Coyago-Cruz, E.; Baldeón, M. Novel List of Ecuadorian Flowers with Antimicrobial Activity. In *Medicinal Plants of Ecuador*; CRC Press: Boca Raton, FL, USA; Taylor Francis Group: Boca Raton, FL, USA, 2022; p. 229, ISBN 9780367775865.
13. Sierra-Vargas, M.P.; Montero-Vargas, J.M.; Debray-García, Y.; Vizuet-de-Rueda, J.C.; Loaeza-Román, A.; Terán, L.M. Oxidative Stress and Air Pollution: Its Impact on Chronic Respiratory Diseases. *Int. J. Mol. Sci.* **2023**, *24*, 853. [CrossRef]
14. Liu, X.; Wang, S.; Cui, L.; Zhou, H.; Liu, Y.; Meng, L.; Chen, S.; Xi, X.; Zhang, Y.; Kang, W. Flowers: Precious Food and Medicine Resources. *Food Sci. Hum. Wellness* **2023**, *12*, 1020–1052. [CrossRef]
15. Al-Hazmi, M.; Noorwali, E. Morning Individuals in Saudi Arabia Have Higher Self-Regulation of Eating Behavior Compared to Evening Types. *Chronobiol. Int.* **2023**, *40*, 223–233. [CrossRef] [PubMed]
16. Azevedo-Meleiro, C.H.; Rodriguez-Amaya, D.B. Qualitative and Quantitative Differences in Carotenoid Composition among *Cucurbita moschata, Cucurbita maxima*, and *Cucurbita pepo*. *J. Agric. Food Chem.* **2007**, *55*, 4027–4033. [CrossRef] [PubMed]
17. Jara-Palacios, M.J.; Hernanz, D.; González-Manzano, S.; Santos-Buelga, C.; Escudero-Gilete, M.L.; Heredia, F.J. Detailed Phenolic Composition of White Grape By-Products by RRLC/MS and Measurement of the Antioxidant Activity. *Talanta* **2014**, *125*, 51–57. [CrossRef] [PubMed]
18. Coyago-Cruz, E.; Coronel, M.; Baldeón, M.; Vera, E. Physicochemical Characteristics and Antioxidant Capacity of Yellow and Orange Andean Floral Species. *Commun. Smart Technol. Innov. Soc.* **2022**, *252*, 1–175.
19. Coyago-Cruz, E.; Guachamin, A.; Vera, E.; Moya, M.; Heredia-Moya, J.; Beltrán, E. Physicochemical Characteristics and Antioxidant Capacity of Ecuadorian Paramo Flowers. *Rev. Bionatura* **2023**, *8*, 21. [CrossRef]
20. Coyago-Cruz, E.; Alarcón, A.; Guachamin, A.; Méndez, G.; Osorio, E.; Heredia-Moya, J.; Zuñiga-Miranda, J.; Beltrán-Sinchiguano, E.; Vera, E. Functional, Antioxidant, Antibacterial, and Antifungal Activity of Edible Flowers. *Antioxidants* **2024**, *13*, 1297. [CrossRef]
21. Periago, M.J.; Martínez-Valverde, I.; Chesson, A.; Provan, G. Phenolic Compounds, Lycopene and Antioxidant Activity in Commercial Varieties of Tomato (*Lycopersicum esculentum*). *J. Sci. Food Agric.* **2002**, *82*, 323–330. [CrossRef]
22. Bobo-García, G.; Davidov-Pardo, G.; Arroqui, C.; Vírseda, P.; Marín-Arroyo, M.R.; Navarro, M. Intra-Laboratory Validation of Microplate Methods for Total Phenolic Content and Antioxidant Activity on Polyphenolic Extracts, and Comparison with Conventional Spectrophotometric Methods. *J. Sci. Food Agric.* **2015**, *95*, 204–209. [CrossRef]
23. Tan, J.; Han, Y.; Han, B.; Qi, X.; Cai, X.; Ge, S.; Xue, H. Extraction and Purification of Anthocyanins: A Review. *J. Agric. Food Res.* **2022**, *8*, 100306. [CrossRef]
24. Tian, Y.; Yang, Y.; Gao, P.; Hong, J.; Qin, Y.; Yuan, Z. Optimization of Ultrasonic-Assisted Extraction of Flavonols and Anthocyanins from Blueberry Using RSM. *Adv. Mater. Res.* **2012**, *4*, 2423–2430. [CrossRef]
25. Yuniati, Y.; Elim, P.; Alfanaar, R.; Kusuma, H.; Mahfud. Extraction of Anthocyanin Pigment from *Hibiscus sabdariffa* L. by Ultrasonic-Assisted Extraction. *IOP Conf. Ser. Mater. Sci. Eng.* **2021**, *1010*, 012032. [CrossRef]
26. Lee, J.; Rennaker, C.; Wrolstad, R. Correlation of Two Anthocyanin Quantification Methods: HPLC and Spectrophotometric Methods. *Food Chem.* **2008**, *110*, 782–786. [CrossRef]
27. Lee, J.; Durst, R.; Wrolstad, R. Determination of Total Monomeric Anthocyanin Pigment Content of Fruit Juices, Beverages, Natural Colorants, and Wines by the PH Differential Method: Collaborative Study. *J. AOAC Int.* **2005**, *88*, 1269–1278. [CrossRef] [PubMed]
28. Chan, Y.; Cheng, N.; Nigam, P.; Owusu-Apenten, R. Effect of PH on the Radical Quenching Capacity of Tea Infusions Using the ABTS•+ Assay. *J. Appl. Life Sci. Int.* **2016**, *6*, 1–8. [CrossRef] [PubMed]
29. Coyago-Cruz, E.; Guachamin, A.; Villacís, M.; Rivera, J.; Neto, M.; Méndez, G.; Heredia-Moya, J.; Vera, E. Evaluation of Bioactive Compounds and Antioxidant Activity in 51 Minor Tropical Fruits of Ecuador. *Foods* **2023**, *12*, 4439. [CrossRef]
30. *CLSI M02*; Performance Standards for Antimicrobial Disk Suspectibility Tests, Approved Standard-Eleventh Edition. Clinical and Laboratory Standards Institue: Malvern, PA, USA, 2018; Volume 38, pp. 2162–2914.

31. *CLSI M44-A2*; Method for Antifungal Disk Diffusion Susceptibility Testing of Yeasts. Approved Guideline—Second Edition. Clinical and Laboratory Standards Institue: Malvern, PA, USA, 2009; Volume 29, p. 29.
32. Balouiri, M.; Sadiki, M.; Ibnsouda, S. Methods for in Vitro Evaluating Antimicrobial Activity: A Review. *J. Pharm. Anal.* **2016**, *6*, 71–79. [CrossRef]
33. Coyago-Cruz, E.; Moya, M.; Méndez, G.; Villacís, M.; Rojas-Silva, P.; Corell, M.; Mapelli-Brahm, P.; Vicario, I.; Meléndez-Martínez, A. Exploring Plants with Flowers: From Therapeutic Nutritional Benefits to Innovative Sustainable Uses. *Foods* **2023**, *12*, 4066. [CrossRef]
34. WFO. The World Flora Online. Available online: https://wfoplantlist.org/plant-list (accessed on 14 June 2024).
35. Ramírez-Cortés, B.; Caro-Velarde, F.; Valdivia-Reynoso, M.; Ramírez-Lozano, M.; Machuca-Sánchez, L. Cambios En Tamaño y Características Químicas de Cálices de Jamaica (*Hibiscus sabdariffa* L.) Durante Su Maduración. *Rev. Chapingo Ser. Hortic.* **2011**, *XVII*, 19–31. [CrossRef]
36. Cho, L.; Yoon, J.; An, G. The Control of Flowering Time by Environmental Factors. *Plant J.* **2017**, *90*, 708–719. [CrossRef]
37. Bhatla, S.; Lal, M. *Plant Physiology. Development and Metabolism*; Springer: Berlin/Heidelberg, Germany, 2023; ISBN 9789811320231.
38. Khristi, V.; Patel, V.H. Therapeutic Potential of Hibiscus Rosa Sinensis: A Review. *Int. J. Nutr. Diet.* **2017**, *4*, 105–123. [CrossRef]
39. Abdullah; Mehmood, F.; Shahzadi, I.; Waseem, S.; Mirza, B.; Ahmed, I.; Waheed, M.T. Chloroplast Genome of *Hibiscus rosa-sinensis* (Malvaceae): Comparative Analyses and Identification of Mutational Hotspots. *Genomics* **2020**, *112*, 581–591. [CrossRef] [PubMed]
40. Slamet, A. The Diversity of *Hibiscus rosa-sinensis* Based on Morphological Approach. *Sci. Educ.* **2018**, *7*, 32. [CrossRef]
41. Hapsari, L.; Trimanto; Wahyudi, D. Species Diversity and Phylogenetic Analysis of *Heliconia* spp. Collections of Purwodadi Botanic Garden (East Java, Indonesia) Inferred by RbcL Gene Sequences. *Biodiversitas* **2019**, *20*, 1266–1283. [CrossRef]
42. Daniel, T.; Acosta, S. *Flora del Bajío y de Regiones Adyacentes*; INECOL Instituto de Ecología, A.C.: Pátzcuaro, Mexico, 2003; pp. 1–158. [CrossRef]
43. Acurio, C. Análisis de la Influencia del Estado de Conservación de los Humedales Altoandinos del Parque Nacional Cajas y la Reserva Ecológica Antisana en el Secuestro de Carbono. Master's Thesis, Universidad UTE, Maestría en Ecoturismo y Manejo de Áreas Naturales Quito, Quito, Ecuador, 2015.
44. Rodríguez, F. Identificación de Procesos de Erosión por Monocultivo en la Parroqui de Pacto y Propuestas de Lineas de Acción. Universidad UTE, Facultad de Ciencias de la Ingeniería e Industrias: Quito, Ecuador, 2019.
45. Domene, M.; Segura, M. *Parámetros de Calidad Interna de Hortalizas y Frutas en la Industria Agroalimentaria*; Cajamar Caja Rural: Almería, Spain, 2014.
46. Li, C.; He, M.; Cai, Z.; Qi, H.; Zhang, J.; Zhang, C. Hyperspectral Imaging with Machine Learning Approaches for Assessing Soluble Solids Content of Tribute Citru. *Foods* **2023**, *12*, 247. [CrossRef] [PubMed]
47. Magri, A.; Adiletta, G.; Petriccione, M. Evaluation of Antioxidant Systems and Ascorbate-Glutathione Cycle in Feijoa Edible Flowers at Different Flowering Stages. *Foods* **2020**, *9*, 95. [CrossRef]
48. De-Lima, F.; Rodrigues, M.; Lidório, H.; Menegaes, J.; Martins, L. Chemical Composition of Rose, Sunflower and Calendula Flower Petals for Human Food Use. *Cienc. Tecnol. Agropecu.* **2019**, *20*, 159–168. [CrossRef]
49. Afify, A.; Hassan, H. Free Radical Scavenging Activity of Three Different Flowers—*Hibiscus rosa-sinensis*, *Quisqualis indica* and *Senna surattensis*. *Asian Pac. J. Trop. Biomed.* **2016**, *6*, 771–777. [CrossRef]
50. Hameed, A.; Nawaz, G.; Gulzar, T. Chemical Composition, Antioxidant Activities and Protein Profiling of Different Parts of Allamanda Cathartica. *Nat. Prod. Res.* **2014**, *28*, 2066–2071. [CrossRef]
51. Gómez-Betancur, I.; Benjumea, D. Traditional Use of the Genus *Renealmia* and *Renealmia alpinia* (Rottb.) Maas (Zingiberaceae)—A Review in the Treatment of Snakebites. *Asian Pac. J. Trop. Med.* **2014**, *7*, S574–S582. [CrossRef]
52. Kumar, N.; Khan, N. Phytochemical Investigation and Antidepressant Activity of *Celosia cristata* Leaves and Flowers in Experimental Animals. *Phytochem. Investig. Antidepressant Act. Celosia Cris.* **2023**, *12*, 298–305.
53. Torres-Guevara, F.; Ganoza-Yupanqui, M.; Mantilla-Rodriguez, E.; Suárez-Rebaza, L.; Bussmann, R. Ethnobotany of Fruit Species Native to Paramos and Cloud Forests of Northern Peru. *Ethnobot. Res. Appl.* **2023**, *25*, 10. [CrossRef]
54. Mu, K.; Liu, Y.; Liu, G.; Ran, F.; Zhou, L.; Wu, Y.; Peng, L.; Shao, M.; Li, C.; Zhang, Y. A Review of Hemostatic Chemical Components and Their Mechanisms in Traditional Chinese Medicine and Ethnic Medicine. *J. Ethnopharmacol.* **2023**, *307*, 116200. [CrossRef] [PubMed]
55. Kaczmarek-Szczepańska, B.; Grabska-Zielińska, S.; Michalska-Sionkowska, M. The Application of Phenolic Acids in the Obtainment of Packaging Materials Based on Polymers—A Review. *Foods* **2023**, *12*, 1343. [CrossRef] [PubMed]
56. Gontova, T.; Mashtaler, V.; Romanova, S.; Maloshtan, L.; Shanaida, M. Phytochemical Analysis of Anthocyanins Extracted from the Flowers of Two Dahlia Cultivars and Their Cytotoxic Properties. *Biomed. Pharmacol. J.* **2024**, *17*, 735–749. [CrossRef]
57. Bal, A.; Shilpa, H.; Debnath, S.; Rastogi, N. Extraction of Anthocyanin from *Hibiscus rosa-sinensis* and Concentration by Forward Osmosis Membrane Process. *Innov. Food Sci. Emerg. Technol.* **2024**, *96*, 103782. [CrossRef]
58. Izquierdo-Vega, J.; Arteaga-Badillo, D.; Sánchez-Gutiérrez, M.; Morales-González, J.; Vargas-Mendoza, N.; Gómez-Aldapa, C.; Castro-Rosas, J.; Delgado-Olivares, L.; Madrigal-Bujaidar, E.; Madrigal-Santillán, E. Organic Acids from Roselle (*Hibiscus sabdariffa* L.)—A Brief Review of Its Pharmacological Effects. *Biomedicines* **2020**, *8*, 100. [CrossRef]
59. Ma, W.-F.; Li, Y.-B.; Nai, G.-J.; Liang, G.-P.; Ma, Z.-H.; Chen, B.-H.; Mao, J. Changes and Response Mechanism of Sugar and Organic Acids in Fruits under Water Deficit Stress. *PeerJ* **2022**, *10*, e13691. [CrossRef]

60. Mignard, P.; Beguería, S.; Giménez, R.; Forcada, C.; Reig, G.; Moreno, M. Effect of Genetics and Climate on Apple Sugars and Organic Acids Profiles. *Agronomy* **2022**, *12*, 827. [CrossRef]
61. Walker, R.P.; Famiani, F. Organic Acids in Fruits: Metabolism, Functions and Contents. In *Horticultural Reviews*; Wiley: Hoboken, NJ, USA, 2018; Volume 45, pp. 371–430, ISBN 9781119431077.
62. Shi, Y.; Pu, D.; Zhou, X.; Zhang, Y. Recent Progress in the Study of Taste Characteristics and the Nutrition and Health Properties of Organic Acids in Foods. *Foods* **2022**, *11*, 3408. [CrossRef]
63. Cano, A.; Maestre, A.; Hernández-Ruiz, J.; Arnao, M. ABTS/TAC Methodology: Main Milestones and Recent Applications. *Processes* **2023**, *11*, 185. [CrossRef]
64. Cano, A.; Arnao, M. ABTS/TEAC (2,2′-Azino-Bis(3-Ethylbenzothiazoline-6-Sulfonic Acid)/Trolox®—Equivalent Antioxidant Capacity) Radical Scavenging Mixed—Mode Assay. In *Measurement of Antioxidant Activity Capacity: Recent Trends and Applications*; Apak, R., Capanuglu, E., Shahidi, F., Eds.; Wiley Online Library: Hoboken, NJ, USA, 2018; pp. 117–139, ISBN 9781119135357.
65. Platzer, M.; Kiese, S.; Herfellner, T.; Schweiggert-Weisz, U.; Miesbauer, O.; Eisner, P. Common Trends and Differences in Antioxidant Activity Analysis of Phenolic Substances Using Single Electron Transfer Based Assays. *Molecules* **2021**, *26*, 1244. [CrossRef] [PubMed]
66. Müller, L.; Fröhlich, K.; Böhm, V. Comparative Antioxidant Activities of Carotenoids Measured by Ferric Reducing Antioxidant Power (FRAP), ABTS Bleaching Assay (ATEAC), DPPH Assay and Peroxyl Radical Scavenging Assay. *Food Chem.* **2011**, *129*, 139–148. [CrossRef]
67. Van-Den-Berg, R.; Haenen, G.; Van-Den-Berg, H.; Bast, A. Applicability of an Improved Trolox Equivalent Antioxidant Capacity (TEAC) Assay for Evaluation of Antioxidant Capacity Measurements of Mixtures. *Food Chem.* **1999**, *66*, 511–517. [CrossRef]
68. Kuskoski, E.; Asuero, A.; Troncoso, A.; Mancini-Filho, J.; Fett, R. Aplicación de Diversos Métodos Químicos Para Determinar Actividad Antioxidante En Pulpa de Frutos. *Ciênc. Tecnol. Aliment.* **2005**, *25*, 726–732. [CrossRef]
69. Arias, A.; Feijoo, G.; Moreira, M. Exploring the Potential of Antioxidants from Fruits and Vegetables and Strategies for Their Recovery. *Innov. Food Sci. Emerg. Technol.* **2022**, *77*, 102974. [CrossRef]
70. Da-Silva, E.; Queiroz, V.; Nascimento, D.; Pereira, M.; Pereira, L.; Dantas, J.; De-Freitas, F.; Gomez, A.; Do-Amaral, F.; De-Barros, E. The Role of Phenolic Compounds in Metabolism and Their Antioxidant. *Res. Soc. Dev.* **2022**, *11*, 1–12. [CrossRef]
71. Luo, Y.; Xu, Y.; Zhang, H.; Zhang, J.; Qin, X.; Jin, S. *Celosia cristata* L.: A Review of Its Traditional Uses, Phytochemistry, Pharmacology, Toxicology, and Quality Control, along with Network Pharmacological Analysis of Its Components and Targets. *J. Ethnopharmacol.* **2024**, *332*, 118325. [CrossRef]
72. Khare, T.; Anand, U.; Dey, A.; Assaraf, Y.; Chen, Z.; Liu, Z.; Kumar, V. Exploring Phytochemicals for Combating Antibiotic Resistance in Microbial Pathogens. *Front. Pharmacol.* **2021**, *12*, 720726. [CrossRef]
73. Dębczak, A.; Grzegorczyk, A.; Świątek, Ł.; Sandomierski, M.; Fekner, Z.; Tyśkiewicz, K.; Milovanovic, S. Green and Sustainable Recovery of Polyphenols, Flavonoids, Alkaloids, and Pigments from Green Tea Leaves: Comparative Analysis of Soxhlet, Accelerated Solvent, and Supercritical Fluid Extraction Techniques. *Sustain. Chem. Pharm.* **2024**, *42*, 101810. [CrossRef]
74. Ondusko, D.; Nolt, D. *Staphylococcus aureus*. *Pediatr. Rev.* **2018**, *39*, 287–298. [CrossRef]
75. Khameneh, B.; Eskin, N.; Iranshahy, M.; Fazly, B. Phytochemicals: A Promising Weapon in the Arsenal against Antibiotic-resistant Bacteria. *Antibiotics* **2021**, *10*, 1044. [CrossRef] [PubMed]
76. Wardoyo, E.; Hildayati, U.; Rachmawati; Kurniatuhadi, R. Phytochemical Analysis and Antifungi Activity of Methanol Extract of *Acalypha hispida* Burm. F. Flower against to *Candida albicans* (Y116). *J. Phys. Conf. Ser.* **2021**, *1940*, 012056. [CrossRef]
77. Narayanan, A.; Palaniappan, S. Evaluation of *Mussaenda erythrophylla* Schumach. & Thonn. for Anthelmintic Activity. *Int. J. Res. Pharm. Sci.* **2021**, *12*, 77–80. [CrossRef]
78. Hajare, C.; Inamdar, F.; Patil, R.; Shete, C.; Wadkar, S.; Patil, K.; Ghosh, J. Antibacterial Activity of the Leaves of *Bougainvillea spectabilis* against *E. coli* NCIM 2832 and *M. aureus* NCIM 5021. *Int. J. Pharm. Sci. Rev. Res.* **2015**, *34*, 194–196.
79. Makerly, H.; Hashim, Z. Antibacterial Activity and Metabolite Content of Water and Methanolic Extracts of Purple and White *Bougainvillea* sp. *J. Bioprocess. Biomass Technol.* **2024**, *3*, 33–40. [CrossRef]
80. Ravikumar, P.; Thangaraj, P. Anti-Bacterial Attributes of Phytochemicals from *Bougainvillaea spectabilis*: Computational Approach. *Pharmacogn. Res.* **2024**, *16*, 384–390. [CrossRef]
81. Dame-Teixeira, N.; El-Gendy, R.; Monici, I.; Holanda, C.; De-Oliveira, A.; Romeiro, L.; Do, T. Sustainable Multifunctional Phenolic Lipids as Potential Therapeutics in Dentistry. *Sci. Rep.* **2022**, *12*, 9299. [CrossRef]
82. Wang, X.; Li, J.; Zhang, S.; Zhou, W.; Zhang, L.; Huang, X. PH-Activated Antibiofilm Strategies for Controlling Dental Caries. *Front. Cell. Infect. Microbiol.* **2023**, *13*, 1130506. [CrossRef]
83. Beladi, S.; Kiridena, S.D.; Wijayaratna, U.; Taylor, C.; Anker, J.; Tzeng, T.-R. PH Variation in Medical Implant Biofilms: Causes, Measurements, and Its Implications for Antibiotic Resistance. *Front. Microbiol.* **2022**, *13*, 1028560. [CrossRef]

foods

MDPI

Article

Inhibition of Amyloid β Accumulation by Protease-Digested Whitebait (Shirasu) in a Murine Model of Alzheimer's Disease

Takahiro Katsuki [1,2], Kayako Ogi [1], Ayaka Kinno [2], Shingo Kasamatsu [2], Hideshi Ihara [2] and Hidenobu Sumitani [1,*]

[1] Toyo Institute of Food Technology, 23-2-4, Minami-Hanayashiki, Kawanishi-shi 666-0026, Hyogo, Japan; takahiro_katsuki@shokuken.or.jp (T.K.)

[2] Department of Biological Chemistry, Graduate School of Science, Osaka Metropolitan University, 1-1 Gakuen-cho, Naka-ku, Sakai-shi 599-8531, Osaka, Japan; su23029l@st.omu.ac.jp (A.K.); kasamatsu@omu.ac.jp (S.K.); iharah@omu.ac.jp (H.I.)

* Correspondence: hidenobu_sumitani@shokuken.or.jp

Abstract: The number of people with dementia is increasing annually worldwide. Alzheimer's disease (AD), which accounts for the highest percentage of dementia-causing diseases, remains difficult to cure, and prevention of its onset is important. We aimed to discover new AD-preventive ingredients and investigate the inhibitory effects of ten different species of seafood digests prepared by protease treatment on β-secretase 1 (BACE1) activity. Substantial inhibition of BACE1 activity was observed in five species of seafood, and protease-digested whitebait (WPD) showed the highest inhibitory effect among the ten marine samples. We further examined the potential of WPD as an AD preventive component using a familial AD strain (5xFAD) murine model. The intraperitoneal administration of WPD for 28 days substantially decreased the insoluble amyloid β_{1-42} content and the expression of glial fibrillary acidic protein, a marker of astrogliosis, in the cerebral cortex of the 5xFAD mice. These results strongly suggest that WPD is a novel functional food-derived ingredient with preventive effects against AD.

Keywords: Alzheimer's disease; β-secretase; whitebait; functional food-derived ingredient

1. Introduction

The increase in the number of patients with dementia is an urgent issue globally. The number of people with dementia worldwide is predicted to increase to approximately 153 million by 2050 [1]. Alzheimer's, Lewy body, vascular, and frontotemporal lobar dementia are well known as the four major causative diseases of dementia, of which Alzheimer's disease (AD) accounts for over 60% of causative diseases [2]. AD is a cerebral disorder that affects memory, thinking, and behavior. Two types of medicines are used to treat patients, cholinesterase (tacrine, donepezil, rivastigmine, galantamine) and *N*-methyl-D-aspartate receptor inhibitors (memantine), which regulate neurotransmitters and temporarily stabilize cognitive functions; however, both are symptomatic therapies [3]. Lecanemab is an antibody medicine that removes amyloid beta (Aβ) from the brain by binding Aβ soluble protofibrils [4]. In 2023, this medicine was approved by the U.S. Food and Drug Administration for the treatment of AD; however, it is important to administer the drug in the early stages of AD [4]. Thus, AD treatment after onset remains difficult and prevention seems important [3].

The most supported hypothesis about the mechanism of AD onset is Aβ plaque formation, resulting in neuronal cell death and memory disorders [5]. Aβ is a 40–42-residue peptide that is enzymatically produced from amyloid precursor protein (APP) by β-secretase (BACE1) and γ-secretase [6]. BACE1 is an aspartic protease involved in the cleavage of membrane proteins. Especially for APP, BACE1 cleaves the extracellular domain and causes Aβ generation [6]. For AD therapy, BACE1 inhibitors have been actively developed [7], and they were reported to inhibit the accumulation of intracerebral Aβ after intraperitoneal administration in a mouse experiment [8]. Some BACE1 inhibitors have been tested in clinical trials; however, they were discontinued due to insufficient therapeutic efficacy and safety [9–12]. In particular, it is difficult to recognize the therapeutic effectiveness after AD onset and immediately before the onset of AD [9]. Although BACE1 inhibitors require early administration, it is impractical to routinely administer medicines before the onset of symptoms. In contrast, functional foods, including bioactive components with BACE1 inhibitory effects, can be consumed via daily diets regardless of AD onset, contributing to the prevention of AD development.

Previous studies have demonstrated the BACE1 inhibitory activity of various food-derived components, including polyphenols, such as catechins [13–15]; polysaccharides, such as glycosaminoglycans [16–18]; bioactive peptides [19–21]; and sulfur-containing components, such as sulforaphane [22]. In addition, several BACE1 inhibitory components have been identified in marine products [23], but most of them are derived from seaweeds or bacteria, and there are only a few reports on components derived from fish [23]. Fish are known for their broad-spectrum health-promoting effects; however, to the best of our knowledge, only a small amount of attention has been paid to the inhibitory effects of fish-derived components on BACE1 activity.

Protease digestion is commonly used to prepare protein-derived bioactive peptides and increase the efficiency of the extraction of functional components from foods, including fish [24,25]. However, little is known about the effects and bioactivity of protease digests of fish in the prevention of AD. In this study, we aimed to evaluate the potential of seafood-derived components prepared by protease digestion as functional ingredients for the prevention of AD onset using a fluorescent in vitro BACE1 assay and animal experiments with a familial AD model (5xFAD, [26]) mouse strain.

2. Materials and Methods

2.1. Protease Digestion of Fish

Ten different types of seafood, sardine (*Sardinops melanostictus*, from Chiba, Japan) squid (*Heterololigo bleekeri*, from Aomori, Japan), horse mackerel (*Decapterus maruadsi*, from Oita, Japan), sea bream (*Pagrus major*, from Hiroshima, Japan), salmon (*Oncorhynchus mykiss*, from Turkey), flounder (*Hippoglossoides dubius*, from Tottori, Japan), shrimp (*Litopenaeus vannamei*, from Indonesia), tuna (*Thunnus albacares*, from Chiba, Japan), mackerel (*Scomber australasicus*, from Kochi, Japan), and boiled whitebait (*Engraulis japonica* fry, from Ehime, Japan), were purchased from a local grocery store. Except for whitebait, the meat parts of the other nine species of seafood were cut into 3–10 mm pieces and used for further protease digestion. The whole whitebait was subjected to protease digestion. Protease solution was prepared by dissolving 9 mg of protease (Amano 3SD, Amano Enzyme, Inc., Nagoya, Japan) in 750 μL of ultrapure water. For protease digestion, seafood samples (900 mg) were mixed with 750 μL of protease solution and incubated at 50 °C for 90 min using a heat block incubator (THB-1, As One Corp., Osaka, Japan). The protease-digested mixtures were then incubated at 95 °C for 10 min to terminate the protease digestion and centrifuged at 7740× *g* at room temperature (approximately 25 °C) for 5 min. The supernatants were filtered using a membrane filter (Watman GD/X 13, 0.45 μm, Cytiva, Tokyo, Japan), lyophilized, and stored at −30 °C until further experiments. Seafood pieces were also prepared in the absence of protease, followed by the same sample procedure, and used as negative controls.

2.2. Fluorescent BACE1 Assay

An evaluation of the inhibitory effects of protease-digests of seafood was carried out using a SensoLyte 520 BACE1 Assay Kit (Anaspec, Fremont, CA, USA) according to the manufacturer's protocol with slight modifications. Lyophilized protease- and non-protease-treated samples were re-dissolved in ultrapure water, and the insoluble fractions were filtered using membrane filters. In a 384-well black plate (Thermo Fisher Scientific,

Waltham, MA, USA), 4 μL/well of sample and 16 μL/well of BACE1 enzymatic solution (diluted 1/200 in the assay buffer) were mixed. After a 10 min incubation at 32 °C, the BACE1 enzymatic reaction was initialized by the addition of 20 μL/well of substrate solution (diluted 1/100 in the assay buffer). Fluorescence intensity (excitation wavelength: 490 nm, emission wavelength: 510 nm) was measured at 2 min intervals for 30 min using a microplate reader (CLARIOstar Plus, BMG Labtech, Aylesbury, UK). BACE1 activity was calculated using the following formula:

$$\text{BACE1 activity} = \text{fluorescence intensity at 30 min} - \text{fluorescence intensity at 0 min} \quad (1)$$

2.3. Evaluation of BACE1 Activity by Fluorescence HPLC

The inhibitory effects of protease-digested seafood on BACE1 activity were also evaluated using a fluorescent high-performance liquid chromatography (HPLC)-based assay. Human recombinant BACE1 (Cat#931-AS, R&D Systems, Minneapolis, MN, USA) was dissolved in ultrapure water, and fluorescent peptide substrate IV (Mca-SEVNLDAEFRK(DNP)RR-NH_2; Cat#ES004, R&D Systems) was dissolved in dimethyl sulfoxide (DMSO). They were diluted with 0.1 M sodium acetate buffer (pH 4.0) before evaluation, respectively.

In a 96-well black plate (OptiPlate-96 F, PerkinElmer, Waltham, MA, USA), 50 μL/well of BACE1 enzyme (final concentration 22 nM), 40 μL/well of substrate solution (final concentration 10 μM), and 10 μL/well of BACE1 inhibitor (LY2886721; Cat#S2156, Selleck Chemicals, Houston, TX, USA; final concentration 200 nM) or the same volume of protease-digest samples were mixed and incubated at 37 °C for 60 min in the dark. One hundred microliters of the reaction mixture was collected into a new 1.5 mL tube containing 100 μL of 2.5% trifluoroacetic acid to terminate the enzymatic reaction. After centrifugation at 4 °C and 15,000× *g* for 10 min, the obtained supernatant was subjected to HPLC analysis. Fifty microliters of sample was injected into an HPLC pump consisting of PU-2089 (JASCO, Tokyo, Japan), autosampler (AS-1550; JASCO), and a fluorescent detector (FP 2020 Plus; JASCO), using Mightysil RP-18 GP (75 × 3.0 mm, φ5 μm; Kanto Chemical Co., Inc., Tokyo, Japan) with an isocratic elution of methanol/0.1% formic acid (50/50) with a flow rate of 0.3 mL/min. The fluorescent product cleaved from the substrate peptide was detected by monitoring the fluorescence excitation wavelength at 320 nm and emission wavelength at 405 nm and quantified by a standard curve obtained using 7-methoxycoumarin-3-carboxylic acid.

2.4. Preparation of Feed Administration

Lyophilized protease-digested samples were dissolved in 1 mL of ultrapure water and separated using a flash automated purifier (Isolera One; Biotage, Uppsala, Sweden) under the following conditions. Ultrapure water and ethanol (liquid chromatography–mass spectrometry grade, FUJIFILM Wako Pure Chemical Corporation, Osaka, Japan) were used as solvents A and B, respectively. The column was composed of SNAP Ultra 25 g (Biotage) and pre-equilibrated with solvent A. Samples were applied to the column at a flow rate of 10 mL/min and separated using a linear gradient of solvents A and B (0% B for 0–3 min, 0–100% B for 3–26 min, and 100% B for 26–30 min). The fractions of 0–10 min were collected, evaporated with ethanol, and lyophilized. The lyophilized powder was stored at −80 °C until animal testing.

2.5. Animal Test

The mouse strain, 5xFAD (B6SJL-Tg6799 strain, [26]) was obtained from The Jackson Laboratory (Bar Harbor, ME, USA), and 9–10-week-old male mice were used. This study was performed in accordance with the Guidelines for Animal Experimentation of Osaka Metropolitan University (Osaka, Japan). All animal experiments were approved by the Animal Ethics Committee of Osaka Prefecture University (protocol code: No. 20-86). The animals were kept at 22–24 °C and with a 12 h light/12 h dark cycle. The animals were fed commercial pellets (CE-2; CLEA Japan Inc., Tokyo, Japan) and provided water ad libitum. The genotype of individual animals was determined using polymerase chain reaction, as recommended by The Jackson Laboratory, using homozygous wild-type (WT) and heterozygous 5xFAD mice.

The mice were divided into four groups: control WT (WT-control, n = 8), WT-medicated (WT-IP, n = 5), control 5xFAD (5xFAD-control, n = 11), and 5xFAD-medicated (5xFAD-IP, n = 5). Body weight was measured immediately before administration. Enzymatic digests were administered intraperitoneally at a dose of 500 mg/kg body weight every 2 days for 28 days. Saline was administered to mice in the control group. After the administration period, mice were anesthetized using isoflurane inhalation and euthanized by cervical dislocation. The brain tissue was immediately removed; the cerebral cortex was harvested, frozen in liquid nitrogen, and stored at −80 °C.

2.6. Quantification of Insoluble Aβ

Extraction of insoluble Aβ from the cerebral cortex was performed as previously reported [27]. In brief, the cerebral cortex was homogenized in 15-fold volumes of ice-cold 20 mM tris(hydroxymethyl)aminomethane (Tris)-HCl buffer (pH 7.4) containing 150 mM NaCl protease inhibitor cocktail (Nacalai Tesque Inc., Kyoto, Japan) with a Teflon homogenizer for 20 strokes, and centrifuged at 100,000× g at 4 °C for 1 h using an ultracentrifuge (Model: CS 120FNX, Eppendorf Himac Technologies Co., Ltd., Ibaraki, Japan) with a S120AT3 rotor (Eppendorf Himac Technologies Co., Ltd.). The resultant supernatant was removed, and the pellet was re-suspended in a 5 M guanidine-HCl (GuHCl, pH 8.0)-containing protease inhibitor cocktail followed by incubation end-over-end overnight at 25 °C. After centrifuging at 23,000× g at 4 °C for 30 min, the supernatant was transferred to a new 1.5 mL tube (referred to as GuHCl extract) and stored at −80 °C until further experiments. Aβ content in the GuHCl extract (referred to as the insoluble Aβ fraction) was measured using a Human β amyloid (1–42) enzyme-linked immunosorbent assay (ELISA) Kit (FUJIFILM Wako Pure Chemical Corporation) in accordance with the manufacturer's protocol. The insoluble Aβ contents were corrected for the protein amount, which was quantified using the Protein Assay BCA Kit (Nacalai Tesque Inc.) with bovine serum albumin as a standard, in accordance with the manufacturer's protocol.

2.7. Western Blotting

Western blotting was performed as previously reported with minor modifications [28]. Cerebral cortex was homogenized in ice-cold 50 mM 4-(2-hydroxyethyl)-1-piperazineethanesulfonic acid buffer (pH 8.0) containing 150 mM NaCl, 10 mM sodium pyrophosphate, 10 mM sodium fluoride, 2% 3-[(3-cholamidopropyl)dimethylammonio]-1-propanesulfonate, 2.5% lithium dodecyl sulfate, 10% glycerol, 2 mM ethylenediaminetetraacetic acid, 2 mM sodium vanadate, 1 mM dithiothreitol, and 1% protease inhibitor cocktail by sonication. After incubation on ice for 30 min, lysates were centrifuged at 23,000× g and 4 °C for 30 min. The supernatant was transferred to a new tube, and proteins in tissue lysates were heat-denatured. Proteins (20 μg/lane) were subjected to 12% acrylamide gel electrophoresis, separated by sodium dodecyl sulfate-polyacrylamide gel electrophoresis, and transferred to nitrocellulose membranes. Nitrocellulose membranes were blocked with Blocking One (Nacalai Tesque Inc.), followed by a reaction with an anti-glial fibrillary acidic protein (GFAP) antibody (Cell Signaling Technology, Danvers, MA, USA) and an anti-β-actin antibody (Santa Cruz Biotechnology, Dallas, TX, USA) at 4 °C overnight. After washing three times with TBST (20 mM Tris-HCl, 150 mM NaCl, 0.1% Tween 20, pH 7.6), the membranes were incubated with horseradish peroxidase-conjugated anti-rabbit (Santa Cruz Biotechnology) or anti-mouse (Cytiva) secondary antibodies for 1 h at room temperature. After washing three times with TBST, immunoreactive bands were detected using a chemiluminescence reagent (ImmunoStar LD, Fujifilm Wako Pure Chemical Corporation) and a luminescent image analyzer (LAS-1000 mini, Fujifilm Corp., Tokyo, Japan). Band intensities were quantified using analysis software (Multigauge, Fujifilm Corp.).

2.8. Statistical Analyses

All experiments were performed at least three times, and values for individual experiments are presented as mean ± standard deviation (SD) or error (SE). Statistical significance was determined using Student's unpaired t-test and the Tukey–Kramer test using the GraphPad Prism 8.1.2 software (GraphPad, Inc., La Jolla, CA, USA); $p < 0.05$ was considered significant.

3. Results

3.1. Screening of BACE1 Inhibitory Activity in Vitro

Protease-treated digests were prepared from ten different species of seafood, including sardine, squid, horse mackerel, sea bream, salmon, flounder, shrimp, tuna, mackerel, and boiled whitebait, and their inhibitory effects on BACE1 activity were evaluated (Figure 1). Compared to the control group, BACE1 activity was substantially inhibited in the presence of protease-digested seafood derivatives (squid, horse mackerel, sea bream, salmon, flounder, shrimp, mackerel, and boiled whitebait), except in the presence of sardine and tuna. Moreover, we confirmed protease digest-dependent BACE1 inhibition in five species of seafood (sea bream, flounder, shrimp, mackerel, and boiled whitebait), and the most remarkable inhibition was observed in the presence of protease-digested whitebait (WPD), which almost completely blocked BACE1 activity (less than 1% activity compared with the control). Based on these results, we further evaluated the potential of WPD.

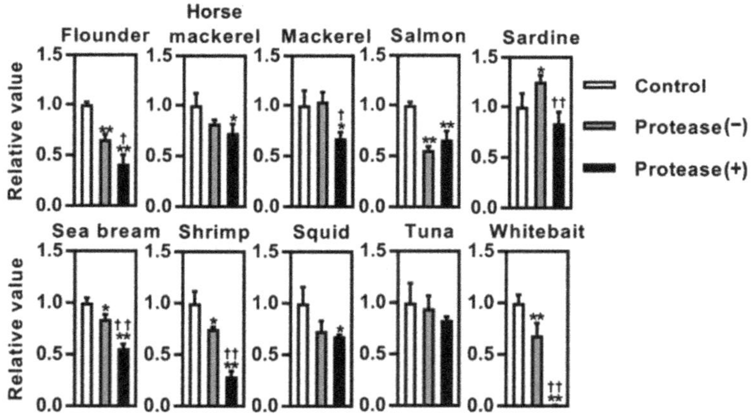

Figure 1. β-secretase 1 (BACE1) inhibitory activity of protease-digested derivatives of seafood. Each lyophilized protease-digested fish powder was dissolved in ultrapure water to 500 µg/mL (final assay concentration, 50 µg/mL). Non-protease-digested samples were dissolved in the same amount of ultrapure water. Ultrapure water was used in the control. White, gray, and black bars represent the control, non-protease-digested, and protease-digested samples, respectively. Data represent the mean ± standard deviation (n = 3 each); * $p < 0.05$, ** $p < 0.01$ versus the control, † $p < 0.05$, †† $p < 0.01$ versus the protease (−).

We performed a different assay to support the effect of WPD on BACE1 inhibitory activity using a fluorescent HPLC system (Figure 2). A strong signal of the BACE1-dependent enzymatic reaction product was observed at a retention time of 5.5 min (Figure 2A, top). Only a modest signal was detected in the presence of the well-established BACE1 inhibitor LY2886721 (Figure 2A, middle). Similarly, the signal disappeared almost completely in the presence of WPD (Figure 2A, bottom). Calculation of the amount of the BACE1-dependent enzymatic reaction product using a standard curve of 7MCA demonstrated that WPD strongly inhibited BACE1 activity, which was similar to that of LY2886721 (Figure 2B).

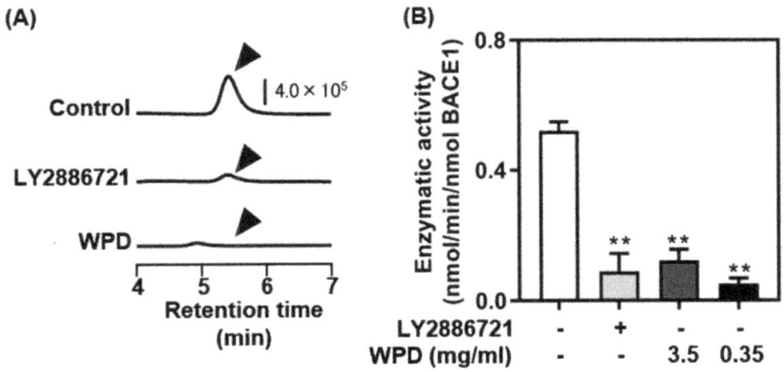

Figure 2. Confirmation of β-secretase 1 (BACE1) inhibitory effects of protease-digested whitebait (WPD) by fluorescent high-performance liquid chromatography (HPLC): (**A**) HPLC chromatograms of the BACE1 enzymatic reaction in the absence (control, **top**) or the presence of LY2886721 (**middle**) or WPD (**bottom**). (**B**) BACE1 enzymatic activities were calculated using a standard curve of 7MCA. Data represent mean ± standard deviation ($n = 3$ each); ** $p < 0.01$ versus the control.

3.2. Evaluation of Effects of WPD Administration on 5xFAD Mice

To evaluate the potential of WPD in the prevention of AD in vivo, we performed animal experiments using AD model 5xFAD-strain mice with the intraperitoneal administration of WPD. A slight and transient decrease in body weight was observed in the WPD-administered groups (both WT-IP and FAD-IP); there was no marked change in body weight among any of the four groups on the final day of the administration period (Figure 3). However, six mice died during days 0–7 in the 5xFAD-IP group.

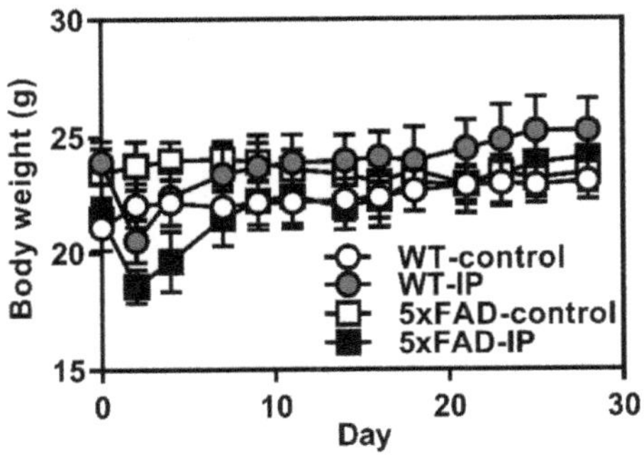

Figure 3. Effects of protease-digested whitebait (WPD) administration on body weight of 5xFAD mice. WPD was administered intraperitoneally every 2 days for 28 days, and the body weight was measured immediately before administration. The white circle shows the WT control group ($n = 8$), the gray circle shows the WT WPD group (day 0: $n = 11$, day 2: $n = 10$, day 4: $n = 7$, days 7–28: $n = 5$), the white square shows the 5xFAD control group ($n = 11$), and the black square shows the 5xFAD WPD group ($n = 5$).

At the end of the administration period, the cerebral cortex was harvested and the insoluble $A\beta_{1-42}$ content was quantified using ELISA (Figure 4). The results indicated that the insoluble $A\beta_{1-42}$ content in the control 5xFAD mice (1231.9 ± 198.3 pmol/mg protein) was substantially higher than that in the control WT, consistent with previous reports [26]; the results suggest that amyloid plaque was formed in the cerebral cortex of 5xFAD mice. In contrast, a marked decrease in the insoluble $A\beta_{1-42}$ content was observed in the WPD-administered 5xFAD mice (206.2 ± 53.11 pmol/mg protein) compared with control 5xFAD mice. In WT mice, insoluble $A\beta_{1-42}$ was not observed regardless of WPD administration.

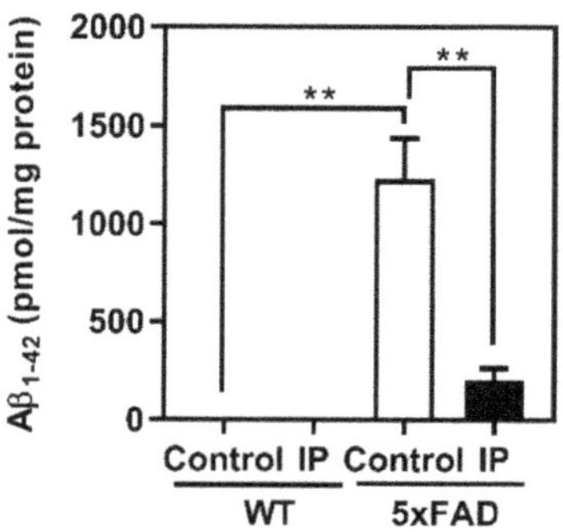

Figure 4. Decrease in the insoluble $A\beta_{1-42}$ content in the cerebral cortex after protease-digested whitebait administration. The insoluble $A\beta_{1-42}$ content in the cerebral cortex was quantified using an enzyme-linked immunosorbent assay. WT control group ($n = 8$), WT WPD group ($n = 5$), 5xFAD control group ($n = 11$), and 5xFAD WPD group ($n = 5$); ** $p < 0.01$.

3.3. Analysis of Astrogliosis

Aggregated forms of Aβ are known to cause aberrant activation of astrocytes (astrogliosis) in the brain [29,30]. Therefore, we examined the effects of WPD administration on the expression of GFAP, a marker of astrogliosis, using Western blotting. A marked increase in GFAP expression was observed in control 5xFAD mice compared to WT mice (Figure 5). In contrast, the WPD-administered 5xFAD mice showed a marked decrease in GFAP expression in the cerebral cortex compared with that in control 5xFAD mice. In WT mice, no difference in GFAP expression was observed upon WPD administration.

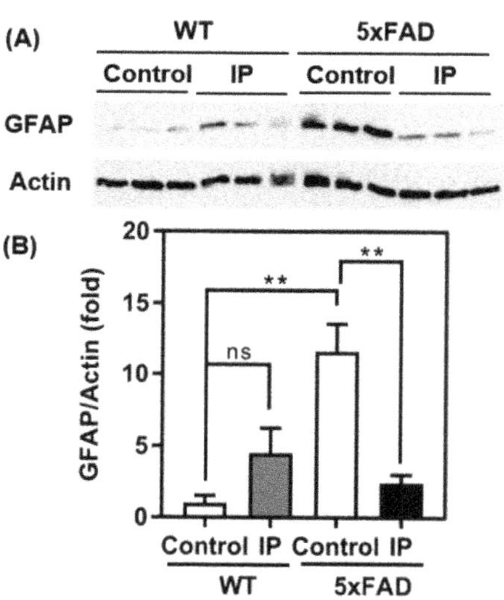

Figure 5. Effect of protease-digested whitebait (WPD) administration on GFAP expression in the cerebral cortex of mice. GFAP expression in the cerebral cortex was detected using Western blotting: (**A**) Representative image of Western blotting using anti-GFAP (**upper**) and anti-β-actin (**lower**) antibodies. (**B**) Relative band intensity of GFAP to β-actin. WT control ($n = 3$), WT WPD ($n = 3$), 5xFAD control ($n = 3$), and 5xFAD WPD group ($n = 3$); ** $p < 0.01$; ns, not significant.

4. Discussion

In this study, we examined the potential of ten different seafood species, including sardine, squid, horse mackerel, sea bream, salmon, flounder, shrimp, tuna, mackerel, and boiled whitebait, using both in vitro and in vivo experiments. Using the in vitro BACE1 assay, a marked inhibitory effect was confirmed in 8 of the 10 species. Furthermore, in sea bream, flounder, shrimp, mackerel, and boiled whitebait, the inhibitory effects were enhanced by protease pretreatment (Figure 1). Consistent with the present results, several previous studies have reported that muscle protein hydrolysates of marine animals, including shrimp [21] and a typical herbivorous gastropod sea hare (*Aplysia kurodai*) [31], exhibit an inhibitory effect on BACE1. In the current study, the results obtained from two different in vitro BACE1 assays revealed that protease-digested derivatives of boiled whitebait (WPD) exhibited the most remarkable effects on BACE1 inhibition (Figures 1 and 2). To the best of our knowledge, this study provides the first evidence that WPD has strong potential to block BACE1 activity.

The results obtained from the animal experiments using 5xFAD mice demonstrated that intraperitoneal administration of WPD strongly suppressed not only the accumulation of insoluble Aβ, but also the expression of GFAP protein, a marker of astrogliosis, in the cerebral cortex (Figures 4 and 5). Astrogliosis is reportedly caused by the accumulation of Aβ [30,31]; thus, the suppression of astrogliosis observed in the WPD-administered 5xFAD group may be because the WPD administration effectively inhibits BACE1 activity in vivo, resulting in the suppression of Aβ accumulation and downstream astrogliosis.

Whitebait is a collective term for the immature fry of fish, including the Japanese anchovy and sardine. WPD potently inhibited BACE1 activity, whereas no such effect was observed in sardines (Figure 1). Based on these results, we predicted two possibilities. First, the bioactive substances contained in whitebait may change during fish development and aging. Peptide fragments prepared by the enzymatic degradation of several foods exhibit BACE1 inhibitory activity [19–21,31] and improve cognitive function in vivo [32,33]. Thus, whitebait may be abundant in these precursor proteins. In the future, mass spectrometry-based analyses will facilitate the identification of peptides present in WPD and the investigation of their pharmacological effects in vitro and in vivo.

Another possibility is that while WPD was prepared by protease digestion of whole shirasu, only the meat was used for protease digestion of sardines, suggesting that bioactive substances with BACE1 inhibitory activity may exist in parts other than the meat. In relation to the latter possibility, it has been reported that chondroitin sulfate, a sulfated glycosaminoglycan that is abundantly present in fish waste [34], has BACE1 inhibitory activity [17,35,36]. Therefore, although further studies are required to identify bioactive substances with BACE1 inhibitory activity and to analyze their pharmacological effects in vivo, functional ingredients with BACE1 inhibitory activity produced from fish waste could lead to the effective utilization of unused resources.

In the animal study, transient weight loss was observed in the early phase of treatment, and six mice died during the administration period in the 5xFAD-IP group. Since the sample used in this animal study was a crude purified product of protease-digested whitebait, various substances other than the active ingredient coexisted, and these ingredients may have contributed to the toxicity. In the future, identification and isolation of the active ingredients in WPD will facilitate a detailed evaluation of the pharmacological activity of the active ingredients in whitebait under low-toxicity dosing conditions.

Several previous studies using 5xFAD mice reported that accumulation of Aβ in the brain was suppressed to approximately half of that in the control group by the administration of natural compounds (i.e., hop-derived iso-α-acids, *Inula britannica*-derived 1,6-O,O-diacetylbritannilactone, whey-derived β-lactolin peptide, and a Chinese natural medicine, Qi-fu-yi) [32,37–39]. In the present study, the insoluble $A\beta_{1-42}$ content in the cerebral cortex of WPD-administered 5xFAD mice markedly decreased to one-sixth of that in control 5xFAD mice, suggesting that WPD can be a functional ingredient derived from food with an AD-prevention effect. In this study, WPD was administered intraperitoneally to mice to eliminate the effects of digestion and absorption in vivo. Therefore, it is essential to evaluate the potential of WPD as a food ingredient in future studies using oral animal experiments.

Although Aβ accumulation and GFAP protein expression were evaluated in this study, cognitive assessment was not performed. 5xFAD mice are reportedly impaired in the Y-maze test at 16–20 weeks [26,40]. The current study terminated administration at 13–14 weeks, which was too early to observe impaired cognitive function. This study aimed to prevent AD onset, and administration was initiated at 9–10 weeks, when Aβ accumulation began. The study was terminated at 13–14 weeks because it was difficult to prepare sufficient amounts of the purified product until the age of impaired cognitive function. Because cognitive assessment is a key factor in AD studies, we plan to extend the duration of this study in the future.

5. Conclusions

In this study, BACE1 activity was evaluated in ten protease-digested fishes. Five protease-digested fishes suppressed BACE1 activity, of which WPD showed the most remarkable inhibition of BACE1. To evaluate the in vivo effects of WPD, WPD was administered to 5xFAD mice, and the accumulation of Aβ and the expression of GFAP protein in the cerebral cortex were found to be suppressed. Thus, WPD was effective in preventing AD both in vitro and in vivo. Moreover, compared with previous studies on natural compounds, WPD may have more powerful AD-preventive effects. Based on these results, WPD may be a novel and effective food for AD prevention.

Author Contributions: Conceptualization, T.K. and H.S.; methodology, T.K., H.S., K.O. and S.K.; formal analysis, T.K., K.O., A.K. and S.K.; investigation, T.K., H.S. and K.O.; resources, H.S.; data curation, T.K., K.O., A.K. and S.K.; writing—original draft preparation, T.K.; writing—review and editing, H.S. and H.I.; supervision, H.S. and H.I.; project administration, H.S.; funding acquisition, H.S. All authors have read and agreed to the published version of the manuscript.

Funding: This research received no external funding.

Institutional Review Board Statement: The study was conducted in accordance with the Declaration of Helsinki, and approved by the Animal Ethics Committee of Osaka Prefecture University (protocol code: No. 20-86).

Informed Consent Statement: Not applicable.

Data Availability Statement: The original contributions presented in the study are included in the article, further inquiries can be directed to the corresponding author.

Conflicts of Interest: The authors declare no conflicts of interest.

References

1. GBD 2019 Dementia Forecasting Collaborators. Estimation of the global prevalence of dementia in 2019 and forecasted prevalence in 2050: An analysis for the Global Burden of Disease Study 2019. *Lancet Public Health* **2022**, *7*, e105–e125. [CrossRef] [PubMed]
2. Crous-Bou, M.; Minguillon, C.; Gramunt, N.; Molinuevo, J.L. Alzheimer's disease prevention: From risk factors to early intervention. *Alzheimers Res. Ther.* **2017**, *9*, 71. [CrossRef] [PubMed]
3. Passeri, E.; Elkhoury, K.; Morsink, M.; Broersen, K.; Linder, M.; Tamayol, A.; Malaplate, C.; Yen, F.T.; Arab-Tehrany, E. Alzheimer's disease: Treatment strategies and their limitations. *Int. J. Mol. Sci.* **2022**, *23*, 13954. [CrossRef] [PubMed]
4. van Dyck, C.H.; Swanson, C.J.; Aisen, P.; Bateman, R.J.; Chen, C.; Gee, M.; Kanekiyo, M.; Li, D.; Reyderman, L.; Cohen, S.; et al. Lecanemab in early Alzheimer's disease. *N. Engl. J. Med.* **2023**, *388*, 9–21. [CrossRef]
5. Hardy, J.A.; Higgins, G.A. Alzheimer's disease: The amyloid cascade hypothesis. *Science* **1992**, *256*, 184–185. [CrossRef]
6. Wolfe, M.S. When loss is gain: Reduced presenilin proteolytic function leads to increased Abeta42/Abeta40. Talking Point on the role of presenilin mutations in Alzheimer disease. *EMBO. Rep.* **2007**, *8*, 136–140. [CrossRef]
7. Hampel, H.; Vassar, R.; De Strooper, B.; Hardy, J.; Willem, M.; Singh, N.; Zhou, J.; Yan, R.; Vanmechelen, E.; De Vos, A.; et al. The beta-secretase BACE1 in Alzheimer's disease. *Biol. Psychiatry* **2021**, *89*, 745–756. [CrossRef]
8. Devi, L.; Tang, J.; Ohno, M. Beneficial effects of the β-secretase inhibitor GRL-8234 in 5XFAD Alzheimer's transgenic mice lessen during disease progression. *Curr. Alzheimer Res.* **2015**, *12*, 13–21. [CrossRef]
9. Egan, M.F.; Kost, J.; Tariot, P.N.; Aisen, P.S.; Cummings, J.L.; Vellas, B.; Sur, C.; Mukai, Y.; Voss, T.; Furtek, C.; et al. Randomized trial of verubecestat for mild-to-moderate Alzheimer's disease. *N. Engl. J. Med.* **2018**, *378*, 1691–1703. [CrossRef]
10. Sperling, R.; Henley, D.; Aisen, P.S.; Raman, R.; Donohue, M.C.; Ernstrom, K.; Rafii, M.S.; Streffer, J.; Shi, Y.; Karcher, K.; et al. Findings of efficacy, safety, and biomarker outcomes of atabecestat in preclinical Alzheimer disease: A truncated randomized phase 2b/3 clinical trial. *JAMA Neurol.* **2021**, *78*, 293–301. [CrossRef]

11. Lo, A.C.; Evans, C.D.; Mancini, M.; Wang, H.; Shcherbinin, S.; Lu, M.; Natanegara, F.; Willis, B.A. Phase II (NAVIGATE-AD study) results of LY3202626 effects on patients with mild Alzheimer's disease dementia. *J. Alzheimers Dis. Rep.* **2021**, *5*, 321–336. [CrossRef] [PubMed]
12. Wessels, A.M.; Tariot, P.N.; Zimmer, J.A.; Selzler, K.J.; Bragg, S.M.; Andersen, S.W.; Landry, J.; Krull, J.H.; Downing, A.M.; Willis, B.A.; et al. Efficacy and safety of lanabecestat for treatment of early and mild Alzheimer disease: The AMARANTH and DAYBREAK-ALZ randomized clinical trials. *JAMA Neurol.* **2020**, *77*, 199–209. [CrossRef] [PubMed]
13. Youn, K.; Park, J.H.; Lee, S.; Lee, S.; Lee, J.; Yun, E.Y.; Jeong, W.S.; Jun, M. BACE1 inhibition by genistein: Biological evaluation, kinetic analysis, and molecular docking simulation. *J. Med. Food* **2018**, *21*, 416–420. [CrossRef] [PubMed]
14. Jeon, S.Y.; Bae, K.; Seong, Y.H.; Song, K.S. Green tea catechins as a BACE1 (beta-secretase) inhibitor. *Bioorg. Med. Chem. Lett.* **2003**, *13*, 3905–3908. [CrossRef]
15. Das, S.; Sengupta, S.; Chakraborty, S. Scope of β-secretase (BACE1)-targeted therapy in Alzheimer's disease: Emphasizing the flavonoid based natural scaffold for BACE1 inhibition. *ACS Chem. Neurosci.* **2020**, *11*, 3510–3522. [CrossRef]
16. Mycroft-West, C.J.; Devlin, A.J.; Cooper, L.C.; Guimond, S.E.; Procter, P.; Guerrini, M.; Miller, G.J.; Fernig, D.G.; Yates, E.A.; Lima, M.A.; et al. Glycosaminoglycans from *Litopenaeus vannamei* inhibit the Alzheimer's disease beta secretase, BACE1. *Mar. Drugs* **2021**, *19*, 203. [CrossRef]
17. Mycroft-West, C.J.; Devlin, A.J.; Cooper, L.C.; Procter, P.; Miller, G.J.; Fernig, D.G.; Guerrini, M.; Guimond, S.E.; Lima, M.A.; Yates, E.A.; et al. Inhibition of BACE1, the beta-secretase implicated in Alzheimer's disease, by a chondroitin sulfate extract from Sardina pilchardus. *Neural Regen. Res.* **2020**, *15*, 1546–1553. [CrossRef]
18. Mycroft-West, C.J.; Cooper, L.C.; Devlin, A.J.; Procter, P.; Guimond, S.E.; Guerrini, M.; Fernig, D.G.; Lima, M.A.; Yates, E.A.; Skidmore, M.A. A Glycosaminoglycan extract from *Portunus pelagicus* inhibits BACE1, the beta secretase implicated in Alzheimer's disease. *Mar. Drugs* **2019**, *17*, 293. [CrossRef]
19. Lee, J.K.; Li-Chan, E.C.Y.; Cheung, I.W.Y.; Jeon, Y.J.; Ko, J.Y.; Byun, H.G. Neuroprotective effect of β-secretase inhibitory peptide from Pacific hake (*Merluccius productus*) fish protein hydrolysate. *Curr. Alzheimer Res.* **2019**, *16*, 1028–1038. [CrossRef]
20. Lee, J.K.; Li-Chan, E.C.Y.; Byun, H.-G. Characterization of β-secretase inhibitory peptide purified from skate skin protein hydrolysate. *Eur. Food Res. Technol.* **2015**, *240*, 129–136. [CrossRef]
21. Li-Chan, E.C.Y.; Cheung, I.W.Y.; Byun, H.-G. Shrimp (*Pandalopsis dispar*) waste hydrolysate as a source of novel β–secretase inhibitors. *Fish. Aquat. Sci.* **2016**, *19*, 11. [CrossRef]
22. Youn, K.; Yoon, J.H.; Lee, N.; Lim, G.; Lee, J.; Sang, S.; Ho, C.T.; Jun, M. Discovery of sulforaphane as a potent BACE1 inhibitor based on kinetics and computational studies. *Nutrients* **2020**, *12*, 3026. [CrossRef] [PubMed]
23. Hafez Ghoran, S.; Kijjoa, A. Marine-derived compounds with anti-Alzheimer's disease activities. *Mar. Drugs* **2021**, *19*, 410. [CrossRef] [PubMed]
24. Nadar, S.S.; Rao, P.; Rathod, V.K. Enzyme assisted extraction of biomolecules as an approach to novel extraction technology: A review. *Food Res. Int.* **2018**, *108*, 309–330. [CrossRef] [PubMed]
25. Kristinsson, H.G.; Rasco, B.A. Fish protein hydrolysates: Production, biochemical, and functional properties. *Crit. Rev. Food Sci. Nutr.* **2000**, *40*, 43–81. [CrossRef]
26. Oakley, H.; Cole, S.L.; Logan, S.; Maus, E.; Shao, P.; Craft, J.; Guillozet-Bongaarts, A.; Ohno, M.; Disterhoft, J.; Van Eldik, L.; et al. Intraneuronal beta-amyloid aggregates, neurodegeneration, and neuron loss in transgenic mice with five familial Alzheimer's disease mutations: Potential factors in amyloid plaque formation. *J. Neurosci.* **2006**, *26*, 10129–10140. [CrossRef]
27. Shankar, G.M.; Li, S.; Mehta, T.H.; Garcia-Munoz, A.; Shepardson, N.E.; Smith, I.; Brett, F.M.; Farrell, M.A.; Rowan, M.J.; Lemere, C.A.; et al. Amyloid-beta protein dimers isolated directly from Alzheimer's brains impair synaptic plasticity and memory. *Nat. Med.* **2008**, *14*, 837–842. [CrossRef]
28. Kinno, A.; Kasamatsu, S.; Akaike, T.; Ihara, H. Reactive sulfur species omics analysis in the brain tissue of the 5xFAD mouse model of Alzheimer's disease. *Antioxidants* **2023**, *12*, 1105. [CrossRef]
29. Itagaki, S.; McGeer, P.L.; Akiyama, H.; Zhu, S.; Selkoe, D. Relationship of microglia and astrocytes to amyloid deposits of Alzheimer disease. *J. Neuroimmunol.* **1989**, *24*, 173–182. [CrossRef]
30. Rogers, J.; Webster, S.; Lue, L.F.; Brachova, L.; Civin, W.H.; Emmerling, M.; Shivers, B.; Walker, D.; McGeer, P. Inflammation and Alzheimer's disease pathogenesis. *Neurobiol. Aging* **1996**, *17*, 681–686. [CrossRef]
31. Lee, J.K.; Kim, S.R.; Byun, H.-G. Purification and characterization of β-secretase inhibitory peptide from sea hare (*Aplysia kurodai*) by enzymatic hydrolysis. *Fish. Aquat. Sci.* **2018**, *21*, 13. [CrossRef]
32. Ano, Y.; Ohya, R.; Takaichi, Y.; Washinuma, T.; Uchida, K.; Takashima, A.; Nakayama, H. Beta-lactolin, a whey-derived lacto-tetrapeptide, prevents Alzheimer's disease pathologies and cognitive decline. *J. Alzheimers Dis.* **2020**, *73*, 1331–1342. [CrossRef] [PubMed]
33. Cheng, L.; Tanaka, M.; Yoshino, A.; Nagasato, Y.; Takata, F.; Dohgu, S.; Matsui, T. A memory-improving dipeptide, Tyr-Pro, can reach the mouse brain after oral administration. *Sci. Rep.* **2023**, *13*, 16908. [CrossRef]
34. Urbi, Z.; Azmi, N.S.; Ming, L.C.; Hossain, M.S. A Concise review of extraction and characterization of chondroitin sulphate from fish and fish wastes for pharmacological application. *Curr. Issues Mol. Biol.* **2022**, *44*, 3905–3922. [CrossRef] [PubMed]
35. Zhang, Q.; Li, J.; Liu, C.; Song, C.; Li, P.; Yin, F.; Xiao, Y.; Li, J.; Jiang, W.; Zong, A.; et al. Protective effects of low molecular weight chondroitin sulfate on amyloid beta (Aβ)-induced damage in vitro and in vivo. *Neuroscience* **2015**, *305*, 169–182. [CrossRef]

36. Zhao, N.; Wu, L.; Zhang, X.; Jiang, W.; Wang, F. Low molecular weight chondroitin sulfate ameliorates pathological changes in 5XFAD mice by improving various functions in the brain. *Neuropharmacology* **2021**, *199*, 108796. [CrossRef] [PubMed]
37. Ano, Y.; Dohata, A.; Taniguchi, Y.; Hoshi, A.; Uchida, K.; Takashima, A.; Nakayama, H. Iso-α-acids, bitter components of beer, prevent inflammation and cognitive decline induced in a mouse model of Alzheimer's disease. *J. Biol. Chem.* **2017**, *292*, 3720–3728. [CrossRef]
38. Tang, J.J.; Huang, L.F.; Deng, J.L.; Wang, Y.M.; Guo, C.; Peng, X.N.; Liu, Z.; Gao, J.M. Cognitive enhancement and neuroprotective effects of OABL, a sesquiterpene lactone in 5xFAD Alzheimer's disease mice model. *Redox Biol.* **2022**, *50*, 102229. [CrossRef]
39. Yang, X.; Ye, T.; He, Y.; Wen, L.; Cheng, X. Qi-fu-yin attenuated cognitive disorders in 5xFAD mice of Alzheimer's disease animal model by regulating immunity. *Front. Neurol.* **2023**, *14*, 1183764. [CrossRef]
40. Devi, L.; Ohno, M. Phospho-eIF2α level is important for determining abilities of BACE1 reduction to rescue cholinergic neurodegeneration and memory defects in 5XFAD mice. *PLoS ONE* **2010**, *5*, e12974. [CrossRef]

foods

MDPI

Article

Nutritional Value, Fatty Acid and Phytochemical Composition, and Antioxidant Properties of Mysore Fig (*Ficus drupacea* Thunb.) Fruits

Hosakatte Niranjana Murthy [1,2,3,*], Guggalada Govardhana Yadav [1], Kadanthottu Sebastian Joseph [4], Sabha Khan H. S. [1], Snehalata M. Magi [1], Yaser Hassan Dewir [5,*] and Nóra Mendler-Drienyovszki [6]

1 Department of Botany, Karnatak University, Dharwad 580003, India; govardhanayadavgs@gmail.com (G.G.Y.); sabakhanhs231999@gmail.com (S.K.H.S.); snehalatam98@gmail.com (S.M.M.)
2 Department of Horticultural Science, Chungbuk National University, Cheongju 28644, Republic of Korea
3 Department of Biotechnology, School of Advanced Sciences, KLE Technological University, Hubballi 580031, India
4 Department of Life Sciences, Christ University, Bengaluru 560029, India; ksjoseph15@gmail.com
5 Plant Production Department, College of Food and Agriculture Sciences, King Saud University, Riyadh 11451, Saudi Arabia
6 Research Institute of Nyíregyháza, Institutes for Agricultural Research and Educational Farm (IAREF), University of Debrecen, P.O. Box 12, 4400 Nyíregyháza, Hungary; mendlerne@agr.unideb.hu
* Correspondence: hnmurthy60@gmail.com (H.N.M.); ydewir@ksu.edu.sa (Y.H.D.)

Abstract: *Ficus drupacea* is a fruit-bearing tree that is distributed in Southeast Asia and Australia. The objective of this research was to ascertain the following with regard to ripened fruits: (i) their nutritional value, (ii) their mineral status, (iii) the fatty acid composition of fruit and seed oil, (iv) their phytochemical makeup, and (v) their antioxidant properties. The ripened fruits contained 3.21%, 3.25%, 0.92%, 1.47%, and 2.20% carbohydrate, protein, fat, ash, and fiber, respectively. Fruits had an energy content of 30.18 kcal/100 g. In terms of mineral content, the fruit was rich in potassium, magnesium, calcium, and nitrogen, with values of 21.03, 13.24, 11.07, and 4.13 mg/g DW. Iron, zinc, manganese, and boron had values of 686.67, 124.33, 114.40, and 35.78 μg/g DW, respectively. The contents of oxalate and phytate were 14.44 and 2.8 mg/g FW, respectively. The fruit and seed oil content were 0.67 and 8.07%, respectively, and the oil's physicochemical properties were comparable to those of fig fruit and seed oils. Omega-3 (α-linolenic acid), omega-6 (linoleic acid), and omega-9 (oleic acid) fatty acids were abundant in the oils. Fruit extracts in acetone, methanol, and water have greater concentrations of phenolics, flavonoids, and alkaloids. The 2,2-diphenyl-2-picrylhydrazyl, total antioxidant activity, and ferric reducing antioxidant power assays demonstrated increased antioxidant activities in close correlation with the higher concentrations of phenolics, flavonoids, and alkaloids. The results of this study demonstrate that the fruits of *F. drupacea* are a strong source of nutrients and phytochemicals, and they merit more investigation and thought for possible uses.

Keywords: antioxidant activity; *Ficus drupacea*; minerals; mysore fig; nutrients; proximate analysis

1. Introduction

Ficus drupacea Thunb. (syn. *Ficus drupacea* Thunb. var. *pubescens* (Roth) Corner, Family: Moraceae), commonly known as 'Mysore fig' or 'brown-woolly fig', is a fruit-bearing tropical tree native to Southeast Asia and Northeast Australia [1]. It possesses a dense woolly pubescence; therefore, it is designated as a 'brown-woolly fig', and it grows to a height of 10–30 m (Figure 1A). Trees possess aerial roots arising in tufts from the stout branches. The bark is greyish brown and exudates milky latex. The syconium is the type of inflorescence that is formed by an enlarged, fleshy, hollow receptacle with numerous female and male flowers on the inside surface, and it subsequently develops into multiple and accessory fruits. Fruits are ovoid to cylindrical, yellow to red, and edible (Figure 1B) [2]. *Ficus drupacea* is known for its medicinal properties and is used for the treatment of various ailments, including malaria and sinusitis [3]. *Ficus drupacea* stem bark has been shown in recent studies to possess antifungal, antibacterial, and anticancer properties. Yessoufou et al. [3] isolated and identified several phytochemicals from the bark, including β-amyrin, β-sitosterol-3-O-β-D-glucopyranoside, 5-O-methyllatifolin, oleanolic acid, epifriedelanol, friedelin, and epilupeol acetate. Manjuprasanna et al. [4] have reported the isolation of a cysteine protease called Drupin from latex, which has exhibited hemostatic characteristics.

Figure 1. Morphology of *Ficus drupacea*. (**A**) Habit; (**B**) Fruits; (**C**) Fruits with seeds; (**D**) Fruit oil; (**E**) Seed oil.

Common fig (*Ficus carica* L.) fruits are highly commercially valuable and provide essential nutrients for human nutrition, including antioxidants, minerals, vitamins, carbohydrates, and amino acids [5]. Additionally, *Ficus carica* has several phytochemicals that aid in the treatment of many diseases, such as obesity, diabetes, cancer, neurodegenerative diseases, and cardiovascular diseases [6,7]. *Ficus drupacea* is collected in the wild and used as a food source, medicinally, and as a fiber source [2]. Even though there is a lot of potential for human use, this species remains underappreciated and ignored. The nutritional, mineral, and phytochemical properties are not known. Consequently, in the current investigation, we aimed to examine the nutritional content of *Ficus drupacea* fruits. We also evaluated the antioxidant properties of fruit extract, conducted mineral and phytochemical analysis, and examined the fatty acid composition of fruit and seed oil.

2. Materials and Methods

2.1. Plant Materials and Sample Preparation

Ficus drupacea fruits were collected at random from five trees in the Haveri district of Karnataka, India, which is located near Shiggavi (14.992726 N, 75.183157 E). The fruits were divided into two groups; from one, only the seeds were removed for study, while from the other, the entire fruit was used. To eliminate the moisture content, the fruits were cut into pieces (Figure 1C) and dried in an oven at 40 ± 2 °C. Following the drying process, the fruits and seeds were ground into a powder using a mechanical grinder and kept at room temperature in airtight polythene bags until additional examination.

2.2. Chemicals

Chemicals, such as Folin–Cicalteau reagent, BF3-methanol, and anthrone, and standard chemicals, such as bovine serum albumin, glucose, and sodium phytate, used in this study were procured from Himedia laboratories, Mumbai, India, whereas heptadecanoic acid was purchased from Sigma-Aldrich, Bengaluru, India. All the other chemicals and solvents used were of analytical grade.

2.3. Proximate Analysis

The fruit's moisture, lipid, ash, and protein levels were examined using AOAC [8] procedures. To put it briefly, the weight difference of the oven-dried sample at 102 °C was recorded for six hours to measure the moisture content gravimetrically. Gravimetric analysis was used to determine the oil content of the sample's fruit and seeds (Figure 1D,E). Using a Soxhlet apparatus at 65 ± 2 °C for 8 h, the finely crushed powder of fruit and seeds was extracted with petroleum ether (40–60 °C) to obtain the oil. The solvent fraction was then evaporated using a rotary evaporator (Buchi, Rotavapor R-100, Flawil, Switzerland). To eliminate any remnants of the solvents, the oil was maintained at 40 ± 2 °C in an oven until its weight stabilized. Additionally, the ash content of the samples was ascertained by igniting the oven-dried samples in the muffle furnace at 750 °C. The oil content was ascertained gravimetrically and kept at −20 °C until additional analysis. Then, 500 mg of defatted samples were ground with 5–10 mL of buffer for the protein content assay. The known volume of extracted material was mixed with 4.5 mL of a 2% sodium carbonate solution. The sodium carbonate solution was made with 0.1 N sodium hydroxide (50 mL), 0.5% copper sulphate (1 mL), and 1% potassium sodium tartrate. Following a 10 min incubation period, tubes were supplemented with 0.5 mL of Folin–Cicalteau reagent and maintained at room temperature in the dark for 30 min to produce a blue color. The spectrophotometric method was used to measure the color developed at 660 nm. The standard utilized was bovine serum albumin. The anthrone reagent method was used to quantify the amount of carbohydrates, and acid and alkali digestion was used to determine the fiber content of the samples. Atwater-specific conversion factors were used to compute the energy value, as reported by the FAO [9].

2.4. Analysis of Mineral Composition

An air or acetylene flame was used with a NOVA 400 atomic absorption spectrophotometer (type Analytic Jena, Jena, Germany) to analyze potassium, phosphorus, sulfur, sodium, calcium, boron, manganese, magnesium, copper, iron, and zinc. Using hollow cathode lamps, the absorbance was measured [8]. To measure nitrogen, a two-step digestion–UV spectrophotometric method was employed [10].

2.5. Physicochemical Estimation of Seed Oil

After four hours of incubation at room temperature, the extracted oil's color and physical condition were assessed. An Abbe refractometer and a specific gravity bottle were used to calculate the oil's density and refractive index, respectively. The AOCS [11] methods were followed to evaluate the free fatty acid (FFA) concentration, peroxide value (PV), iodine value, and unsaponification values. Spectrophotometric methods were employed to determine the content of carotenoids and lignans, as stated by Manasa et al. [12].

2.6. Fatty Acid Profiling

The esterification procedure was used to produce fatty acid methyl esters (FAMEs) following AOCS [11] recommendations. In this process, 1 mL of BF3-methanol was mixed with 15 mg of the oil sample, and the mixture was incubated for 30 min at 60 °C. Without delay, the reaction tubes were placed in an ice bath and kept there for five minutes. The mixture was then vortexed after 1 mL of hexane and distilled water were added. After that, the undisturbed methyl esters that made up the top layer were transferred to GC vials. The internal standard was heptadecanoic acid. The identification of FAMEs was accomplished, as reported by Manasa et al. [12], using a GC-MS (PerkinElmer, Turbo-mass Gold, mass spectrometer, Waltham, MA, USA) equipped with a flame ionization detector (FID) and a fused silica Rtx-2330 column (Restek manufactured, 30 m, 90.32 mm ID, and 0.20 mm film thickness, Bellefonte, PA, USA). Nitrogen was used as the carrier gas, the injector port was kept at 230 °C, and the detector temperature was set at 250 °C. The column temperature was 120 °C at first, then it rose gradually over the course of 20 min to 220 °C, and it remained there for an additional 10 min. The fragmentation pattern and retention time were compared to industry norms and the NIST library to identify FAMEs.

2.7. Analysis of Anti-Nutritional Factors

2.7.1. Estimation of Phytate

Next, 10 mL of 2.4% HCl was used to extract the defatted seed cake (0.5 g) for 16 h with continuous agitation. The mixture was then filtered. After adding 1 g of NaCl to the filtrate, it was vigorously agitated for 20 min. After centrifuging the mixture for 20 min at 10 °C at 1000× *g*, the known volume of the supernatant was diluted to 3 mL with distilled water, and Wade's reagent (0.03% $FeCl_3 \cdot 6H_2O$ + 0.3% sulfosalicylic acid) was added. With a UV-Vis spectrophotometer, the color-developed absorbance was measured at 500 nm. A control was created without any sample additions. The standard utilized was sodium phytate [13]. The linear equation for the analysis was $y = 0.0012x + 0.0165$.

2.7.2. Estimation of Oxalate

Oxalate was determined using the method of Salgado et al. [14]. After adding 2 g of defatted seed cake to 190 mL of distilled water and 10 mL of 6 N HCl, the mixture was heated for 4 h at 90 °C in a water bath. After filtering the mixture and adjusting its volume to 250 mL, 50 mL of this solution was titrated against concentrated ammonia in the presence of a methyl orange indicator. It was then heated to 95 °C, and 10 mL of 5% $CaCl_2$ was added. For the calcium oxalate precipitation, 6 N NH_4OH was added after 10 min, and the color change was monitored and maintained overnight. After filtering and dissolving the precipitate in hot sulfuric acid, 125 mL of the filtrate was prepared, heated to 95 °C, and titrated against 0.05 N $KMnO_4$. To calculate oxalate, the following formula was used:

$$\text{Oxalate (\%)} = (\text{mL } KMnO_4)\,(0.05)\,(45.02) \times (100)\,(5)/(1000)\,(\text{wt. of the sample in grams})$$

2.8. Quantitative Phytochemical Analysis

2.8.1. Extraction Procedure

Three separate solvents were used to extract the defatted fruit powder during an eight-hour period using a Soxhlet apparatus: acetone, methanol, and water, in increasing order of their polarity (acetone < methanol < water). The next solvent was made from the residue left over from the previous solvent extraction. Following the extraction process, the solvents were removed using a rotary evaporator (Buchi, Rotavapor R-100, Flawil, Switzerland). The extracts were then stored at 4 °C until they were needed, and any leftover solvents were removed by heating them in an oven at 40 ± 2 °C.

2.8.2. Quantification of Phenolics

A slightly modified version of the Murthy et al. [15] method was used to calculate the total phenolics. In summary, 0.5 mL of the extract (0.33 mg/mL concentration) was diluted to 3 mL using distilled water, and 0.1 mL of 2 N Folin–Ciocalteau reagent was added. Following a 6 min incubation period, 0.5 mL of 20% sodium carbonate (Na_2CO_3) was added. A UV-Vis (ultraviolet-visible) spectrophotometer (Hitachi U-3310, Ibaraki, Japan) was used to measure the developed color absorbance at 760 nm after the tubes had been left in a warm water bath for 30 min. The standard was gallic acid. The linear equation for the analysis was y = 0.0208x + 0.0155.

2.8.3. Quantification of Flavonoids

The flavonoid concentration in each of the extracts was estimated using the method suggested by Dalawai et al. [16]. In summary, 0.5 mL of extract (1 mg/mL concentration) was diluted to 3 mL using distilled water, then 0.15 mL of $NaNO_3$ was added and the mixture was incubated for 5 min at room temperature. After adding 0.3 mL of 10% $AlCl_3$ and 2 mL of 1 M NaOH, the solutions were vortexed, and the absorbance at 510 nm was measured. The standard was quercetin. The linear equation for the analysis was y = 0.0066x + 0.0003.

2.8.4. Quantification of Alkaloids

The Murthy et al. [15] approach was used to quantitatively measure the alkaloid content: 6.98 mg of bromocresol green powder was dissolved in 0.3 mL of NaOH to prepare the bromocresol green solution, which was then finally diluted with distilled water to a final volume of 100 mL. Then, 5 mL of the previously prepared bromocresol green solution was added to 1.0 mL (5 mg/mL concentration) of a known quantity of material. Next, 5 mL of phosphate buffer (2 M sodium phosphate and 0.2 M citric acid with pH adjusted to 4.7) was added. After adding and vigorously shaking 5 mL of chloroform, the absorbance at 470 nm was measured, and the chloroform layer was recovered. The standard practice was to use atropine. The linear equation for the analysis was y = 0.0039x + 0.0032.

2.9. Antioxidant Activity

2.9.1. Determination of Antioxidant Activity Using 2,2′-Diphenyl-1-Picrylhydrazyl (DPPH) Radical Scavenging Method

The 0.1 mL extracts of various extract concentrations (3 mg/mL for acetone and methanol extracts and 0.5 mg/mL for water extract) were mixed with 1.9 mL of a 0.1 mM DPPH solution made in methanol. After shaking the tubes well, they were dark-incubated for fifteen minutes. Using a spectrophotometer set at 517 nm, the color intensity of the DPPH solution was determined. Gallic acid was used as the standard, and activity was expressed in milligrams of gallic acid equivalent (GAE)/gram of extract, according to Yadav et al. [17]. The linear equation for the analysis was y = 0.0618x − 0.0023.

2.9.2. Total Antioxidant Activity (TAA)

The phosphomolebdenum method was used to carry out the total antioxidant assay [17]. The 0.15 mL extracts at different concentrations (3 mg/mL for acetone and methanol extracts and 0.1 mg/mL for water extract) were combined with 1.5 mL of the reagent solution, which comprised 0.6 M sulfuric acid, 28 mM sodium phosphate, and 4 mM ammonium molybdate. The hue developed in the tubes was measured at 695 nm following 90 min of incubation at 95 °C. Ascorbic acid was used as the reference, and the activity was expressed in milligrams of ascorbic acid equivalent per gram (mg AAE/g) of extract. The linear equation for the analysis was y = 0.0078x − 0.0007.

2.9.3. Ferric Reducing Antioxidant Power (FRAP)

Using the Murthy et al. [18] methodology, the FRAP assay was conducted. First, 0.1 mL of the extract (3 mg/mL for acetone and methanol extracts and 0.1 mg/mL for water extract) was mixed with 3 mL of FRAP reagent, which contained 300 mM acetate buffer (pH 3.6), 10 mM of TPTZ (2,4,6-tripyridyl-s-triazine) in 40 mM HCl, and 20 mM $FeCl_3 \cdot 6H_2O$ (10:1:1). The absorbance at 593 nm was measured in relation to a blank solution after the sample and FRAP reagent-filled tubes were vortexed and allowed to stand for six minutes at room temperature. The activity of the extracts is represented as mg ascorbic acid equivalent (AAE)/g extract, with ascorbic acid serving as the standard. The linear equation for the analysis was y = 0.0526x + 0.0373.

2.10. Analysis of Data and Statistical Treatment

Each experiment was conducted three times, and the results were reported as mean values with standard errors. Descriptive statistics, including the mean and standard error, were calculated using Microsoft Excel 2019. The statistical significance of the differences between mean values was assessed using Duncan's multiple range test at $p < 0.05$. All the statistical analyses were performed using SAS 9.4 software (SAS Institute Inc., Cary, NC, USA).

3. Results and Discussion

3.1. Nutritional Value

One of the most prevalent fruits in the Mediterranean region, the common fig (*F. carica*), is a significant crop grown all over the world [19]. According to Alzaharni et al. [5], the approximate composition of a common fresh fig is as follows: 79.9–88.1% moisture, 7.60–20.0% carbohydrate, 0.53–1.30% fat, 0.60–4.00% ash, 2.10–2.20% fiber, and an overall energy value of 37.0 kcal/100 g. According to the current investigation, the *F. drupacea* fruits' approximate composition was 87.99 ± 0.20% moisture, 3.21 ± 0.15% carbohydrate, 3.25 ± 0.26% protein, 0.92 ± 0.15 fat, 1.47 ± 0.04% ash, 2.20 ± 0.25% fiber, and 30.18 kcal/100 g of energy, which is similar to that of common figs (Table 1). The fruits of *F. auriculata* were found to have the following nutritional values: 15.22 ± 3.4% ash, 1.82 ± 1.0% fiber, 0.01 ± 0% fat, 3.19 ± 1.0% protein, 35.42 ± 4.5% carbohydrate, and 141.68 kcal/100 g energy, respectively [20]. On the other hand, the approximate contents of *F. hispida* fruits were 1.71 ± 1.0 fiber, 14.94 ± 3.1% ash, 3.11 ± 1.0% protein, 36.33 ± 5.6% moisture, 43.86 ± 5.3% carbohydrate, and 175.44 kcal/100 g, in that order. According to Rusmadi et al. [20], fruits of *F. fistulosa* have 44.0 ± 5.3 moisture, 33.66 ± 4.4% carbohydrate, 2.90 ± 1.0% protein, 0.02 ± 0% fat, 15.91 ± 3.6% ash, 1.61 ± 1.0% fiber, and 142.29 kcal/100 g energy. The *F. amplissima* fruits had nutritional values of 83.29%, 15.45%, 1.81%, 0.63%, 0.82%, and 0.81% for moisture, protein, fat, ash, and fiber, respectively, and 53.09 kcal/100 g for energy [21]. Variations in the environmental conditions of agriculture and species diversity may have contributed to the variation in the nutritional qualities of different fruits.

Table 1. Proximate composition of *Ficus drupacea* fruit.

Component	Composition (%)
Moisture	87.99 ± 0.20 [z]
Carbohydrate	3.21 ± 0.15
Protein	3.25 ± 0.26
Fat	0.92 ± 0.15
Ash	1.47 ± 0.04
Fiber	2.20 ± 0.25
Energy (kcal/100 g)	30.18

[z] Each value represents the mean ± standard error of three replicates.

3.2. Mineral Composition

Fruits are an excellent source of minerals and are essential for maintaining many physiological processes in the human body, such as the growth of bones, muscles, and nerves, as well as the control of the body's water balance [22]. *F. carica* is rich in minerals, which help the body function normally by supplying calcium, iron, phosphorus, potassium, and sodium [5]. Among the several minerals contained in *F. carica* that support healthy bone development are iron and strontium [23]. Fruits of many *Ficus* species are well known for having a mineral-rich makeup, and *F. drupacea* fruits are no exception (Table 2). Particularly, fruits of *F. drupacea* have high concentrations of potassium, magnesium, calcium, and nitrogen (21.03, 13.24, 11.07, and 4.13 mg/g DW, respectively). Furthermore, the concentrations of phosphorus, sulphur, and sodium were 1.64, 1.02, and 0.54 mg/g DW, respectively.

Different microelements with notable concentrations include iron, zinc, manganese, boron, and copper (686.67, 124.33, 114.40, 35.78, and 13.98 μg/g DW, respectively) (Table 2).

Table 2. Mineral composition of *Ficus drupacea* fruit.

Mineral	Composition
Microelements (mg/g DW)	
Nitrogen	4.13 ± 0.06 [z]
Phosphorous	1.64 ± 0.02
Potassium	21.03 ± 0.12
Sulphur	1.02 ± 0.03
Sodium	0.54 ± 0.01
Calcium	11.07 ± 0.12
Magnesium	13.24 ± 0.09
Microelements (μg/g DW)	
Boron	35.78 ± 0.44
Zinc	124.33 ± 0.88
Iron	686.67 ± 8.82
Manganese	114.40 ± 0.29
Copper	13.93 ± 0.20

[z] Each value represents the mean ± standard error of three replicates.

Ficus begalensis, *F. recemosa*, *F. religosa*, *F. palmata*, *F. microcorpa*, *F. johannis*, *F. sarmentosa*, *F. hispida*, and *F. auriculata* were among the fruits of various *Ficus* species that Khan et al. [24] analyzed elementally. The range of potassium, magnesium, and calcium in these fruits was 57.3–11.29, 7.26–3.83, and 24.9–24.08 mg/g DW in various species. In contrast, the ranges for iron, manganese, copper, and zinc in different species were 2.9–0.51, 32.53–10.01, 6.96–1.00, and 20.49–5.05 mg/g DW. Such variation is evident in the variety of *Ficus* species, and it also depends on the mineral nutrition levels and soil conditions in the plant's growing medium. Many elements, including potassium, calcium, manganese, iron, copper, and zinc, are essential for the synthesis of secondary metabolites, which give plants their medicinal and pharmacological properties and enable the plants to fight disease. Studies show that calcium helps prevent osteoporosis and the fractures that result from it. It takes magnesium to prevent cardiovascular diseases. Furthermore, phosphorus is required for the development and maintenance of bodily structures and cells [22]. The fruits of *F. drupacea* have the highest concentration of iron (686.67 μg/g DW). Iron is linked to angiotensin-converting enzyme inhibitors and is a necessary mineral for the prevention of anemia [25]. In contrast, several trace elements can both prevent and treat disease [26]. The aforementioned findings demonstrate that *F. drupacea* fruits are edible and a rich source of the necessary mineral components for human health.

3.3. Antinutrient Composition

Table 3 lists the antinutritional components of *F. drupacea* fruits, including their oxalate and phytate contents. The phytate concentration was 2.8 mg/g FW, which is lower than the quantity reported in *Rourea minor* [27] but equivalent to phytate values found in fruits of the *Balanitis aegyptica* species (0.06–1.82 mg/g) [28]. Phytate decreases the digestibility of amino acids and forms complexes with phosphorous, calcium, iron, and zinc to obstruct their absorption. As a result, the body can easily no longer access these nutrients. The phytate levels found in this investigation, however, were less than the 10–60 mg/g that have been linked to issues with mineral bioavailability [29]. The fruit of *F. drupacea* contained 14.44 mg/g FW of oxalate. It is well known that oxalate, particularly at concentrations of roughly 45 g/100 g, inhibits the renal absorption of calcium [30]. Nonetheless, the values found in this investigation are significantly lower than the values that are thought to be hazardous. This implies that consuming the fruits might not provide any issues with mineral absorption.

Table 3. Anti-nutritional factors of *Ficus drupacea* fruit.

Factor	Composition (mg/g FW)
Phytate	2.80 ± 0.01 [z]
Oxalate	14.44 ± 0.06

[z] Each value represents the mean ± standard error of three replicates.

3.4. Characterization of Fruit and Seed Oil

For human health, edible oil is a vital dietary resource. Edible oil can be extracted from plants using their fruits or seeds. The acquired oils are either discovered to have medicinal and cooking uses or they are utilized for edible purposes. Due to rising demand for vegetable oils in the chemical industry, animal feed, pharmaceutical, and other fields, oil-yielding plants have garnered increased attention recently. Rapid population growth and resource depletion present problems for the entire planet. Investigating substitute sources of edible oil is crucial [31]. Different plant species or plant parts differ significantly in terms of their oil concentration, composition, and qualities [32]. To extract oils from fruits and seeds, we investigated the oil content of *F. drupacea* in the current study. Fruits of *F. drupacea* have less oil (0.67%) (Figure 1D), but seeds have a substantial amount of oil (8.07%) (Figure 1E). Table 4 lists the fruit and seed oil's physicochemical characteristics. In contrast to the seed oil, which was crimson yellow and liquid, the fruit oil had a blood-red hue and was viscous at room temperature. The refractive index of *F. drupacea* fruit oil is 1.510, while that of the seed oil is 1.498. These values are similar to those of common edible oils, such as corn oil (1.473) and soybean oil (1.472), and are correlated with the molecular weight, degree of unsaturation, and length of the fatty acid chain of the oil [33]. *F. drupacea* fruit and seed oils have densities of 0.921 and 0.931 g/cm^3, respectively, which are similar to corn oil's value of 0.917 g/cm^3 [33]. Essential characteristics of edible seed oils are their peroxide and free fatty acid levels. Edible natural oils with less than 5% free fatty acid concentration are a possibility [34,35]. *F. drupacea* fruit and seed oil had a lower free fatty acid concentration (1.03 and 1.55%, respectively) than mustard oil, but it was nevertheless comparable to several commercial oils like sesame oil [36]. The fruit and seed oils of *F. drupacea* have peroxide values of 19.79 and 19.92 meq O_2/kg (Table 4), which are lower than those of various commercial edible oils, including sunflower, peanut, olive, mustard, and rape seed oils [37] and *Balanites roxhbergii* crude oil [17]. As a measure of an oil's unsaturation level, the iodine value for *F. drupacea* fruit and seed oil was 88.21 and 154.03 I_2/100 g, respectively. These values are higher than those of edible refined oils like olive oil (80.03 I_2/100 g) and mustard oil (94.95 I_2/100 g) [37]. Table 4 shows that the fruit and seed oils of *F. drupacea* had unsaponification values of 15.28% and 2.80%, respectively, indicating the presence of nutraceuticals in the oils besides fatty acids. The results are greater than those of mustard oil (1.01%) and olive oil (1.23%) [37]. The formation of cataracts and age-related macular degeneration are both slowed down by the vital natural pigments called carotenoids. Additionally, they have antioxidant properties that enhance cognitive and cardiovascular health [38]. The fruit and seed oils from *F. drupacea* had a greater carotenoid concentration (1048.73 and 36.64 mg/kg, respectively), higher than apple, apricot, and blackberry fruit oils (15.80, 66.80, and 13.30 mg/kg, respectively) [39]. Lignans are polyphenols with biological activities that include preventive properties against diseases including hormone-dependent cancers and cardiovascular disorders. They also exhibit prooxidant and antioxidant actions [40]. *F. drupacea* fruit and seed oils had lignan contents of 0.59 and 0.83%, respectively, which are lower than *Erythrina stricta* seed oil (165.47 mg/100 g) [35].

Table 5 shows the fatty acid content of *F. drupacea*'s fruit and seed oils, and Figure 2 displays the GC-MS chromatograms analyzed for fatty acid profiling. Fruit oil contained the following main fatty acids: oleic acid (7.07%), linoleic acid (28.61%), α-linolenic acid (11.91%), and palmitic acid (46.08%), and 50.80:8.68:40.52% was the overall concentration of monounsaturated, polyunsaturated, and saturated fatty acids (Table 5). These results contrast from the fatty acid composition of fruit oil from *F. carica*, which contains oleic acid

(10.0%), palmitic acid (14.0%), linoleic acid (21.0%), and α-linolenic acid (53%) [5]. On the other hand, *F. exasperata* fruit oil had a relatively lower amount of saturated fatty acids, such as stearic acid (9.10%) and palmitic acid (7.32%), and a higher amount of unsaturated fatty acids, such as linoleic acid (54.54%) and oleic acid (18.89%) [41].

Table 4. Physicochemical characteristics of *Ficus drupacea* fruit and seed oil.

Parameter	Fruit Oil [z]	Seed Oil [z]
Oil yield (%)	0.67 ± 0.15 [b]	8.07 [a]
Color	Blood red	Crimson yellow
State at room temperature	Solid	Liquid
Refractive index	1.510 ± 0.01 [a]	1.498 ± 0.01 [a]
Density (g/cm^3)	0.921 ± 0.01 [a]	0.931 ± 0.01 [a]
Free fatty acid content (%)	1.03 ± 0.02 [a]	1.55 ± 0.08 [a]
Peroxide value (meq O_2/kg)	19.79 ± 1.46 [a]	19.92 ± 0.77 [a]
Iodine value (I_2/100 g)	88.21 ± 0.12 [b]	154.03 ± 2.30 [a]
Unsaponification value (%)	15.28 ± 0.10 [a]	2.80 ± 0.06 [b]
Carotenoids (mg/kg)	1048.73 ± 39.87 [a]	36.64 ± 3.32 [b]
Lignans (% SE)	0.59 ± 0.02 [a]	0.83 ± 0.10 [a]

[z] Each value represents the mean ± standard error of three replicates. Mean values followed by different letters in their superscript are significantly different from each other (p = 0.05) in the respective row according to Duncan's multiple range test.

Table 5. Fatty acid composition of *Ficus drupacea* pulp and fruit oil.

Fatty Acid	Chain Length	Pulp Oil		Seed Oil	
		RT (in min)	% Composition [z]	RT (in min)	% Composition [z]
Lauric acid	12:0	10.733	0.25 ± 0.02 [a]	ND	ND
Myristic acid	14:0	12.251	0.61 ± 0.02 [a]	12.251	0.12 ± 0.01 [b]
Pentadecanoic acid	15:0	13.082	0.46 ± 0.01 [a]	13.072	0.08 ± 0.01 [b]
Palmitic acid	16:0	13.954	46.08 ± 0.29 [a]	13.964	18.16 ± 0.12 [b]
Palmitelaidic	16:1n9	14.384	0.52 ± 0.01 [a]	14.384	0.14 ± 0.01 [b]
Hexadecanoic acid, 14-methyl	14Me-16:0	14.877	0.59 ± 0.02 [a]	ND	ND
Stearic acid	18:0	15.923	2.81 ± 0.06 [b]	15.943	3.73 ± 0.09 [a]
Oleic acid	18:1n9	16.395	7.07 ± 0.09 [b]	16.426	12.63 ± 0.17 [a]
Vaccenic acid	18:1n11	16.497	1.09 ± 0.02 [a]	16.518	0.99 ± 0.06 [a]
Linoleic acid	C18:2n9,12	17.256	28.61 ± 0.17 [a]	17.297	26.81 ± 0.17 [a]
Linolenic acid	18:3n9,12,15	18.426	11.91 ± 0.12 [b]	18.487	36.87 ± 0.18 [a]
Paullinic acid	20:1n13	ND	ND	19.195	0.11 ± 0.01 [a]
Behenic acid	22:0	ND	ND	21.236	0.37 ± 0.01 [a]
Total saturated fatty acids		NRV	50.80	-	22.46
Total monounsaturated fatty acids		NRV	8.68	-	13.87
Total polyunsaturated fatty acids		NRV	40.52	-	63.68

[z] Each value represents the mean ± standard error of three replicates. ND = Not detected. NRV = No retention value. Mean values followed by different letters in their superscript are significantly different from each other (p = 0.05) in the respective row according to Duncan's multiple range test.

The major fatty acids found in the seed oil *F. drupacea* in the present study were oleic acid (12.63%), palmitic acid (18.16%), α-linolenic acid (36.87%), and linoleic acid (26.81%) (Table 5; Figure 3). The proportions of saturated fat, monounsaturated fat, and polyunsaturated fat were 22.46%, 13.87%, and 66.68%, respectively. These findings were similar to those of seed oil *F. carica*, where the primary fatty acids α-linolenic acid (41.80–37.87%), linoleic acid (37.95–31.80%), and palmitic acid (7.40–3.58%) were found in several accessions [42]. Consistent with the aforementioned findings, Baygeldi et al. [43] detected noteworthy concentrations of omega-3 (α-linolenic acid, 41.75%), omega-6 (linoleic acid, 30.9%), and omega-9 (oleic acid,15.98%). The essential fatty acids α-linolenic and linoleic can be found in the seed oils of *F. carica* and *F. drupacea*, which can be employed in nutraceutical products.

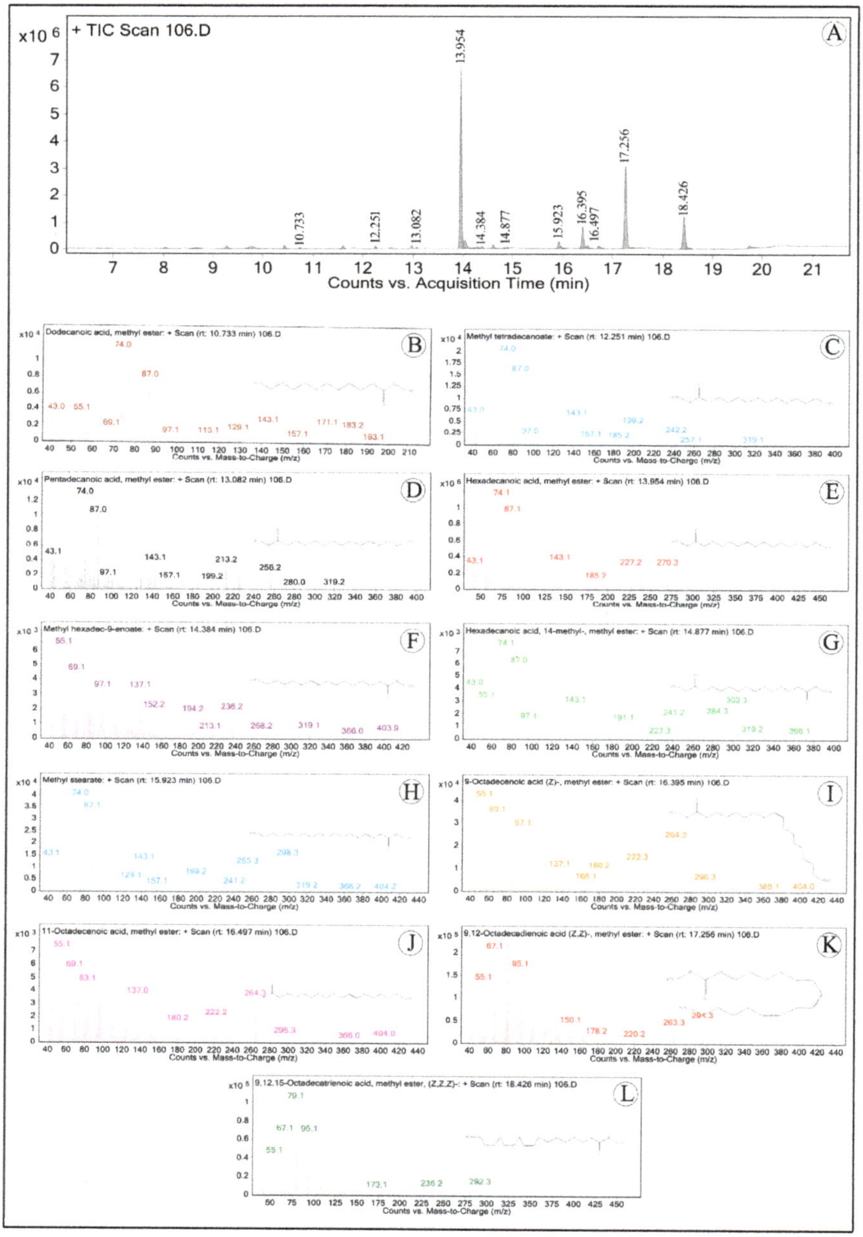

Figure 2. (**A**) GC-MS chromatograms of *Ficus drupacea* pulp oil fatty acid methyl esters (FAMEs); B-L. Mass spectra of major FAMEs of analyzed fatty acids. (**B**) Methyl laurate; (**C**) Methyl myristate; (**D**) Pentadecanoic acid, methyl ester; (**E**) Methyl palmitate; (**F**) Methyl hexadic-9-enoate; (**G**) Hexadecanoic acid, 14-methyl-, methyl ester; (**H**) Methyl stearate; (**I**) Methyl oleate; (**J**) Methyl vaccinate; (**K**) Methyl linoleate; (**L**) Methyl linolenate.

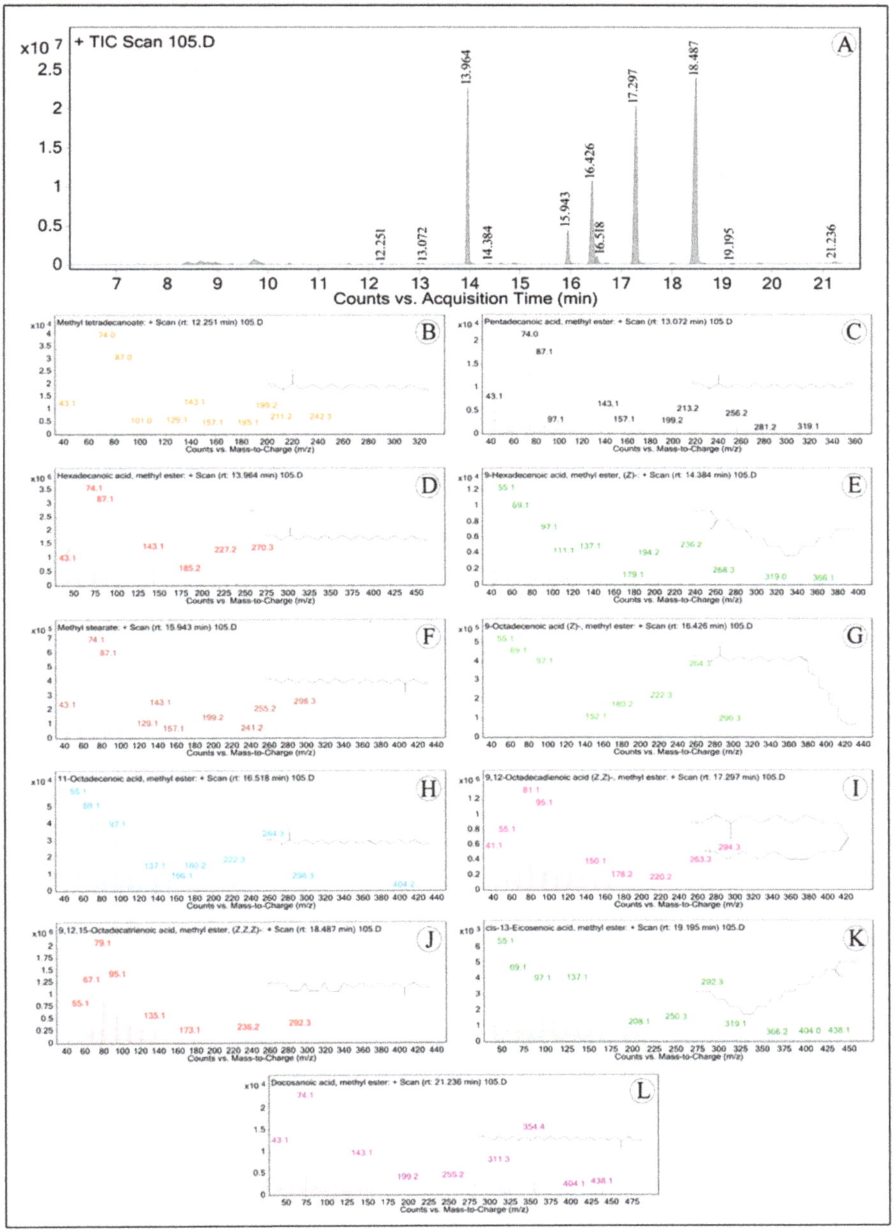

Figure 3. (**A**) GC-MS chromatograms of *Ficus drupacea* seed oil fatty acid methyl esters (FAMEs); B-L. Mass spectra of major FAMEs of analyzed fatty acids. (**B**) Methyl myristate; (**C**) Pentadecanoic acid, methyl ester; (**D**) Methyl palmitate; (**E**) Methyl hexadic-9-enoate; (**F**) Methyl stearate; (**G**) Methyl oleate; (**H**) Methyl vaccinate; (**I**) Methyl linoleate; (**J**) Methyl linolenate; (**K**) Methyl paullinate; (**L**) Methyl behenate.

3.5. Phytochemical Composition

Three important phytochemical groups—phenolics, flavonoids, and alkaloids—were examined in the fruits of *F. drupacea* through extraction using various solvents. The findings are shown in Table 6. Methanol yielded the highest extract (259.98 mg/100 g) of all the extracts, compared to acetone (20.86 mg/100 g) and water extract (3.96 mg/100 g). Fruits of *F. drupacea* are better suited for the methanol extraction method, which has the highest yield. Several researchers employed a similar approach and discovered that it was helpful to extract phytochemicals from plant material using a variety of solvents [18].

Table 6. Phytochemical composition of *Ficus drupacea* fruit extracts.

Activity	Acetone (mg/g Extract) ^{z}	Methanol (mg/g Extract) ^{z}	Water (mg/g Extract) ^{z}
Extract yield (mg/100 g DW)	20.86	259.98	3.96
Total phenolics (GAE)	60.12 ± 2.69^{b}	25.78 ± 1.46^{c}	70.50 ± 2.04^{a}
Flavonoids (QE)	16.53 ± 1.03^{a}	5.88 ± 0.53^{b}	7.12 ± 0.37^{b}
Alkaloids (AE)	3.76 ± 0.62^{a}	1.37 ± 0.08^{b}	0.04 ± 0.01^{c}

^{z} Each value represents the mean ± standard error of three replicates. GAE—Gallic acid equivalent; QE—Quercetin equivalent; AE—Atropine equivalent. Mean values followed by different letters in their superscript are significantly different from each other ($p = 0.05$) in the respective row according to Duncan's multiple range test.

The antioxidant activity of phenolics is responsible for their anti-inflammatory, antitumor, anticancer, antidiabetic, antibacterial, and anti-osteoporosis characteristics [44]. The total phenolic content of six underutilized fruits (*Carrisa carandas*, *Dovyalis hebecarpa*, *Flocourtia indica*, *Malphighia emerginata*, *Slacia chinensis*, and *Syzygium indica*) was examined by Perera et al. [45] and the results ranged from 6.77 to 10.29 mg GAE/g on a fresh weight basis. The total phenolic content of *F. drupacea* fruit extracts was 60.12, 25.78, and 70.50 mg GAE/g extract in acetone, methanol, and water extract, respectively, indicating the fruits' high power for preventing free radicals that cause health problems in the body. According to Ullah et al. [46], flavonoids are an additional class of antioxidants that help regulate inflammatory, cardiovascular, and carcinogenic issues in humans. The *Rourea minor* fruit extracts were tested with acetone, methanol, water, and 70% methanol extract. The total flavonoid content was 1.37, 1.27, 3.25, and 1.61 mg QE/g DW, respectively [27]. Total flavonoid concentrations in the fruits of *F. drupacea* were 16.53, 5.88, and 7.12 mg QE/g extract, respectively. These values were greater than those of the following wild fruits: *Baccaurea sapida*, *Docynia indica*, *Elaeagnus latifolia*, *Elaeagnus pyriformis*, *Haematocarpus validus*, *Myrica esculanta*, *Myrica nagi*, *Prunus nepalensis*, *Pyrus pashia*. In these fruits, the flavonoid content ranged from 2.77 mg QE/g DW in *Pyrus pashia* to 5.46 mg QE/g DW in *Haematocarpus validus* [47]. Another significant class of secondary metabolites found in plants is called the alkaloids, and they are widely used as antibacterial, anticancer, cardioprotective, antidiabetic, and neuroprotective medicines [48]. The total alkaloid concentrations of longan fruits were estimated by Tang et al. [49], who reported 1.67, 6.44, and 7.40 mg HE/g for the pulp, pericarp, and seed, respectively. The acetone, methanol, and water extract of *F. drupacea* fruits yielded alkaloid contents of 3.76, 1.37, and 0.04 mg AE/g extract, respectively. These values were significantly lower than those of longan fruits. These findings show that the content of alkaloids present in different fruits varies by species.

3.6. Antioxidant Activities

Antioxidants are essential for preserving human health because they protect the body from oxidative stress and prevent a number of illnesses, including cancer, autoimmune disorders, Parkinson's disease, and Alzheimer's. Bioactive chemicals obtained from plants protect cells against oxidative damage by their inhibition or interaction with reactive oxygen species and free radicals [44,46]. The fruit extract of *F. drupacea* exhibited a noteworthy phytochemical presence, as evidenced by the antioxidant activities evaluated using three conventional in vitro methods: FRAP assay, total antioxidant activity, and DPPH radical scavenging activity. The findings are displayed in Table 7.

Table 7. Antioxidant activities of *Ficus drupacea* fruit extracts.

Activity	Acetone (mg/g Extract) [z]	Methanol (mg/g Extract) [z]	Water (mg/g Extract) [z]
DPPH (mg GAE)	3.81 ± 0.87 [a]	2.54 ± 1.09 [a]	4.82 ± 0.59 [a]
TAA (mg AAE)	3103 ± 3.20 [a]	97.99 ± 13.68 [b]	139.15 ± 0.95 [b]
FRAP (mg AAE)	965.21 ± 25.20 [a]	10.66 ± 0.35 [b]	38.14 ± 0.85 [b]

[z] Each value represents the mean ± standard error of three replicates. GAE—Gallic acid equivalent; AAE—Ascorbic acid equivalent. Mean values followed by different letters in their superscript are significantly different from each other ($p = 0.05$) in the respective row according to Duncan's multiple range test.

Acetone extract had the highest activity among the studied extracts, including methanol and water, as determined by TAA and FRAP methods. The DPPH radical scavenging activity of the water extract was 4.82 mg GAE/g, but the acetone and methanol extracts showed 3.81 and 2.54 mg GAE/g extract, respectively. TAA was measured as follows: 139.5 and 97.99 mg AAE/g extract in the case of water and methanol extracts, and 3103 mg AAE/g extract in the case of acetone extract. The acetone extract had a FRAP activity of 965.21 mg AAE, while the methanol and water extracts had FRAP activities of 10.66 and 38.14 mg AAE/g, respectively. Similar to the current findings, Arunachalam et al. [21] found that *Ficus amplissima* extracts prepared in hot water, petroleum ether, chloroform, acetone, and methanol have extremely strong antioxidant activity when tested utilizing TAA and FRAP assay techniques. On the other hand, Rajesh et al. [50] have suggested that the high phenolic content in the *Ficus racemosa* samples may be the cause of their ability to scavenge free radicals. Numerous other studies using a range of antioxidant assays have effectively established the antioxidant qualities of wild and underutilized fruits. For example, the unripe pulp outperformed the matured one in *Diospyros chloroxylon* when it came to scavenging DPPH radicals, while acetone extracts outperformed methanol and water extracts [15]. As demonstrated by the DPPH and FRAP tests, the antioxidant activity of black chokeberry, raspberry, blackberry, and mulberry was attributed to their phenolic components [51].

4. Conclusions

This study showed that *Ficus drupacea* fruits are nutrient-dense, containing ash, fiber, protein, fat, and important minerals. The fruits have comparatively low levels of antinutrients. α-linolenic acid, oleic acid, and linoleic acid were the main fatty acids in the fruit and seed oil, which have properties similar to those of figs. The fruits also exhibit strong free radical scavenging properties due to their abundance of phytochemicals, which include flavonoids, phenolics, alkaloids in acetone, methanol, and water extracts of fruits. The aforementioned findings demonstrate that *Ficus drupacea* fruits are valuable for food uses. The current study's merits include providing information on the nutritional value, oil content, fatty acid composition, phytochemical status, and antioxidant status of the underutilized fruit, *F. drupacea*. However, further investigation would be required to examine the additional beneficial phytochemicals, particularly pigments, and to precisely quantify the specific phenolic, flavonoid, and associated antioxidant activities of these compounds.

Author Contributions: Conceptualization, methodology, H.N.M.; formal analysis and data curation, H.N.M., G.G.Y. and K.S.J.; validation, G.G.Y. and K.S.J.; visualization, G.G.Y., K.S.J., Y.H.D. and N.M.-D.; investigation, S.K.H.S. and S.M.M.; writing, review and editing, H.N.M., G.G.Y., Y.H.D. and N.M.-D. All authors have read and agreed to the published version of the manuscript.

Funding: HNM was supported by the Brain Pool Program of the National Research Foundation of Korea (Grant No. 2022H1D3A2A02056665). YHD was supported by the Researchers Supporting Project number (RSP-2024R375), King Saud University, Riyadh, Saudi Arabia.

Data Availability Statement: The original contributions presented in the study are included in the article, further inquiries can be directed to the corresponding author.

Acknowledgments: The authors acknowledge Researchers Supporting Project number (RSP-2024R375), King Saud University, Riyadh, Saudi Arabia.

Conflicts of Interest: The authors declare no conflicts of interest.

References

1. POWO Plants of the World Online. Facilitated by the Royal Botanic Gardens, Kew. Available online: https://powo.science.kew.org (accessed on 13 August 2024).
2. Tropical Plants Database, Ken Fern. Available online: http://tropical.theferns.info/viewtropical.php?id=Ficus+drupacea (accessed on 13 August 2024).
3. Yessoufou, K.; Elansary, H.O.; Mahmoud, E.A.; Skalicka-Wozniak, K. Antifungal, antibacterial and anticancer activities of *Ficus drupacea* L. stem bark extract and biologically active isolated compounds. *Ind. Crops Prod.* **2015**, *74*, 752–758. [CrossRef]
4. Manjuprasanna, V.N.; Rudresha, G.V.; Urs, A.P.; Milan Gowda, M.D.; Rajaiah, R.; Vishwanath, B.S. Drupin, a cysteine protease from *Ficus durpacea* latex accelerates excision would healing n mice. *Int. J. Biol. Macromol.* **2020**, *165*, 691–700. [CrossRef]
5. Alzaharni, A.Y.; Alshaikhi, A.I.; Hazzazi, J.S.; Kurdi, J.R.; Ramadan, M.F. Recent insight on nutritional value, active phytochemicals, and health-enhancing characteristics of fig (*Ficus carica*). *Food Safe Health* **2024**, *2*, 179–195. [CrossRef]
6. Mawa, S.; Husain, K.; Jantan, I. *Ficus carica* L. (Moraceae): Phytochemistry, traditional uses and biological activities. *Evid. Based Complement. Altern. Med.* **2013**, *2013*, 974256. [CrossRef]
7. Wojdyło, A.; Nowicka, P.; Carbonell-Barrachina, Á.A.; Hernández, F. Phenolic compounds, antioxidant and antidiabetic activity of different cultivars of *Ficus carica* L. fruits. *J. Funct. Foods* **2016**, *25*, 421–432. [CrossRef]
8. AOAC. *Official Methods of Analysis of the Association of Analytical Chemists*, 22nd ed.; Association of Official Analytical Chemists, Inc.: Rockville, MD, USA, 2023.
9. FAO. *Food Energy-Methods of Analysis and Conversion Factors*; Food and Agriculture Organization of the United Nations: Rome, Italy, 2003.
10. Koistinen, J.; Sjoblom, M.; Spilling, K. Total nitrogen determination by spectrophotometric method. *Methods Mol. Biol.* **2020**, *1980*, 81–86. [CrossRef]
11. AOCS. *Official Methods and Recommended Practices of the American Oil Chemist's Society*; American Oil Chemist's Society: Champaign, IL, USA, 2017.
12. Manasa, V.; Vaishnav, S.R.; Tumaney, A.W. Physicochemical characterization and nutraceutical compounds of the selected spice fixed oils. *J. Food Sci. Technol.* **2021**, *58*, 3094–3105. [CrossRef] [PubMed]
13. Kahriman, F.; Songur, U.; Serment, M.; Akbulut, S.; Egesel, C.O. Comparision of colorimetric methods for determination of phytic acid content in raw and oil extracted flour samples of maize. *J. Food Compos. Anal.* **2020**, *86*, 103380. [CrossRef]
14. Salgado, N.; Silva, M.A.; Figueira, M.E.; Costa, H.S.; Albuquerque, T.G. Oxalate in foods: Extraction conditions, analytical methods, occurance, and health implications. *Foods* **2023**, *12*, 3201. [CrossRef]
15. Murthy, H.N.; Dalawai, D.; Arer, I.; Karadakatti, P.; Hafiz, K. Nutritional value of underutilized fruit: *Diospyros chloroxylon* Roxb. (green ebony persimmon). *Int. J. Fruit Sci.* **2022**, *22*, 249–263. [CrossRef]
16. Dalawai, D.; Murthy, H.N.; Dewir, Y.H.; Sebastian, J.K.; Nag, A. Phytochemical composition, bioactive compounds, and antioxidant properties of different parts of *Andrographis macrobotrys* Nees. *Life* **2023**, *13*, 1166. [CrossRef] [PubMed]
17. Yadav, G.G.; Murthy, H.N.; Dewir, Y.H. Nutritional composition and in vitro antioxidant activities of seed kernel and seed oil of *Balanites roxburghii*: An underutilized species. *Horticulturae* **2022**, *8*, 798. [CrossRef]
18. Murthy, H.N.; Yadav, G.G.; Kadapati, S.S.; Lamani, S.; Desai, A.S.; Sumbad, M.M.; Aslshahrami, T.S.; Deweri, Y.H. Bioacitve constituents and antioxidant activities of *Erythrina strica* Roxb. seeds. *Appl. Ecol. Environ. Res.* **2024**, *22*, 3275–3284. [CrossRef]
19. Viuda-Martos, M.; Barber, X.; Perez-Alvarez, J.A.; Fernandez-Lopez, J. Assessment of chemical, physico-chemical, techno-functional and antioxidant properties of fig (*Ficus carica* L.) powder co-products. *Ind. Corps Prod.* **2015**, *69*, 472–479. [CrossRef]
20. Rusmadi, N.N.N.N.; Shahari, R.; Amri, C.N.A.C.; Tajudin, N.S.; Mispan, M.R. Nutritional value of selected edible *Ficus* fruits in Kuntan. *J. Trop. Life Sci.* **2020**, *10*, 11–14.
21. Aurnachalam, K.; Murugan, R.; Primelazhagan, T. Evaluation of antioxidant activity, and nutritional and chemical composition of *Ficus amplissima* Smit fruit. *Int. J. Food Prop.* **2014**, *17*, 454–468. [CrossRef]
22. Weyh, C.; Krüger, K.; Peeling, P.; Castell, L. The role of minerals in the optimal functioning of the immune system. *Nutrients* **2022**, *14*, 644. [CrossRef]
23. Pal, I. A comprehensive review on *Ficus carica* L. an unexploited medicinal plant. *Int. J. Adv. Res.* **2020**, *8*, 876–881. [CrossRef]
24. Khan, K.K.; Khan, M.A.; Niamat, R.; Munir, M.; Fazal, H.; Mazari, P.; Seema, N.; Bashir, T.; Kanwal, A.; Ahamed, S.N. Element content analysis of plants of genus *Ficus* using atomic absorption spectrophotometer. *Afr. J. Pharm. Pharmacol.* **2011**, *5*, 317–321. [CrossRef]
25. Man, Y.; Xu, T.; Adhikari, B.; Zhou, C.; Wang, Y.; Wang, B. Iron supplementation and iron-fortified foods: A review. *Crit. Rev. Food Sci. Nutr.* **2022**, *62*, 4504–4525. [CrossRef]
26. Mehri, A. Trace elements in human nutrition (II)—An update. *Int. J. Prev. Med.* **2020**, *11*, 2. [CrossRef] [PubMed]
27. Murthy, H.N.; Yadav, G.G.; Kadapatti, S.S.; Pote, A.H.; Jagali, R.; Yarashi, V.; Dewir, Y.H. Evaluation of nutritional, phytochemical, and antioxidant potential of *Rourea minor* fruits: An underutilized species. *Horticulturae* **2023**, *9*, 606. [CrossRef]

28. Achaglinkame, M.A.; Aderibigbe, R.O.; Hensel, O.; Sturm, B.; Korese, J.K. Nutritional characteristics of four underutilized edible wild fruits of dietary interest in Ghana. *Foods* **2019**, *8*, 104. [CrossRef]
29. Pujol, A.; Sanchis, P.; Grases, F.; Masmiquel, L. Phytate intake, health and disease: "Let thy food be thy medicine and medicine be thy food. *Antioxidants* **2023**, *12*, 146. [CrossRef]
30. Huynh, N.K.; Nguyen, D.H.M.; Nguyen, H.V.H. Effects of processing on oxalate contents in plant foods: A review. *J. Food Compos. Anal.* **2022**, *112*, 104685. [CrossRef]
31. Choe, U.; Childs, H.; Zeng, M.; Zheng, W.; Zhu, H.; Zhu, L.; Xie, Z.; Gao, B.; Yu, L. Value-added utilization of fruit seed oils for improving human health: A progress review. *ACS Food Sci. Technol.* **2023**, *3*, 528–538. [CrossRef]
32. Zhou, Y.; Zhao, W.; Lai, Y.; Zhang, B.; Zhang, D. Edible plant oil: Global status, health issues, and perspectives. *Front. Plant Sci.* **2020**, *11*, 1315. [CrossRef]
33. Endo, Y. Analytical methods to evaluate the quality of edible fats and oils: The JOCS standard methods for analysis of fats, oils and related materials (2013) and advanced methods. *J. Oleo Sci.* **2018**, *67*, 1–10. [CrossRef] [PubMed]
34. Lamani, S.; Anu-Appaiah, K.A.; Murthy, H.N.; Dewir, Y.H.; Rihan, H.Z. Fatty acid profile, tocopherol content of seed oil, and nutritional analysis of seed cake of wood apple (*Limonia acidissima* L.), an underutilized fruit-yielding tree species. *Horticulturae* **2021**, *7*, 275. [CrossRef]
35. Murthy, H.N.; Yadav, G.G.; Kadapatti, S.S.; Lamani, S.; Desai, A.S.; Sumbad, M.M.; Dewir, Y.S.; Magyar-Tabori, K. Nutritional and oil characterization of *Erythrina stricta* Roxb. seeds: A potential resource for functional foods. *Cogent Food Agric.* **2024**, *10*, 2337770. [CrossRef]
36. Neme, K.; Tola, Y.B.; Mohammed, A.; Tadesse, E.; Shaheen, F.; Ahmed, S.; Jahan, H.; Qaiser, S.; Muller, F. Effect of seed processing treatments on oil quality of Ethiopian sesame varieties. *CyTA-J. Food* **2023**, *21*, 31–40. [CrossRef]
37. Konuskan, D.B.; Arslan, M.; Oksuz, A. Physicochemical properties of cold pressed sunflower, peanut, rapeseed, mustard and olive oils grown in the Eastern Mediterranean region. *Saudi J. Biol. Sci.* **2019**, *26*, 340–344. [CrossRef]
38. Eggersdorfer, M.; Wyss, A. Carotenoids in human nutrition and health. *Arch. Biochem. Biophys.* **2018**, *652*, 18–25. [CrossRef]
39. Kanopka, I.; Tanska, M.; Dabrowski, G.; Ogrodowska, D.; Czaplicki, S. Edibile oils from selected unconventional sources-a comprehensive review of fatty acid composition and phytochemical content. *Appl. Sci.* **2023**, *13*, 12829. [CrossRef]
40. Soleymani, S.; Habtemariam, S.; Rahimi, R.; Nabavi, S.M. The what and who of dietary lignans in human health: Special focus on prooxidant and antioxidant effects. *Trends Food Sci. Technol.* **2020**, *106*, 382–390. [CrossRef]
41. Bello, M.O.; Abdul-Hammed, M.; Adepoju, A.J.; Esan, O.A.; Tiamiyu, A.A. Nutritional composition and fatty acid profile of *Ficus exasperata* fruit and fruit oil. *J. Nat. Sci. Res.* **2014**, *4*, 25–29.
42. Tas, E.N. Biochemical characteristics of fig (*Ficus carica* L.) seeds. *J. Agric. Sci.* **2019**, *25*, 232–237.
43. Baygeldi, N.; Kucukerdonmez, O.; Akder, R.N.; Cagindi, O. Medicinal and nutritional analysis of fig (*Ficus carica*) seed oil: A new gamma-tocopherol and omega-3 source. *Prog. Nutr.* **2021**, *23*, e2021052.
44. Rahman, M.M.; Rahman, M.S.; Islam, M.R.; Rahman, F.; Mithi, F.M.; Alqahtani, T.; Almikhlafi, M.A.; Alghamdi, S.Q.; Alruwali, A.S.; Hossain, M.S.; et al. Role of phenolic compounds in human disease: Current knowledge and future prospects. *Molecules* **2022**, *27*, 233. [CrossRef]
45. Perera, S.; Silva, A.B.G.; Amarathung, Y.; De Silva, S.; Jayatissa, R.; Gamage, A.; Merah, O.; Madhujith, T. Nutritional composition and antioxidant activity of selected underutilized fruits grown in Sri Lanka. *Agronomy* **2022**, *12*, 1073. [CrossRef]
46. Ullah, A.; Munir, S.; Badshah, S.L.; Khan, N.; Ghani, L.; Poulson, B.G.; Emwas, A.H.; Jaremko, M. Important flavonoids and their roles as a therapeutic agent. *Molecules* **2020**, *25*, 5243. [CrossRef] [PubMed]
47. Rymbai, H.; Verma, V.K.; Talang, H.; Assumi, S.R.; Devi, M.B.; Vanlalruati; Sangma, R.H.C.H.; Biam, K.P.; Chanu, L.J.; Makdoh, B.; et al. Biochemical and antioxidant activity of wild edible fruits of eastern Himalaya, India. *Front. Nutr.* **2023**, *10*, 1039965. [CrossRef] [PubMed]
48. Debnath, B.; Singh, W.S.; Das, M.; Goswami, S.; Singh, M.K.; Maiti, D.; Manna, K. Role of plant alkaloids on human health: A review of biological activities. *Mater. Today Chem.* **2018**, *9*, 56–72. [CrossRef]
49. Tang, Y.Y.; He, X.M.; Sun, J.; Li, C.B.; Li, L.; Sheng, J.F.; Xin, M.; Li, Z.C.; Zheng, F.J.; Liu, G.M.; et al. Polyphenols and alkaloid byproducts of Longan fruits (*Dimocarpus longan* Lour.) and their bioactivities. *Molecules* **2019**, *24*, 1186. [CrossRef]
50. Rajesh, M.; Anusuya, N.; Siddhuraj, P.; Manian, S. The antioxidant activity and free radical scavenging potential of two different solvent extracts of *Cammelia sinensis* (L.) O. Kuntz, *F. bengalensis* L. and *F. racemosa* L. *Food Chem.* **2008**, *107*, 1000–1007.
51. Zhang, M.Q.; Zhang, J.; Zhang, Y.T.; Sun, J.Y.; Prieto, M.A.; Simal-Gandara, J.; Putnik, P.; Li, N.Y.; Liu, C. The link between phenolic composition and the antioxidant activity in different small berries: A metabolic approach. *LWT-Food Sci. Technol.* **2023**, *182*, 114853. [CrossRef]

foods

MDPI

Article

Qualitative and Quantitative Analyses of Sialyl *O*-Glycans in Milk-Derived Sialylglycopeptide Concentrate

Junichi Higuchi [1,†], Masaki Kurogochi [2,†,‡], Toshiyuki Yamaguchi [1], Noriki Fujio [1], Sho Mitsuduka [1], Yuko Ishida [1], Hirofumi Fukudome [1], Noriko Nonoyama [1], Masayuki Gota [1], Mamoru Mizuno [2] and Fumihiko Sakai [1,*]

1 Milk Science Research Institute, Megmilk Snow Brand Co., Ltd., 1-1-2 Minamidai, Kawagoe-shi 350-1165, Saitama, Japan; junichi-higuchi@meg-snow.com (J.H.); toshiyuki-yamaguchi@meg-snow.com (T.Y.); noriki-fujio@meg-snow.com (N.F.); s-mitsuduka@meg-snow.com (S.M.); y-ishida@meg-snow.com (Y.I.); h-fuku@meg-snow.com (H.F.); noriko-adachi@meg-snow.com (N.N.); m-gota@meg-snow.com (M.G.)

2 Laboratory of Glyco-Organic Chemistry, The Noguchi Institute, 1-9-7 Kaga, Itabashi-ku, Tokyo 173-0003, Japan; kuroko@noguchi.or.jp or kurogochi.masaki.d2@f.mail.nagoya-u.ac.jp (M.K.); mmizuno@noguchi.or.jp (M.M.)

* Correspondence: f-sakai@meg-snow.com; Tel.: +81-49-242-8118

† These authors contributed equally to this study.

‡ Current address: Institute for Glyco-core Research (iGCORE), Nagoya University, Furo-cho, Chikusa-ku, Nagoya 464-8601, Aichi, Japan.

Abstract: Sialyl glycans have several biological functions. We have previously reported on the preparation and bifidogenic activity of milk-derived sialylglycopeptide (MSGP) concentrate containing sialyl *O*-glycans. The current study qualitatively and quantitatively analyzed the sialyl *O*-glycans present in the MSGP concentrate. Notably, our quantitative analysis indicated that a majority of *O*-glycopeptides in the MSGP concentrate were derived from glycomacropeptides. The concentrate was found to contain mainly three types of sialyl core 1 *O*-glycans, with the disialyl core 1 *O*-glycan being the most abundant. We successfully quantified three types of sialyl core 1 *O*-glycans using a meticulous method that used homogeneous *O*-glycopeptides as calibration standards. Our results provide valuable insights into assessment strategies for the quality control of *O*-glycans in dietary products and underscore the potential applications of MSGP concentrate in the food industry and other industries.

Keywords: sialylglycopeptide; *O*-glycan; homogeneous *O*-glycopeptide

1. Introduction

Milk, a rich source of nutritionally valuable protein, offers a promising avenue for health research. Its proteins, which boast a well-balanced amino acid profile, excellent digestibility, and numerous bioactive peptides (antioxidant, antihypertensive, immunomodulatory, and antimicrobial peptides), have been the focus of recent proteomics studies [1,2]. In fact, such studies have revealed that various milk proteins, including lactoferrin, α-lactalbumin, immunoglobulin, κ-casein, and several milk fat globule membrane proteins, are extensively glycosylated [3–5]. The glycan moieties of these glycoproteins play crucial roles in protein folding, biological recognition, and protection from digestion [6]. They have also been shown to protect infants against pathogens [7,8] and serve as substrates for bifidobacterial growth in the infant's gut [9,10]. The potential of the glycan moiety of milk glycoproteins to become a new functional material for maintaining and improving health inspires hope for the future of nutrition research. In particular, sialic acid attached to the non-reducing ends of these glycopeptides has shown potential in promoting brain development, immunomodulation, and even stress and inflammation relief [11,12].

Glycomacropeptide (GMP), which is a hydrophilic peptide released into whey from κ-casein during the production of cheese, is one of the most heavily glycosylated peptides in whey proteins. Studies over the past decades have characterized the *O*-glycan structures and the site-specific profiling of *O*-glycosylation in GMP. A total of eleven distinct glycan structures have been identified in bovine GMP. Of these, five are present in both colostrum and mature milk, whereas the remaining six are exclusively found in colostrum [6,13,14]. Moreover, Kurogochi et al. reported *O*-acetylation of sialic acid in GMP [5]. A total of seven potential sites of *O*-glycosylation in GMP have been reported, including ^{121}T, ^{131}T, ^{133}T, ^{136}T, ^{141}S, ^{142}T, and ^{165}T [15,16]. In a recent study, intact GMP in commercial GMP powders was identified through a top-down approach based on liquid chromatography (LC)–mass spectrometry (MS) analysis [15,17]. The glycosylation of intact GMP and the presence of fragment GMP were identified in commercial GMP powders [17]. These glycans, especially Neu5Ac, are of great importance in the biological activities of GMP, including the inhibition of influenza virus binding to oligosaccharide receptors on host cells [18], the promotion of *Bifidobacterium* growth [19], and the facilitation of early brain development in young piglets [20].

In our previous study, we developed a unique "milk-derived sialylglycopeptide (MSGP)" concentrate from GMP-rich whey protein concentrate (G-WPC) [21]. The glycan types and structures in the MSGP concentrate, prepared from G-WPC, were consistent with those previously reported for GMP, mainly sialyl core 1 *O*-glycans. The concentrate has been demonstrated to possess higher bifidogenic properties than those of GMP in vitro. Given that no food ingredient has ever utilized sialyl *O*-glycan, the MSGP concentrate represents a potential game changer in the field of nutraceuticals and functional foods. We anticipate that the MSGP concentrate will be prepared from any cheese whey-derived whey protein concentrate (WPC) or whey protein isolate (WPI) containing GMP. It should be noted, however, that the *O*-glycans present in WPC or WPI are not limited to those derived from GMP. For quality control, when materials other than G-WPC are used as raw materials, it is necessary to analyze the chemical structure and quantity of sialyl *O*-glycans in the concentrate.

To accurately identify the structure of a glycopeptide, the glycosylation sites on the peptide chain need to be determined in addition to conducting a structural analysis of the peptide chain and glycopeptide. Unlike *N*-glycosylation, which features a consensus motif (NXS/T), no consensus motif exists for *O*-glycosylation. The use of LC–MS/MS with collision-induced dissociation (CID) and electron transfer dissociation (ETD) is indispensable when identifying the glycosylation sites of *O*-glycopeptides contained in the MSGP concentrate [5], underscoring the need for advanced techniques in our research.

O-glycopeptides are significantly more difficult to quantify than *N*-glycopeptides. Although *N*-glycans attached to asparagine residues can be released from proteins or peptides by commercially available enzymes, such as peptide-*N*-glycosidase F (PNGase F), and then identified or quantified through LC–MS techniques [22], no enzyme can release intact *O*-glycans attached to threonine or serine residues. *O*-glycans can be released from proteins or peptides via β-elimination under mild alkaline conditions; however, some of the released glycans are degraded under alkaline conditions, referred to as peeling reaction, thereby complicating the structural analysis and quantification of *O*-glycans. Kameyama et al. developed eliminative oximation that enables β-elimination of *O*-glycans from glycoproteins using hydroxylamine and 1,8-diazabicyclo[5.4.0]undec-7-ene (DBU) to suppress *O*-glycan degradation [23]. *O*-glycans released following this procedure were then analyzed using LC with a fluorescence detector (LC–FLD) after labeling with 2-aminobenzoate (2-AA). Yamaguchi et al. relatively quantified *O*-glycans in human milk using LC–MS without labeling, taking advantage of the fact that the *N*-acetylgalactosamine (GalNAc) at the reducing end of the *O*-glycan immediately forms an oxime upon release from the glycoprotein using Kameyama's method; the oximated *O*-glycans are to be more ionized by electrospray ionization than native *O*-glycans [24]. These advanced methods are valuable for identifying and quantifying *O*-glycans in MSGP concentrate.

The current study aimed to qualitatively and quantitatively characterize the sialyl *O*-glycans in an MSGP concentrate prepared from whey protein isolates. We demonstrated that most of the *O*-glycopeptides in the MSGP concentrate were derived from GMP. To quantify *O*-glycans, we purified homogeneous *O*-glycopeptides from MSGP concentrate to be used as standards for calibration and then successfully quantified three types of sialyl core 1 *O*-glycans. These meticulous methods ensure the accuracy and reliability of our findings.

2. Materials and Methods

2.1. Reagents

Whey protein isolate (WPI) from cheese whey was purchased from a commercial manufacturer (Glanbia Nutritionals, Twin Falls, ID, USA). Alcalase 2.4 L FG (EC 3.4.21.62, 2.4 Anson unit/g, produced by *Bacillus licheniformis*, with endoprotease activity) and Flavorzyme 1000 L (EC 3.4.11.1, 1000 leucine amino-peptidase unit/g, produced by *Aspergillus oryzae*, with both endoprotease and exopeptidase activities) were obtained from Novozymes (Copenhagen, Denmark). The EZGlyco O-glycan Prep Kit was purchased from Sumitomo Bakelite (Tokyo, Japan). Other reagents were purchased from standard vendors.

2.2. Preparation of the MSGP Concentrate from WPI

The MSGP concentrate was prepared from WPI according to previously reported methods [21] with slight modifications. Briefly, WPI dissolved in water (10% [wt/wt]) was heated to 55 °C and adjusted to pH 7.0 using a KOH solution. Alcalase and Flavorzyme (0.45% [wt/wt], respectively) were then added to the solution simultaneously. Enzyme reaction was performed at 55 °C for 8 h and stopped by heating to 85 °C. We used a molecular weight cutoff 1000 ultrafiltration membrane (XT-3B-3838, Synder Filtration, CA, USA) to concentrate the glycopeptides from the WPI hydrolysate. The hydrolysate was then concentrated through ultrafiltration. The retentate was lyophilized using a freeze dryer (Nissei Limited, Tokyo, Japan). The resulting lyophilized powder, that is, the MSGP concentrate, was stored at −30 °C until further use.

2.3. Gross Chemical Composition of WPI and the MSGP Concentrate

The moisture, protein, fat, and ash contents of the MSGP concentrate were determined at the Japan Food Research Laboratories (Tokyo, Japan). Moisture content was determined using the oven-drying method under normal pressure. Protein content was determined using the Kjeldahl method with a nitrogen-to-protein conversion factor of 6.38. Fat content was determined using the Rose Gottlieb method. Ash content was determined using the dry ashing method. Carbohydrate content was calculated by subtracting the moisture, protein, fat, and ash contents from the total weight.

2.4. Determination of Sialic Acid and Galacto-N-Biose

To measure sialic acid (Neu5Ac) and galacto-*N*-biose (GNB), WPI and MSGP concentrates were treated with neuraminidase derived from *Arthrobacter ureafaciens* (EC 3.2.1.18, one unit is defined as the amount of enzyme required to liberate 1 μmol of Neu5Ac per minute at pH 5.0 at 37 °C, Nacalai Tesque Inc., Kyoto, Japan) and *O*-glycosidase (EC 3.2.1.97, one unit is defined as the amount of enzyme required to liberate 0.68 nmol of *O*-linked disaccharides from 5 mg of neuraminidase-treated fetuin in 1 h at 37 °C, New England Biolabs, MA, USA). A reaction mixture containing 5 mg/mL of WPI or 500 μg/mL of the MSGP concentrate, 0.6 U/mL of neuraminidase, and 800,000 U/mL of *O*-glycosidase in 50 mmol/L sodium phosphate buffer (pH 5.0) was incubated at 37 °C for 2 h. The Neu5Ac and GNB released by the enzymes were analyzed using high-performance anion-exchange chromatography with pulsed amperometric detection (HPAE-PAD). Enzymatic hydrolysates were applied to a Dionex CarboPac PA1 column (4 mm × 250 mm; 10 μm particle size; Thermo Fisher Scientific, Waltham, MA, USA). Standards for Neu5Ac and GNB were obtained as the purest available grades (Sigma-Aldrich, St. Louis, MO, USA).

For Neu5Ac analysis, the mobile phase consisted of water (solvent A), 200 mmol/L NaOH solution (solvent B), and 600 mmol/L sodium acetate in 100 mmol/L NaOH solution (solvent C) with the following gradient elution: from 0 to 10 min 45% B and 10% C; from 10 to 25 min 37.5% B and 25% C; and from 25 to 30 min 45% B and 10% C at a flow rate of 1 mL/min. For GNB analysis, the mobile phase consisted of solvents A, B, and C with the following gradient elution: from 0 to 9 min 50% B and 0% C; from 9 to 12 min 0% B and 100% C; and from 12 to 30 min 50% B and 0% C at a flow rate of 1 mL/min. The chromatograms were analyzed using Chromeleon software version 7.3 (Thermo Fisher Scientific).

2.5. Enzymatic Digestion of the MSGP Concentrate

A solution of MSGP concentrate (5 mg/mL) in 50 mmol/L phosphate buffer (pH 5.0) was treated with 1.2 units/mL of neuraminidase from *Arthrobacter ureafaciens* and 1,600,000 units/mL of *O*-glycosidase at 37 °C for 2 h. Subsequently, an equal amount of 2 U/mL of Proteinase K (EC 3.4.21.64, one unit is defined as the amount of enzyme required to produce a peptide equivalent to 1 μmol/min of tyrosine as the colorant of the Folin & Ciocalteu phenol reagent, Fujifilm Wako Pure Chemical Corporation, Osaka, Japan) was added to the sample followed by treatment at 37 °C for 1 h.

2.6. Size-Exclusion Chromatography (SEC)

To estimate the molecular weight of WPI, MSGP, and enzyme-digested MSGP, SEC analysis was performed using the Waters Alliance e2695 Separations Module (Waters Corporation, Milford, MA, USA). Samples were applied to a tandem combination of Inertsil Diol (4.6 mm × 250 mm; 5 μm particle size; GL Sciences, Inc., Tokyo, Japan) and Inertsil WP300 Diol (4.6 mm × 250 mm; 5 μm particle size; GL Sciences, Inc.) columns and eluted with a mobile phase consisting of 40% (vol/vol) acetonitrile containing 0.01% (vol/vol) trifluoroacetic acid at a flow rate of 0.3 mL/min. Absorbance was monitored at 210 nm using a 2489 UV detector (Waters Corporation). β-Lactoglobulin (molecular weight: 18,277), α-lactoalbumin (molecular weight: 14,146), aprotinin (molecular weight: 6512), sialylglycopeptide (from egg yolk, molecular weight: 2866), and oxytocin (molecular weight: 1007) were used as molecular weight markers.

2.7. Analysis of O-Glycan Composition in the MSGP Concentrate

Labeled MSGP *O*-glycans were prepared using an EZGlyco O-glycan Prep Kit. Briefly, 10 μL of MSGP solution (10 mg/mL in H_2O) was mixed with 5 μL of Glycan Released Reagent A and 10 μL of Glycan Released Reagent B and incubated at 50 °C for 20 min. Subsequently, released glycans were captured using Glycan Capturing Beads, which were washed with acetonitrile. Thereafter, 4 mg of 2-aminobenzeamide (2-AB) and 0.04 mg of reducing reagent in 50 μL of methanol/acetic acid/H_2O (9/2/9) were added to the beads. The *O*-glycan-containing solution was recovered via centrifugation at 3000× *g* for 1 min. The solution was incubated at 50 °C for 2.5 h and washed with acetonitrile to remove excess reagent with a Cleanup Column. The 2-AB-labeled *O*-glycans were recovered by adding H_2O and analyzed using a Q-Exactive mass-spectrometer and an RS Fluorescence (FL) Detector coupled with an UltiMate 3000 (Thermo Fisher Scientific). For analysis, a labeled *O*-glycan solution (10 μL) was applied to a Glycanpac AXH-1 column (2.1 mm × 150 mm; 3 μm particle size; Thermo Fisher Scientific) at 40 °C. The mobile phase comprised acetonitrile (solvent A) and 50 mmol/L ammonium formate (pH 4.4) (solvent B) in a gradient elution of from 0 to 55 min at 10–35% and then from 55 to 65 min at 35% B. The flow rate was 0.4 mL/min. The electrospray voltage and heat capillary temperature were 3.5 kV and 275 °C, respectively. Nitrogen (99.5% purity) was used as sheath gas (set to 35), auxiliary gas (set to 10), and collision gas. Full-scan mass spectra were acquired in positive ion mode from m/z 400 to 2000 and a resolution of 70,000. MS2 spectra were acquired in the automatic data-dependent mode using one precursor scan, followed by five MS2 scans with a higher energy collisional dissociation. The stepped normalized collision energies were set to 10, 15, and 20. FL detector excitation and emission wavelengths were set to 330 and 430 nm, respectively. The spectra and chromatograms were analyzed using Xcalibur version 4.0 (Thermo Fisher Scientific). The structures of labeled *O*-glycans were identified through manual inspection.

2.8. Analysis of O-Glycopeptides in the MSGP Concentrate Using LC ESI-MS and ESI-MS/MS

LC ESI-MS and ESI-MS/MS were used to analyze the MSGP concentrate on an LC system (Dionex, Sunnyvale, CA, USA) with a C-18 reverse phase column (Inertsustain ODS-3, 1.0 × 150 mm, 3.5 μm particle size; GL Sciences, Inc.). The sample was loaded with 25 mmol/L ammonium formate solution (0–4 min), eluted with a gradient of 0–15% acetonitrile with 0.1% formic acid (4–34 min), washed with 100% acetonitrile with 0.1% formic acid (34–40 min), and then equilibrated with 25 mmol/L ammonium formate solution (40–60 min) at a constant flow of 50 μm/min (column oven at 37 °C). Electrospray MS data were collected using the HESI-II probe ion source on an LTQ-Velos Pro instrument (Thermo Fisher Scientific) in the positive mode. The spray voltage was set at 3.5 kV, with a temperature of 150 °C. The sheath gas flow and auxiliary gas flow rates were set at 20 and 5 arb, respectively. MS/MS data were collected using a data-dependent acquisition method, which was implemented as a "top 15" experiment: one precursor scan covering an *m/z* range of 400–2000, followed by CID spectra targeting the top 15 most intense ions in the precursor spectrum. CID event parameters were as follows: isolation width, 3 Da; normalized collision energy, 30%. The peaks and masses were integrated using the annotation procedures (Qual Browser) on Xcalibur 2.2 SP 1.48 software (Thermo Fisher Scientific).

The CID fragmentation normalized energy of the selective ion was set to 30%. ETD fragmentation for *O*-glycopeptides was performed using supplemental collisional activation (35%). MS3 measurements (CID-MS3) were performed through CID fragmentation of peptide-only ions (those with complete loss of the sugar chain from the CID-MS/MS data of *O*-glycopeptides), and the second CID fragmentation normalized energy was set to 35%.

2.9. Preparation of Homogeneous O-Glycopeptides as Standards for Quantification

O-glycopeptides were separated using a preparative HPLC system (PLC761; GL Sciences Inc., Tokyo, Japan) equipped with a UV detector (210 nm) and fraction collector on a HILIC amino column (NH2P-90 20F, 20.0 × 300 mm, 9 μm particle size; Resonac, New York, NY, USA) or a C18 RP column (YMC-Pack ODS-A, 20 × 150 mm, 5 μm particle size; YMC Co., LTD, Kyoto, Japan).

For the separation of *O*-glycopeptide with a disialyl core 1 *O*-glycan, a 5 mL solution of the MSGP concentrate (10 mg/mL) was loaded onto a HILIC amino column equilibrated with 30 mmol/L NaH_2PO_4 solution. Elution was performed with 30 mmol/L (0–5 min), a gradient of 30–165 mM (5–35 min), and 165–300 mM (35–45 min) NaH_2PO_4 solution at a flow of 10.0 mL/min at 40 °C. Each fraction was collected every 40 s. Fractions containing *O*-glycopeptide with a disialyl core 1 *O*-glycan were desalted using a graphite carbon cartridge (InertSep GC column; GL Sciences). The *O*-glycopeptide was eluted with 50% acetonitrile with 0.1% formic acid from the cartridge. The acetonitrile was then removed using a centrifugal evaporator (CVE-3100D; Tokyo Rikakikai Co., Ltd., Tokyo, Japan) after neutralization by adding a 28% ammonia solution. The desalted sample was lyophilized as a semi-purified *O*-glycopeptide with a disialyl core 1 *O*-glycan using EYELA FDU-2200 (Tokyo Rikakikai Co., Ltd.). To obtain a homogeneous *O*-glycopeptide with a disialyl core 1 *O*-glycan, a sample containing the semi-purified *O*-glycopeptide with a disialyl core 1 *O*-glycan dissolved in water was loaded onto a C18 RP column equilibrated with 2.0% acetonitrile with 0.1% formic acid. Elution was performed with a gradient of 2.0–10.8% (0–15 min) and 10.8–90.0% (15–25 min) and an isocratic mode of 90% (25–30 min) at a flow of 10.0 mL/min at 40 °C. Each fraction was collected every 30 s. Fractions containing *O*-glycopeptide with a disialyl core 1 *O*-glycan were desalted and lyophilized as described above. The lyophilized sample was used as the standard for *O*-glycopeptide with a disialyl core 1 *O*-glycan.

To prepare *O*-glycopeptide with a branched monosialyl core 1 *O*-glycan, 10 mL of homogeneous *O*-glycopeptide having a disialyl core 1 *O*-glycan (2 mg/mL) in 50 mmol/L of sodium phosphate buffer (pH 6.0) was treated with five units of α2,3-neuraminidase (EC 3.2.1.18, from *Salmonella typhimurium* LT2; Takara Bio Inc., Shiga, Japan) at 37 °C for 2 h to selectively release one molecule of Neu5Ac from *O*-glycopeptide with a disialyl core 1 *O*-glycan. Subsequently, α2,3-neuraminidase was inactivated through treatment at 100 °C for 5 min. The sample treated with α2,3-neuraminidase was loaded onto a HILIC amino column equilibrated with 15 mmol/L NaH_2PO_4. Elution was performed with 15 mmol/L NaH_2PO_4 (0–5 min) and with a gradient of 15–75 mmol/L (5–35 min), 75–150 mmol/L (35–45 min), and 150 mmol/L (45–50 min) at a constant flow of 10.0 mL/min at 40 °C. Each fraction was collected every 40 s. Fractions containing *O*-glycopeptide with a branched monosialyl core 1 *O*-glycan were desalted and lyophilized as described above. The lyophilized powder was used as the standard for *O*-glycopeptide with a branched monosialyl core 1 *O*-glycan.

For the preparation of *O*-glycopeptide with a linear monosialyl core 1 *O*-glycan, 10 mL of homogeneous *O*-glycopeptide with a disialyl core 1 *O*-glycan (2 mg/mL) in 50 mmol/L sodium phosphate buffer (pH 5.0) was treated with 0.005 units/mL of neuraminidase (from *Arthrobacter ureafaciens*) at 37 °C for 16 h to release one molecule of Neu5Ac from the *O*-glycopeptide with a disialyl core 1 *O*-glycan. Subsequently, neuraminidase was inactivated via treatment at 100 °C for 5 min. The reaction mixture was loaded onto a C18 RP column equilibrated with 2.0% acetonitrile with 0.1% formic acid. Elution was then performed with a gradient of 2.0–10.8% (0–15 min) and 10.8–90.0% (15–25 min) and an isocratic mode of 90% (25–30 min) at a flow of 10.0 mL/min at 40 °C. Each fraction was collected every 30 s. Fractions containing *O*-glycopeptide with a linear monosialyl core 1 *O*-glycan were lyophilized as described above. The lyophilized sample was used as the standard for *O*-glycopeptide with a linear monosialyl core 1 *O*-glycan.

The amino acid sequence and *O*-glycan structure of homogeneous *O*-glycopeptides, as well as the *O*-glycan binding sites of *O*-glycopeptides with a disialyl or monosialyl core 1 *O*-glycan, were identified through LC–MS analysis as described above. The amount of homogeneous *O*-glycopeptides with a disialyl or monosialyl core 1 *O*-glycan was determined based on the Neu5Ac content. Neu5Ac in the homogeneous *O*-glycopeptides with a disialyl or monosialyl core 1 *O*-glycan was released through neuraminidase treatment, with the released Neu5Ac being determined using HPAE-PAD as described above.

2.10. Quantification of Sialyl Core 1 O-Glycans in the MSGP Concentrate

To quantify sialyl core 1 *O*-glycans, 20 μL of the MSGP concentrate (300 μg/mL in H_2O) or 20 μL of purified homogeneous *O*-glycopeptide (2–100 μg/mL in H_2O) containing 10 μg/mL of 3′-sialyllewis x (as internal standards) was mixed with 10 μL of Glycan Released Reagent A and 20 μL of Glycan Released Reagent B (provided in the EZGlyco O-glycan Prep Kit) and incubated at 50 °C for 20 min [24]. Subsequently, 100 μL of 1.33 mol/L acetic acid was added for neutralization. The reaction solution was diluted two times with acetonitrile. An aliquot (10 μL) of a released *O*-glycan solution was applied to a Glycanpac AXH-1 column (2.1 mm × 150 mm; 3 μm particle size) and analyzed using a Q-Exactive mass spectrometer coupled with an UltiMate 3000. The mobile phase comprised acetonitrile (solvent A) and 50 mmol/L ammonium formate (pH 4.4) (solvent B) in a gradient elution of 0 to 38.5 min at 10–27.5% B and then 38.5 to 45 min at 27.5% B. The flow rate was 0.4 mL/min. The electrospray voltage and heat capillary temperature were 3.5 kV and 275 °C, respectively. Nitrogen (99.5% purity) was used as sheath gas (set to 35), auxiliary gas (set to 10), and collision gas. Single ion monitoring (SIM) chromatograms were acquired in negative mode at m/z 489.16 (corresponding to $[M-2H]^{2-}$ of oximated disialyl core 1 *O*-glycan), 688.24 ($[M-H]^{-}$ of oximated monosialyl core 1 *O*-glycans), and 834.30 ($[M-H]^{-}$ of oximated 3′-sialyllewis x). Chromatograms were analyzed using Xcalibur version 4.0 to obtain the SIM area of each sialyl core 1 *O*-glycan. The SIM area of each sialyl *O*-glycan was divided by that of the internal standard, respectively, and each glycan was quantified using a calibration curve generated from that of the purified *O*-glycopeptide.

3. Results

3.1. Chemical Composition of the MSGP Concentrate

The MSGP concentrate was prepared from commercially available WPI derived from cheese whey. After comparing the gross chemical composition of MSGP concentrate with that of WPI, a two-thirds decrease in protein and a fifteenfold increase in carbohydrates were observed (Table 1). Neu5Ac and GNB (Galβ1-3GalNAc), parts of carbohydrates, were more than 10 times higher in MSGP concentrate than in WPI (Table 1).

Table 1. Chemical specifications of WPI and the MSGP concentrate prepared from WPI (wt/wt).

	WPI	MSGP Concentrate
Moisture	4.5%	2.0%
Protein	90.0%	60.8%
Fat	0.5%	2.9%
Ash	3.0%	4.2%
Carbohydrate	2.0%	30.1%
Neu5Ac	1.4%	16.4%
GNB	0.9%	10.1%

3.2. SEC Analysis of the MSGP Concentrate

The molecular weight distribution of the MSGP concentrate was analyzed using SEC. The main peak of the MSGP concentrate was observed in the molecular weight range from 1000 to 3500, which was confirmed to have shifted to a lower molecular weight range than that of WPI (Figure 1a,b). The peak top of MSGP (molecular weight: 1399) treated with neuraminidase and *O*-glycosidase capable of releasing core 1 *O*-glycans from glycopeptides shifted to a lower molecular region than that of MSGP (molecular weight: 1904) (Figure 1c). Although MSGP treated with proteinase K showed no change in molecular weight distribution, MSGP treated with proteinase K after treatment with neuraminidase and *O*-glycosidase demonstrated a shift to a much lower molecular region than that of MSGP treated with neuraminidase and *O*-glycosidase (Figure S1), indicating that the main peak of MSGP determined via SEC consisted of glycopeptides with sialyl core 1 *O*-glycans.

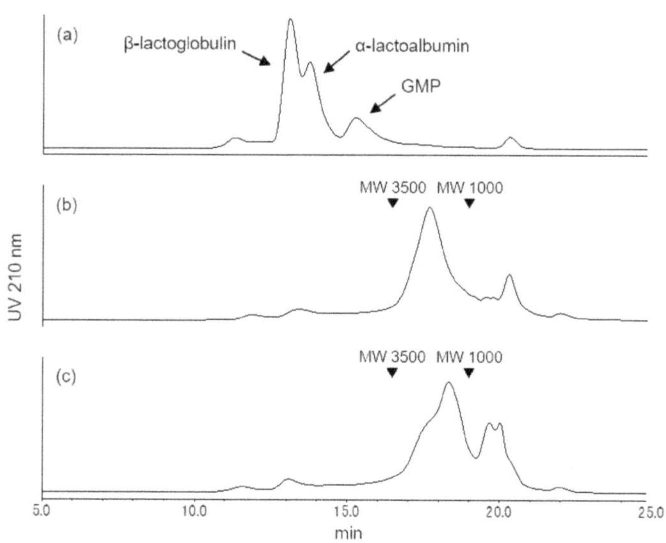

Figure 1. Size-exclusion chromatograms of WPI, MSGP concentrate, and MSGP concentrate treated with glycosidases. WPI (**a**), MSGP concentrate (**b**), MSGP concentrate treated with neuraminidase,

and *O*-glycosidase (c) were applied to a tandem combination of Inertsil Diol and Inertsil WP300 Diol columns (GL Sciences) and eluted with 40% (vol/vol) acetonitrile containing 0.01% trifluoroacetic acid at a flow rate of 0.3 mL/min. Absorbance was monitored at 210 nm. The inverted triangle symbol indicates the retention time of the molecular weight (MW) calculated from β-lactoglobulin (MW: 18,277), α-lactalbumin (MW: 14,146), aprotinin (MW: 6512), sialylglycopeptide (from egg yolk, MW: 2866), and oxytocin (MW: 1007).

3.3. LC–MS/MS Analysis of Glycopeptides in the MSGP Concentrate

Glycopeptides in the MSGP concentrate were analyzed using LC–MS/MS with CID and ETD. Mass spectra were examined using Proteome Discoverer (version 2.5), Byonic software (version 2.13.17), and manual inspection. A total of 141 glycopeptides were detected, among which 131 were *O*-glycopeptides derived from the GMP and 10 were *N*- or *O*-glycopeptides derived from osteopontin and glycosylation-dependent cell adhesion molecule 1 (Table S1). These results demonstrate that GMP-derived *O*-glycopeptides were predominantly enriched in the MSGP concentrate. The 131 types of GMP-derived *O*-glycopeptides comprised combinations of 43 types of peptides and 7 types of glycans (Figure 2). The peptide chain consisted of ≤12 amino acid residues. The disialyl *O*-glycan (glycan **5** in Figure 2B) was the most frequently detected (Table S1). Parts of the disialyl *O*-glycans were *O*-acetylated at Neu5Ac binding to GalNAc (glycans **6** and **7** in Figure 2B). MS/MS analysis with ETD identified eight amino acid residues as glycosylation sites (^{121}T, ^{131}T, ^{133}T, ^{136}T, ^{141}S, ^{142}T, ^{165}T, and ^{167}T shown in red letters in Figure 2A; residue numbering is based on Swiss–Prot entries for the mature form of bovine κ-casein; accession number P02668). In addition, the detection of glycosylated EA^{149}SPE and A^{149}SPE (Nos. 121 and 122 in Table S1) indicated the glycosylation of ^{149}S, whereas the detection of glycosylated ^{155}SPPEIN^{161}TVQ and ^{155}SPPEIN^{161}TVQVT (Nos. 123 and 124 in Table S1) indicated the glycosylation of ^{155}S and/or ^{161}T.

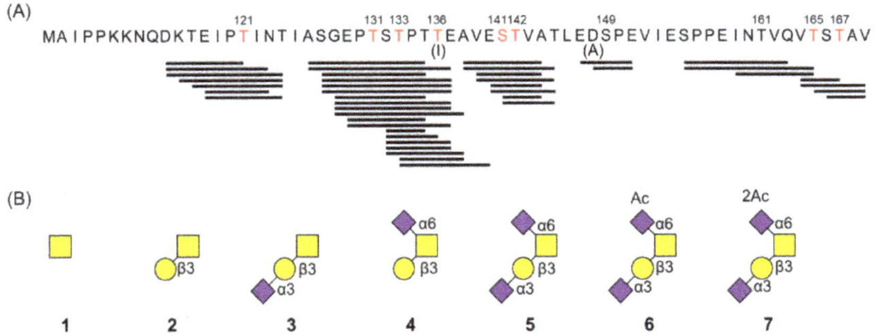

Figure 2. Glycopeptides detected in WPI-derived MSGP concentrate using LC–MS/MS with CID and ETD. (**A**) The sequence of the GMP is shown at the top (residue numbering is based on Swiss–Prot entries for the mature form of bovine κ-casein; accession number P02668). Black bars below the GMP sequence indicate peptides detected via LC–MS, whereas red letters in the GMP sequence indicate glycosylated residues characterized by ETD measurements. (**B**) Proposed structures of glycans. Yellow squares, yellow circles, and purple diamonds indicate GalNAc, galactose (Gal), and Neu5Ac, respectively. *O*-Acetylated Neu5Ac or *O*, *O′*-diacetylated Neu5Ac was represented by the addition of Ac or 2Ac on top of Neu5Ac.

3.4. Composition of O-Glycans in the MSGP Concentrate

LC–FLD was used to analyze the composition of *O*-glycans in the MSGP concentrate after being chemically released and fluorescently (2-AB) labeled using a commercially available kit. The proportion of each 2-AB-labeled *O*-glycan was determined from the signal area of the FL chromatogram (Figure 3). One type of asialyl core 1, three types of sialyl core 1 *O*-glycans, and one type of sialyl core 2 *O*-glycan were detected. The signal area for asialyl core 1 (GNB) could not be evaluated correctly due to overlapping lactose signals. Regarding the proportion of sialyl *O*-glycans excluding peeling by-product in the MSGP concentrate, disialyl core 1 *O*-glycan (glycan **5** in Figure 2B) was the most abundant (72.4%), whereas sialyl core 2 *O*-glycan was the scantest (2.2%) among the sialyl *O*-glycans in the MSGP concentrate (Table 2).

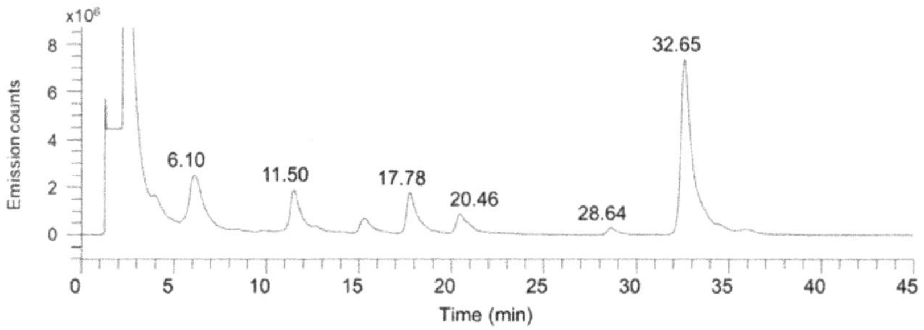

Figure 3. HPLC fluorescence chromatogram of 2-AB-labeled *O*-glycans from the MSGP concentrate. The labeled *O*-glycans were applied to a Glycanpac AXH-1 Column at 40 °C and eluted with a gradient of acetonitrile and 50 mmol/L ammonium formate solution (pH 4.4) in a gradient of 10–35% ammonium formate for 0 to 55 min. The flow rate was maintained at 0.4 mL/min. The excitation and emission wavelengths were set at 330 and 430 nm, respectively. The peaks at 17.78, 20.46, 28.64, and 32.65 min correspond to linear monosialyl core 1 *O*-glycan **3**, branched monosialyl core 1 *O*-glycan **4**, monosialyl core 2 *O*-glycan, and disialyl core 1 *O*-glycan **5**, respectively. The peaks at 6.10 and 11.50 min correspond to GNB overlapped with lactose and peeling by-product, respectively. Each peak was assigned based on MS2 analysis.

Table 2. Composition of sialyl *O*-glycans of the MSGP concentrate.

O-Glycan	Proportion
Linear monosialyl core 1 *O*-glycan **3**	16.5%
Branched monosialyl core 1 *O*-glycan **4**	8.9%
Disialyl core 1 *O*-glycan **5**	72.4%
Monosialyl core 2 *O*-glycan	2.2%

3.5. *Quantification of Sialyl Core 1 O-Glycan in the MSGP Concentrate*

We attempted to quantify sialyl core 1 *O*-glycans (monosialyl and disialyl glycans), which were the most significant glycans in the MSGP concentrate, through eliminative oximation [14] with some modifications [15]. To adjust the efficiency of glycan release and rate of peeling reaction of the released glycans, homogeneous glycopeptides with monosialyl or disialyl core 1 *O*-glycans were prepared as standards. First, homogeneous glycopeptide with disialyl core 1 *O*-glycan was purified from the MSGP concentrate via two-step HPLC purification (Figure S2A,B). The structure of the purified *O*-glycopeptide was determined to be 8-aa *O*-glycopeptide (GEPTSTPT) with a disialyl core 1 *O*-glycan (Neu5Acα2,3Galβ1,3(Neu5Acα2,6)GalNAc; glycan **5** in Figure 2B) based on LC–MS with CID- and ETC-MS/MS analysis (Figure 4A,D). Thereafter, two types of monosialyl glycopeptides were prepared through the digestion of homogeneous glycopeptides with a disialyl core 1 *O*-glycan using α2,3-specific or non-specific neuraminidase and HPLC purification (Figure S2C,D). The structures of the purified monosialyl glycopeptides were determined to be 8-aa *O*-glycopeptide (GEPTSTPT) with a linear (Neu5Acα2,6Galβ1,3GalNAc, glycan **3**; Figure 4B,E) and branched (Galβ1,3(Neu5Acα2,6)GalNAc, glycan **4**; Figure 4C,F) monosialyl core 1 *O*-glycan based on LC–MS. The *O*-glycans in homogeneous glycopeptides and MSGP concentrate were released via eliminative oximation, after which the oximated *O*-glycans were detected using LC–MS without further labeling reaction. The amount of sialyl core 1 *O*-glycans in the MSGP concentrate was calculated using calibration curves generated by homogeneous glycopeptides with sialyl core 1 *O*-glycans (Figure 4G–I). Linear monosialyl *O*-glycan **3**, branched monosialyl *O*-glycan **4**, and disialyl *O*-glycan **5** accounted for 1.3%, 0.4%, and 19.7% (wt/wt) of the MSGP concentrate, respectively. Disialyl *O*-glycan **5** was the major glycan, accounting for 92% of the sialyl core 1 *O*-glycans in the MSGP concentrate.

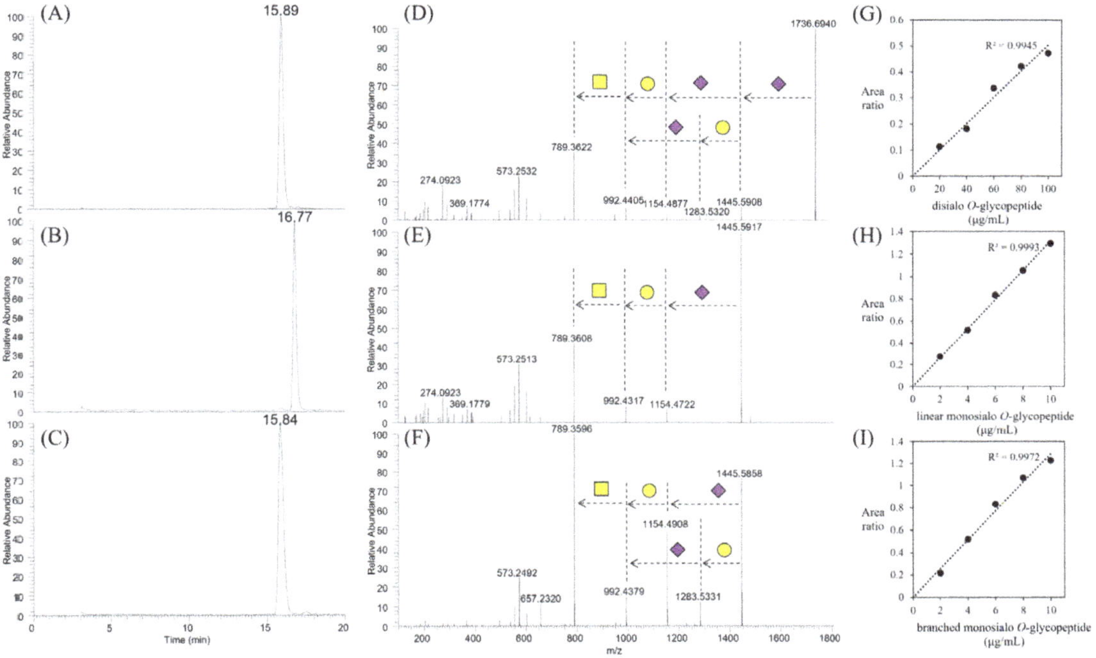

Figure 4. Preparation of homogeneous *O*-glycopeptides with sialyl core 1 *O*-glycans for use as standards. TIC chromatograms (*m*/*z* 400–2000) of the purified glycopeptides with disialyl (**A**), linear monosialyl (**B**), and branched monosialyl (**C**) core 1 *O*-glycans. MS2 spectra of the purified *O*-glycopeptides with disialyl (**D**), linear monosialyl (**E**), and branched monosialyl (**F**) core 1 *O*-glycans. Yellow squares, yellow circles, and purple diamonds indicate GalNAc, Gal, and Neu5Ac, respectively. Calibration curves for the LC–MS area of disialyl (**G**), linear monosialyl (**H**), and branched monosialyl (**I**) *O*-glycans released from the purified *O*-glycopeptides through eliminative oximation.

4. Discussion

In the current study, we prepared an MSGP concentrate to enrich the sialyl *O*-glycan moiety found in whey protein and used SEC, LC–FLD, and LC–MS with CID-MS/MS and ETD for both qualitative and quantitative analyses of the same. Notably, SEC and LC–MS analyses revealed that the molecular weight distribution of the sialylglycopeptides contained in the MSGP concentrate ranged from 1000 to 3500. Moreover, LC–FLD determined that the sialyl *O*-glycans in the MSGP concentrate consisted of disialyl core 1 *O*-glycans, linear monosialyl core 1 *O*-glycans, branched monosialyl core 1 *O*-glycans, and sialyl core 2 *O*-glycans. The total amount of the three types of sialyl core 1 *O*-glycans in the MSGP concentrate was 21.4% (wt/wt), with the disialyl core 1 *O*-glycan being the major glycan identified using the eliminative oximation method and a calibration curve generated from homogeneous glycopeptides. Considering their potential to become a new functional material, these sialyl *O*-glycans certainly hold promise for maintaining and improving health. Therefore, controlling for the quality and quantity of these sialyl *O*-glycans is crucial when using the MSGP concentrate as a nutraceutical food. The results of the current study offer a new approach and technology for quality control checks of glycopeptides in the MSGP concentrate to ensure the reliability and safety of this potential health product.

The present study tackled the challenging task of quantifying *O*-glycans accurately. To quantify *O*-glycans, chemically releasing them from glycopeptides is necessary. In general, although mild reaction conditions are often employed to suppress side reactions, this frequently promotes insufficient glycan release and poor quantification. Recently, three groups have described a novel and versatile method for analyzing *O*-glycans through β-elimination in the presence of a pyrazolone analog [25–27]. These methods combine release under alkaline conditions and *O*-glycan labeling using 1-phenyl-3-methyl 5-pyrazolone, which minimizes the peeling reaction. Alternatively, Kameyama et al. reported a method for releasing *O*-glycans from glycoproteins using hydroxylamine and an organic superbase (DBU) and tagging of released *O*-glycans using 2-AA [23]. Their method involves reacting hydroxylamine with the released glycans to form oximes, which are relatively stable under alkaline conditions. We selected the eliminative oximation method for the release reaction considering its ability to efficiently release *O*-glycans while suppressing decomposition. In the present study, *O*-glycans were analyzed in the oximated form without any further labeling reaction through LC–MS to avoid loss of quantification due to glycan degradation associated with the labeling reaction [24]. Furthermore, accurate quantification of *O*-glycans requires *O*-glycopeptide standards, which are not commercially available. We addressed this issue by preparing three types of homogeneous *O*-glycopeptide standards with different *O*-glycans from the MSGP concentrate through purification and enzymatic conversion. Ultimately, the combination of the eliminative oximation method and the preparation of *O*-glycopeptide standards allowed for the successful quantification of *O*-glycans. Our results showed that sialyl core 1 *O*-glycans accounted for 21.4% (wt/wt) of the MSGP concentrate. The Neu5Ac content bound to sialyl core 1 *O*-glycans is estimated to be 13.4%, a calculation derived from the molecular weights of the sialyl core 1 *O*-glycans (glycans **3** and **4**: 674.61, glycan **5**: 965.87) and of Neu5Ac (309.27), indicating that around 82% of the total Neu5Ac content (16.4%) contained in the MSGP concentrate is present as a sialyl core 1 *O*-glycan. This result seems reasonable considering that some Neu5Ac molecules in the MSGP concentrate exist as sialyl *N*-glycans and sialyl core 2 *O*-glycans. To the best of our knowledge, this is the first report of the absolute quantification of *O*-glycans. Further studies are definitely required given the constant evolution of methods for quantifying *O*-glycan.

We detected 141 glycopeptides in the MSGP concentrate, among which 131 were *O*-glycopeptides derived from GMP. This result is understandable considering that the WPI used as a raw material contains approximately 15% GMP, the most abundant glycoprotein in WPI. Of these *O*-glycopeptides, 84 *O*-glycopeptides were commonly detected in both the WPI-derived MSGP concentrate in this study and the G-WPC-derived MSGP concentrate in our previous study [21]. The seven types of *O*-glycans were completely identical in both MSGP concentrates. These results demonstrated that an equivalent MSGP concentrate could be prepared from different ingredients. This finding has significant implications for the versatility of the MSGP concentrate preparation process. Interestingly, Kurogochi et al. reported that several types of *O*-glycopeptides derived from GMP can be prepared from trypsin-digested WPC or WPI from cheese whey [5]. *O*-glycopeptides are an attractive material for numerous purposes, such as identifying glycosylation sites using MS and exploring the substrate specificity of endoglycosidases and endoproteases. GMP is a good material for preparing *O*-glycopeptides.

Previous studies have reported that GMP has seven glycosylation sites (^{121}T, ^{131}T, ^{133}T, ^{136}T, ^{141}S, ^{142}T, and ^{165}T) [15,16]. In line with this, the current study detected the glycopeptides in which these seven sites were glycosylated, as well as the glycopeptides in which threonine/serine residues other than these seven sites (^{149}S, ^{155}S/^{161}T, and ^{167}T) may be glycosylated. Unfortunately, given the low abundance of glycopeptides that could be new glycosylation sites, we were unable to analyze the binding sites through

ETD. To confirm whether new binding sites exist, further enrichment of the glycopeptides containing them is necessary. Although several studies have analyzed glycans in GMP over the years, several unknown variables still remain.

The function of the WPI-derived MSGP concentrate has yet to be verified. Koh J. et al. reported that the intact GMP is digested in the human jejunum to produce glycopeptides [28]. The glycopeptides derived from GMP they detected in jejunal fluids are structurally similar to the glycopeptides contained in the MSGP concentrate. If the GMP fragments digested in the jejunum have biological activities [29], then the glycopeptides contained in the MSGP concentrate should also have activities similar to those of the GMP fragments. Further studies are needed to elucidate the function of the MSGP concentrate.

While the physical properties of GMP have been the subject of investigation, as well as its effects on the physical properties of other milk proteins, specific experiments addressing these aspects of the MSGP concentrate have yet to be conducted. A comprehensive understanding of the MSGP concentrate is essential to elucidate its potential utilization in functional foods. This understanding should encompass the stability, solubility, and emulsifying properties of the MSGP concentrate itself, as well as the interaction with other components in the food matrix. To fully grasp the possible applications of the MSGP concentrates, these investigations are included in our future research plan.

Aside from WPI from cheese whey, edible bird's nest, breast milk, and eggs have been found to be rich in sialic acid. However, the methods presented in the current study cannot be applied to all of these foods. Considering that the structures and amount of sialyl *O*-glycans vary depending on the food, analytical methods suitable for each food may need to be considered.

In conclusion, we performed qualitative and quantitative analyses of sialyl *O*-glycans in the MSGP concentrate prepared through proteolytic digestion and ultrafiltration of WPI. LC–FLD analysis showed that the sialyl *O*-glycans in the MSGP concentrate mainly consisted of three types of sialyl core 1 *O*-glycans. These three *O*-glycans accounted for a total of 21.4% (wt/wt) of the MSGP concentrate, among which the disialyl core 1 *O*-glycan accounted for over 90%. These results revealed that MSGP concentrate is a food material with high sialyl core 1 *O*-glycan content. This study provides new insights into assessment methods for the quality control of *O*-glycan in food products.

Supplementary Materials: The following supporting information can be downloaded at: https://www.mdpi.com/article/10.3390/foods13172792/s1, Figure S1. Size-exclusion chromatograms of milk-derived sialylglycopeptide (MSGP) concentrate and MSGP concentrate treated with enzymes; Figure S2. HPLC chromatogram of the preparation of homogeneous *O*-glycopeptide with sialyl core 1 *O*-glycans; Table S1. Glycopeptides detected from the MSGP concentrate.

Author Contributions: Conceptualization, J.H., T.Y., H.F. and F.S.; methodology, J.H., T.Y., M.K., M.M. and F.S.; validation, J.H., M.K., T.Y., N.F., S.M., Y.I., H.F. and N.N.; investigation, J.H., T.Y., M.K., N.F., S.M., Y.I., H.F. and N.N.; writing—original draft preparation, J.H., M.K., T.Y. and F.S.; writing—review and editing, M.K., T.Y., M.M. and F.S.; visualization, J.H., T.Y., and N.F.; supervision, M.G., M.M. and F.S.; project administration, M.M. and F.S. All authors have read and agreed to the published version of the manuscript.

Funding: This research received no external funding.

Institutional Review Board Statement: Not applicable.

Informed Consent Statement: Not applicable.

Data Availability Statement: The original contributions presented in the study are included in the article/Supplementary Material, further inquiries can be directed to the corresponding author.

Conflicts of Interest: Higuchi, J., Yamaguchi, T., Fujio, N., Mitsuduka, S., Ishida, Y., Fukudome, H., Nonoyama, N., Gota, M., and Sakai, F. were employees of Megmilk Snow Brand Co., Ltd. Megmilk Snow Brand manufactures and sells dairy foods in Japan and some Asian and Oceanian countries. Kurogochi, M. and Mizuno, M. received support from a cooperative research fund provided by

Megmilk Snow Brand Co., Ltd. The company was not involved in the study design, collection, analysis, interpretation of data, the writing of this article or the decision to submit it for publication.

References

1. Maity, S.; Bhat, A.H.; Giri, K.; Ambatipudi, K. BoMiProt: A Database of Bovine Milk Proteins. *J. Proteom.* **2020**, *215*, 103648. [CrossRef] [PubMed]
2. Dayon, L.; Macron, C.; Lahrichi, S.; Núñez Galindo, A.; Affolter, M. Proteomics of Human Milk: Definition of a Discovery Workflow for Clinical Research Studies. *J. Proteome Res.* **2021**, *20*, 2283–2290. [CrossRef]
3. Wilson, N.L.; Robinson, L.J.; Donnet, A.; Bovetto, L.; Packer, N.H.; Karlsson, N.G. Glycoproteomics of Milk: Differences in Sugar Epitopes on Human and Bovine Milk Fat Globule Membranes. *J. Proteome Res.* **2008**, *7*, 3687–3696. [CrossRef]
4. Valk-Weeber, R.L.; Deelman-Driessen, C.; Dijkhuizen, L.; Eshuis-De Ruiter, T.; Van Leeuwen, S.S. In Depth Analysis of the Contribution of Specific Glycoproteins to the Overall Bovine Whey N-Linked Glycoprofile. *J. Agric. Food Chem.* **2020**, *68*, 6544–6553. [CrossRef]
5. Kurogochi, M.; Matsuda, A.; Mizuno, M. Preparation of O-Glycopeptides from Commercial Bovine Whey Proteins Using Offline Liquid Chromatography–Mass Spectrometry. *Carbohydr. Res.* **2020**, *491*, 107981. [CrossRef] [PubMed]
6. O'Riordan, N.; Kane, M.; Joshi, L.; Hickey, R.M. Structural and Functional Characteristics of Bovine Milk Protein Glycosylation. *Glycobiology* **2014**, *24*, 220–236. [CrossRef]
7. Wang, W.L.; Wang, W.; Du, Y.M.; Wu, H.; Yu, X.B.; Ye, K.P.; Li, C.B.; Jung, Y.S.; Qian, Y.J.; Voglmeir, J.; et al. Comparison of Anti-Pathogenic Activities of the Human and Bovine Milk N-Glycome: Fucosylation Is a Key Factor. *Food Chem.* **2017**, *235*, 167–174. [CrossRef] [PubMed]
8. Yue, H.; Han, Y.; Yin, B.; Cheng, C.; Liu, L. Comparison of the Antipathogenic Effect toward Staphylococcus Aureus of N-Linked and Free Oligosaccharides Derived from Human, Bovine, and Goat Milk. *J. Food Sci.* **2020**, *85*, 2329–2339. [CrossRef] [PubMed]
9. Wang, W.L.; Du, Y.M.; Wang, W.; Conway, L.P.; Cai, Z.P.; Voglmeir, J.; Liu, L. Comparison of the Bifidogenic Activity of Human and Bovine Milk N-Glycome. *J. Funct. Foods* **2017**, *33*, 40–51. [CrossRef]
10. Karav, S.; Le Parc, A.; Leite Nobrega de Moura Bell, J.M.; Frese, S.A.; Kirmiz, N.; Block, D.E.; Barile, D.; Mills, D.A. Oligosaccharides Released from Milk Glycoproteins Are Selective Growth Substrates for Infant-Associated Bifidobacteria. *Appl. Environ. Microbiol.* **2016**, *82*, 3622–3630. [CrossRef]
11. Wang, Y.; Ze, X.; Rui, B.; Li, X.; Zeng, N.; Yuan, J.; Li, W.; Yan, J.; Li, M. Studies and Application of Sialylated Milk Components on Regulating Neonatal Gut Microbiota and Health. *Front. Nutr.* **2021**, *8*, 766606. [CrossRef] [PubMed]
12. Li, D.; Lin, Q.; Luo, F.; Wang, H. Insights into the Structure, Metabolism, Biological Functions and Molecular Mechanisms of Sialic Acid: A Review. *Foods* **2024**, *13*, 145. [CrossRef] [PubMed]
13. Saito, T.; Itoh, T. Variations and Distributions of O-Glycosidically Linked Sugar Chains in Bovine κ-Casein. *J. Dairy Sci.* **1992**, *75*, 1768–1774. [CrossRef] [PubMed]
14. Abd El-Salam, M.H.; El-Shibiny, S.; Buchheimh, W. Characteristics and Potential Uses of the Casein Macropeptide. *Int. Dairy J.* **1996**, *6*, 327–341. [CrossRef]
15. Guerrero, A.; Lerno, L.; Barile, D.; Lebrilla, C.B. Top-down Analysis of Highly Post-Translationally Modified Peptides by Fourier Transform Ion Cyclotron Resonance Mass Spectrometry. *J. Am. Soc. Mass Spectrom.* **2015**, *26*, 453–459. [CrossRef]
16. Hua, S.; Nwosu, C.C.; Strum, J.S.; Seipert, R.R.; An, H.J.; Zivkovic, A.M.; German, J.B.; Lebrilla, C.B. Site-Specific Protein Glycosylation Analysis with Glycan Isomer Differentiation. *Anal. Bioanal. Chem.* **2012**, *403*, 1291–1302. [CrossRef] [PubMed]
17. Qu, Y.; Kim, B.J.; Koh, J.; Dallas, D.C. Analysis of Bovine Kappa-Casein Glycomacropeptide by Liquid Chromatography–Tandem Mass Spectrometry. *Foods* **2021**, *10*, 2028. [CrossRef] [PubMed]
18. Kawasaki, Y.; Isoda, H.; Shinmoto, H.; Tanimoto, M.; Dosako, S.; Idota, T.; Nakajima, I. Inhibition by κ-Casein Glycomacropeptide and Lactoferrin of Influenza Virus Hemagglutination. *Biosci. Biotechnol. Biochem.* **1993**, *57*, 1214–1215. [CrossRef]
19. Idota, T.; Kawakami, H.; Nakajima, I. Growth-Promoting Effects of N-Acetylneuraminic Acid-Containing Substances on Bifidobacteria. *Biosci. Biotechnol. Biochem.* **1994**, *58*, 1720–1722. [CrossRef]
20. Wang, B.; Yu, B.; Karim, M.; Hu, H.; Sun, Y.; Mcgreevy, P.; Petocz, P.; Held, S.; Brand-Miller, J. Dietary Sialic Acid Supplementation Improves Learning and Memory in Piglets. *Am. J. Clin. Nutr.* **2007**, *85*, 561–569. [CrossRef] [PubMed]
21. Fukudome, H.; Yamaguchi, T.; Higuchi, J.; Ogawa, A.; Taguchi, Y.; Li, J.; Kabuki, T.; Ito, K.; Sakai, F. Large-Scale Preparation and Glycan Characterization of Sialylglycopeptide from Bovine Milk Glycomacropeptide and Its Bifidogenic Properties. *J. Dairy Sci.* **2021**, *104*, 1433–1444. [CrossRef] [PubMed]
22. Wang, Y.; Zhao, H.; Tao, J.; Li, M.; Liu, G.; Dong, W. A New Method for Purifying N-Glycans Released from Milk Glycoprotein. *J. Proteom.* **2021**, *245*, 104283. [CrossRef] [PubMed]
23. Kameyama, A.; Thet Tin, W.W.; Toyoda, M.; Sakaguchi, M. A Practical Method of Liberating O-Linked Glycans from Glycoproteins Using Hydroxylamine and an Organic Superbase. *Biochem. Biophys. Res. Commun.* **2019**, *513*, 186–192. [CrossRef] [PubMed]
24. Yamaguchi, T.; Fukudome, H.; Higuchi, J.; Takahashi, T.; Tsujimori, Y.; Ueno, H.M.; Toba, Y.; Sakai, F. Label-Free Liquid Chromatography–Mass Spectrometry Quantitation of Relative N- and O-Glycan Concentrations in Human Milk in Japan. *Int. J. Mol. Sci.* **2024**, *25*, 1772. [CrossRef] [PubMed]

25. Furukawa, J.I.; Fujitani, N.; Araki, K.; Takegawa, Y.; Kodama, K.; Shinohara, Y. A Versatile Method for Analysis of Serine/Threonine Posttranslational Modifications by β-Elimination in the Presence of Pyrazolone Analogues. *Anal. Chem.* **2011**, *83*, 9060–9067. [CrossRef]
26. Zauner, G.; Koeleman, C.A.M.; Deelder, A.M.; Wuhrer, M. Mass Spectrometric O-Glycan Analysis after Combined O-Glycan Release by Beta-Elimination and 1-Phenyl-3-Methyl-5-Pyrazolone Labeling. *Biochim. Biophys. Acta Gen. Subj.* **2012**, *1820*, 1420–1428. [CrossRef] [PubMed]
27. Wang, C.; Yuan, J.; Wang, Z.; Huang, L. Separation of One-Pot Procedure Released O-Glycans as 1-Phenyl-3-Methyl-5-Pyrazolone Derivatives by Hydrophilic Interaction and Reversed-Phase Liquid Chromatography Followed by Identification Using Electrospray Mass Spectrometry and Tandem Mass Spectrometry. *J. Chromatogr. A* **2013**, *1274*, 107–117. [CrossRef]
28. Koh, J.; Kim, B.J.; Qu, Y.; Huang, H.; Dallas, D.C. Top-Down Glycopeptidomics Reveals Intact Glycomacropeptide Is Digested to a Wide Array of Peptides in Human Jejunum. *J. Nutr.* **2022**, *152*, 429–438. [CrossRef] [PubMed]
29. Ebrahimi, A.; Andishmand, H.; Huo, C.; Amjadi, S.; Khezri, S.; Hamishehkar, H.; Mahmoudzadeh, M.; Kim, K.H. Glycomacropeptide: A Comprehensive Understanding of Its Major Biological Characteristics and Purification Methodologies. *Compr. Rev. Food Sci. Food Saf.* **2024**, 23. [CrossRef] [PubMed]

foods

MDPI

Article

Characterization of Antioxidant Bioactive Compounds and Rheological, Color and Sensory Properties in 3D-Printed Fruit Snacks

Anica Bebek Markovinović [1], Tomislav Bosiljkov [1], Tibor Janči [1], Marko Kostić [2], Nebojša Dedović [2], Ela Lučić [1], Katarina Bavrka [1], Branimir Pavlić [3] and Danijela Bursać Kovačević [1,*]

[1] Faculty of Food Technology and Biotechnology, University of Zagreb, Pierottijeva 6, 10000 Zagreb, Croatia
[2] Faculty of Agriculture, University of Novi Sad, Trg Dositeja Obradovića 8, 21102 Novi Sad, Serbia
[3] Faculty of Technology, University of Novi Sad, Blvd. Cara Lazara 1, 21000 Novi Sad, Serbia
* Correspondence: danijela.bursac.kovacevic@pbf.unizg.hr

Abstract: The influence of wheat starch (6%, 8% and 10%, *w*/*w*) and a 3D printing program (program 1 vs. program 2) on the content of bioactive compounds, antioxidant capacity, color parameters and rheological and sensory properties was investigated in 3D strawberry and strawberry tree fruit snacks. Increasing the starch content led to a decrease in the content of almost all the bioactive compounds, while it had no effect on the antioxidant capacity. The printing program had no significant effect on the bioactive compounds (except hydroxycinnamic acids), antioxidant capacity and color parameters. A higher starch content improved the strength of the sample but had no effect on the mechanical properties. Smaller particles with a higher starch content improved the stability of the sample. In contrast to the programs, varying the starch content had a significant effect on all the color parameters except the a* values. Eight different sweeteners in two different concentrations were used for the sensory evaluation of the 3D-printed snacks. The variations in sweetener content only affected the sweet and harmonious taste. In summary, this study confirms the great potential of fruit bases for the production of 3D-printed snacks with excellent biological and rheological properties, which can be a step toward personalized food with the addition of sweeteners.

Keywords: strawberry; strawberry tree fruit; *Arbutus unedo* L.; 3D printing; quality

1. Introduction

Due to growing consumer awareness and increased interest in nutritionally valuable foods, novel plant materials that could improve the nutritional, functional or sensory properties of foods are in high demand [1]. Of particular importance are plants that have not yet been sufficiently researched but show great potential for processing, such as the Mediterranean plant strawberry tree fruit (*Arbutus unedo* L.), which has been shown to have strong biological effects thanks to its antioxidant bioactive compounds [2]. It has already been shown that phytochemicals from *A. unedo* have the ability to slow down the oxidative process by inhibiting the harmful effects of free radicals, thus protecting the body against the development of numerous chronic diseases, such as cardiovascular and neurodegenerative diseases, diabetes and tumor diseases [3]. In addition, these fruits are characterized by an impressive amount of crude fiber, containing between 7.04 and 22.20 g of total dietary fiber per 100 g of fresh ripe strawberry tree fruit [4–7]. From a technological point of view, they could be highlighted as an interesting raw material for the production of functional food.

In response to solving global crises with a focus on sustainability in food production and processing, the fourth industrial revolution or Industry 4.0 was initiated by a combination of information and communication solutions that is now visible in numerous sectors where, thanks to digitalization, it is significantly changing the way new products are designed or manufactured [8]. The guidelines for Industry 4.0 particularly emphasize the use of non-thermal processing technologies together with additive technologies, with three-dimensional printing (3DP) leading the way in food technology [9]. Three-dimensional printing is a relatively fast process of additive, layer-by-layer production in which computer models enable the production of 3D products in various shapes. This technology is already being explored for functional food design, so its application to various raw materials such as broccoli and carrots [10]; a blend of calcium caseinate powder, starch and medium-chain triglyceride powder [11]; a betaine-enriched oat-based blend [12]; orange by-products [13,14]; a gluten-free cereal blend [15,16]; and a pumpkin blend [17] has been investigated. However, fruit matrices for the 3DP are particularly challenging as it is difficult to produce a functional product that meets the nutritional, biological, textural and sensory requirements [18].

In a previous work, the influence of different amounts (10%, 15% and 20%) and types of starch (wheat vs. corn) and the 3DP programs on the stability of bioactive compounds, antioxidant capacity and textural properties of the strawberry-based 3D-printed product was investigated [19]. A similar study was also conducted with 3D-printed *A. unedo* products to optimize the 3DP technology for this fruit material [2]. Both studies showed a significant influence of the amount and type of starch as well as the 3DP processing parameters on the stability of the bioactive compounds, antioxidant capacity and textural properties. However, due to its chemical and textural properties, *A. unedo* proved to be an excellent raw material that can serve as an excellent basis for the development of various formulations of 3D functional products—in contrast to strawberry, which poses a major challenge due to its high water content.

Based on all the above, the aim of this work was to investigate the possibility of combining two fruit bases, *A. unedo* and strawberry, in the production of 3D-printed functional snacks. The idea is to improve the nutritional, biological, rheological and sensory properties of innovative functional 3D snack products by combining these two fruits. To this end, the influence of different amounts of wheat starch (6%, 8% and 10%) and the type of 3D printing program on the content of bioactive compounds (total phenolic content, total hydroxycinnamic acids, total flavonols and condensed tannins), pigments (monomeric anthocyanins, total carotenoids, chlorophyll A and chlorophyll B), antioxidant capacity (DPPH and FRAP), color and rheological properties of the 3D-printed snacks was investigated. The 3D-printed snack sample characterized by its best bioactive potential and rheological properties was selected for the continuation of the sensory acceptability study, where the addition of eight different sweeteners in two different concentrations was tested.

2. Materials and Methods

2.1. Fruit Material

The samples for the 3D snacks were produced from the fruits of strawberries (*Fragaria ananassa* × Duch., cv. 'Albion') and strawberry trees (*Arbutus unedo* L.). The strawberries were supplied by the company Jagodar-HB d.o.o. (Donja Lomnica, Zagreb County, Croatia). The fruits of *A. unedo* were collected in the southern part of the island of Lošinj (Primorsko-Goranska County, Croatia). After delivery to the laboratory, the fruits were washed, cleaned, dried and stored in plastic bags at −18 °C until the experiments. Wheat starch (Denes Natura Kft., Pécs, Hungary) was used as a hydrocolloid carrier to prepare the fruit mixture for 3D printing. All the sweeteners used for the sensory evaluation were purchased from the local market.

2.2. Preparation of Fruit Material for 3D Printing

Each of the fruits was thawed, homogenized with a Cordys SB-1 blender (MS Industrial Ltd., Hong Kong, China) and then mixed in a 1:1 ratio (*w*/*w*). The wheat starch (6%, 8% and 10% *w*/*w*) was added to the prepared fruit mixture to achieve an appropriate viscous texture. The mixture was then heated to 65 °C with constant stirring on an LLG-uniSTIRRER

7 magnetic stirrer (Lab Logistics Group GmbH, Meckenheim, Germany) to obtain a mixture with suitable viscosity for 3D printing.

2.3. Three-Dimensional Printing of Fruit Snacks

A Foodini 3D printer (Natural Machines, Barcelona, Spain) with a nozzle diameter of 4 mm was used for the 3DP of the functional snacks. A 3D heart shape (Figure 1A) with three layers (Figure 1B) was designed using the Foodini Creator computer program. The 3DP was performed with two different programs (P1 and P2), differing in the printing speed (8000 mm min^{-1} vs. 14,000 mm min^{-1}), printing line thickness (3.5 mm vs. 3.4 mm), mixture flow rate (1.4 vs. 1.65 (dimensionless)) and nozzle height of the first layer (6 mm vs. 4.5 mm). The dimensions of the 3D-printed objects were 53 mm (length) × 51 mm (width) × 12 mm (height).

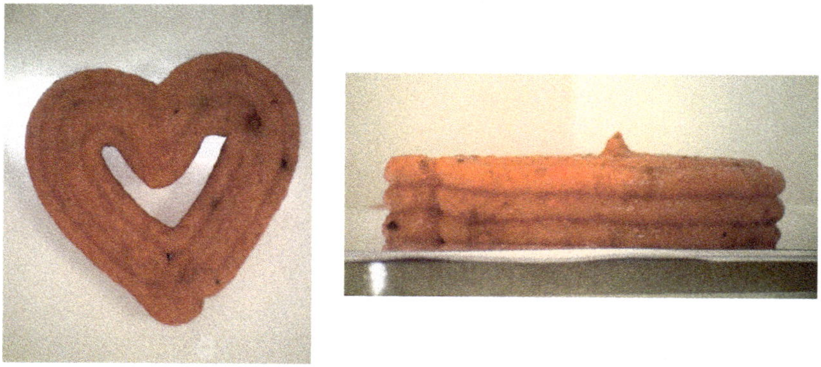

(A) (B)

Figure 1. Designed heart shape for 3D printing: top view (**A**) and side view (**B**).

The design of the 3DP experiments is presented in Table 1.

Table 1. Experimental plan for 3DP of fruit snacks.

Sample ID	Starch Content (%)	3D Program
1	0	P1
2	0	P2
3	6	P1
4	6	P2
5	8	P1
6	8	P2
7	10	P1
8	10	P2

2.4. Extraction of Antioxidant Bioactive Compounds

Ultrasound-Assisted Extraction (UAE) was performed to isolate the bioactive compounds from all the analyzed samples using a processor (UP400St Hielscher Ultrasound Technology, Teltow, Germany) equipped with a DN22 titanium sonotrode (546 mm^2) [20]. In brief, 1% formic acid in 80% methanol (v/v) was used as the extraction solvent. A total of 10 g of the sample was placed in an Erlenmeyer flask and 40 mL of the extraction solvent was poured over it. The UAE was performed at 50% amplitude and 100% pulse for 5 min. The extract was then filtered into a 50 mL flask and made up with the extraction solvent.

The extracts obtained were used in the procedures for the spectrophotometric determination of the total phenols, total flavonoids, hydroxycinnamic acids, flavonols, monomeric anthocyanins and condensed tannins as well as for the analysis of the antioxidant activities using the DPPH and FRAP methods. All the measurements were carried out in duplicate.

2.5. Determination of Polyphenolic Compounds

All the 3D-printed snacks were spectrophotometrically analyzed for the polyphenol content, pigments and antioxidant capacity using a UV–vis spectrophotometer (LLG-uniSPEC 2 spectrophotometer, Buch and Holm, Meckenheim, Germany).

2.5.1. Determination of Total Phenolic Content (TPC)

A total of 400 μL of the extract, 400 μL of the Folin Ciocalteu reagent and 4 mL of a 7.5% sodium carbonate solution were pipetted successively into the test tubes. The reaction mixture was allowed to stand at room temperature for 20 min. The absorbance was measured at 725 nm. The calibration curve was prepared with the solutions of different concentrations of gallic acid (10–250 mg L^{-1}). The equation of the calibration curve used to determine the total phenolic content was:

$$y = 0.0078x - 0.0032,$$

where

y—the absorbance of the sample at 725 nm;

x—the concentration of gallic acid (mg L^{-1}). The TPC results were expressed as mg gallic acid equivalents (GAEs) per 100 g sample [21].

2.5.2. Determination of Total Flavonoids (TF)

Briefly, 0.5 mL of the extract, 1.5 mL of 96% ethanol, 0.1 mL of 10% aluminum chloride, 0.1 mL of 1 M potassium acetate and 2.8 mL of distilled water were added into a glass tube. A blank sample was prepared in the same way, but an extraction solvent was used instead of the extract and the same volume of distilled water (0.1 mL) was added instead of 10% aluminum chloride. The reaction mixture was then allowed to stand for 30 min and the absorbance was measured at 415 nm. Quercetin standard solutions (10–200 mg L^{-1}) were used to generate a calibration curve and the results were expressed as mg quercetin equivalents (QEs) per 100 g of sample [22].

2.5.3. Determination of Total Hydroxycinnamic Acids (HCAs) and Total Flavonols (FLs)

A total of 250 μL of the extract, 250 μL of 1 g L^{-1} HCl in 96% ethanol and 4.55 mL of 2 g L^{-1} HCl were pipetted into a glass tube. The absorbance was then measured at 320 nm for the HCA and at 360 nm for the FL. The HCA content was calculated from the calibration curve obtained from the solutions of different concentrations of chlorogenic acid (10–600 mg L^{-1}), and the results were expressed as the mg chlorogenic acid equivalent (CAE) per 100 g sample. The FL content was calculated from the calibration curve obtained from the solutions of different concentrations of quercetin (10–600 mg L^{-1}), and the results were expressed as the mg quercetin equivalent (QE) per 100 g of sample [23].

2.5.4. Determination of Condensed Tannins (CTs)

In total, 2.5 mL of 1% vanillin, 2.5 mL of 25% H_2SO_4 solution and 1 mL of the extract were pipetted into a glass tube. The mixture was mixed and allowed to stand at room temperature for 10 min. Then, the absorbance was measured at 500 nm. The calibration curve was prepared from the catechin solutions of different concentrations (10–120 mg L^{-1}) and the results were expressed as the mg catechin equivalent (CE) per 100 g of sample [24].

2.6. Determination of Pigments

2.6.1. Determination of Total Monomeric Anthocyanins (ANTs)

The anthocyanin content was determined using the pH differential method [25]. In total, 1 mL of the extract was mixed with 4 mL of potassium chloride buffer pH 1.0 (0.025 M) and also 1 mL of the extract was mixed with 4 mL of sodium acetate buffer pH 4.5 (0.4 M). After 20 min, the absorbance of the reaction mixtures was measured at 520 nm and 700 nm. The ANT content was expressed as the mg cyanidin-3-glucoside equivalent (Cy-3-Glc) per 100 g of sample.

2.6.2. Determination of Total Carotenoids (CARs), Chlorophyll A (CHL A) and Chlorophyll B (CHL B)

The determination of the CAR, CHL A and CHL B was performed using the previously established method [26]. In total, 5 g of the sample was placed in an Erlenmayer flask and 25 mL of the extraction solvent (80% acetone, v/v) was added. The prepared mixture was then extracted in an ultrasonic bath (DT 514 H SONOREX DIGITEC, 13.5 L, 860 W, 40 kHz, Bandelin electronic, Berlin, Germany) at 50 °C for 30 min. After the extraction, the samples were filtered into 25 mL volumetric flasks and filled up to the mark with 80% acetone. The absorbance was measured at 470 nm, 646.8 nm and 663.2 nm. The concentrations of the CAR, CHL A and CHL B were calculated according to the formula from the literature [26] and expressed in mg 100 g^{-1} of sample.

2.7. *In Vitro Antioxidant Capacity (AOC)*

2.7.1. DPPH (2,2-Diphenyl-1-picrylhydrazyl) Scavenging Activity Assay

The antiradical activities of the bioactive antioxidants were determined using the DPPH method [27]. In brief, 1.5 mL of the extract and 3 mL of 0.5 mM DPPH solution were pipetted into a test tube and kept at room temperature in the dark for 20 min. The absorbance was then measured at 517 nm. A calibration curve was constructed from different concentrations of Trolox solutions (10–150 µM) and the results were expressed as the µmol Trolox equivalent (TE) per 100 g of sample.

2.7.2. FRAP (Ferric-Reducing Antioxidant Power) Assay

The FRAP method was performed according to the literature protocol [28]. A total of 600 µL of extract and 4500 µL of FRAP reagent (prepared from acetate buffer (0.3 M), 2.5 mL of TPTZ reagent (2,4,6-tris-2-pyridyl-s-triazine; 10 mM) and 2.5 mL of iron (III) chloride (20 mM) in a 10:1:1 ratio) were pipetted into glass tubes, mixed and thermostatted at 37 °C for 10 min. The absorbance was then measured at 593 nm. A calibration curve was constructed from different concentrations of Trolox solutions (10–150 µM) and the results were expressed as the mmol Trolox equivalent (TE) per 100 g of sample.

2.8. *Determination of Rheological Properties of 3DP Snacks*

2.8.1. Texture Analysis

An evaluation of the rheological properties was conducted using the TA.HD plus Texture Analyser (Stable Micro System, Godalming, UK), applying two tests: the forward extrusion test and penetration test.

Forward Extrusion Test

The testing was conducted using an extrusion set (cylindrical sample container and a piston disc). The base disk is set at the bottom of the sample container with a central opening of 3 mm in diameter. The parameters are as follows: test speed, 1 mms^{-1}; outgoing speed, 10 mms^{-1}; and extrusion distance, 20 mm with trigger force 10 g.

The testing measures the compression force required for the piston to extrude the 3D-printed sample through the opening of the disk. Each sample batch was tested three times, and all the tests were performed at room temperature. The results are expressed as the mean extrusion force (F) (firmness) and work (W) required for extruding the samples.

Penetration Test

A spherical probe with a diameter of 4 mm was utilized during the test. The parameters were set to the following operating speeds: test speed, 0.5 mms^{-1}; outgoing speed, 10 mms^{-1}; and deformation distance, 6 mm with trigger force 2 g. Three parallel measurements were conducted at room temperature, and the results are presented as the maximum force (F_p) (hardness) and work (W_p).

2.8.2. Dimension Measurements

The dimensions of the samples are expressed in terms of the geometric parameters (length × width × height). The differences in the measured values were determined using a digital caliper with an accuracy of 0.01 mm. The minimal differences in the sample dimensions using different programs without starch content are program 1: 58 mm × 58 mm × 9.4 mm and program 2: 62 mm × 59 mm × 9 mm (length × width × height).

2.8.3. Particle Size Distribution

The particle size distribution of the samples in the Hydro 2000S system was determined using the laser diffraction method (Malvern Masterseizer 2000, Malvern Instruments Ltd., Worcestershire, UK). Within the cylinder, 10 g of dissolved sample was dispersed in 30 mL of distilled water, ensuring the homogeneity of the solution for the measurement and achieving a minimal degree of obscuration. The particle diameters were expressed over D (3.2); D (4.3); d (0.1); d (0.5); and d (0.9).

2.9. Determination of Instrumental Color

Color measurements were carried out for each trial using a Konica Minolta Spectrophotometer (CM-700d, Konica Minolta, Tokyo, Japan), which featured a D65 10° standard observer light source and a target mask CM-A183 with an 8 mm aperture, and a glass-covered cone. In each trial, the 3D-printed sample was compressed under the target mask of the spectrophotometer and the colorimetric parameters (L*, a* and b*) were measured. The color change (ΔE), chroma (C) and hue (H*) were calculated using the provided formulas:

$$\Delta E^*_{ab} = \sqrt{\Delta L^{*2} + \Delta a^{*2} + \Delta b^{*2}} \quad (1)$$

$$C^* = \sqrt{a^{*2} + b^{*2}} \quad (2)$$

$$H^* = \tan^{-1}\left(\frac{b^*}{a^*}\right) \quad (3)$$

where all ΔL^{*2}, Δa^{*2} and Δb^{*2} were calculated on the differences between the control and 3D-printed samples. All the measurements were conducted in triplicate.

2.10. Sensory Evaluation of 3D-Printed Snacks

All the 3D-printed snacks were sensory-evaluated using the Quantitative Descriptive Analysis (QDA) method [29]. Based on the results obtained in the determination of the stability of the bioactive antioxidants, pigments and rheological properties by 3DP technology, a sample characterized by the best results was selected and a sensory evaluation was carried out with it. For this purpose, the addition of 8 different sweeteners was tested in 2 concentration levels (Table 2), which were determined in a preliminary sensory evaluation.

A team of 18 sensory panelists rated the sensory attributes using a line intensity scale, with the scores assigned on a scale of 0–7 to indicate the relative intensity of each attribute, with 0 indicating the complete absence of the sensory attribute and 7 indicating a very pronounced attribute. The samples were served in coded Petri dishes. A total of 12 sensory descriptors were evaluated, which included the following attributes: (i) color—intensity of orange color; (ii) odor—strawberry odor, off-odor; (iii) aroma—strawberry flavor, strawberry tree fruit flavor, off-flavor; (iv) taste—sweet taste, sour taste, harmony taste, off-taste; and (v) texture—homogeneity, glossy appearance.

Table 2. Experimental design for the sensory evaluation of 3D-printed snacks.

Simpe ID	Sweeteners	Sweetener Content (%)
A	Control sample	Without sweetener
B1	Saccharose	6.1
B2	Saccharose	9.1
C1	Fructose	7.1
C2	Fructose	8.9
D1	Birch sugar (xylitol)	5.6
D2	Birch sugar (xylitol)	8.5
E1	Erythritol	3.2
E2	Erythritol	4.7
F1	Maple syrup	5.5
F2	Maple syrup	8.7
G1	Date syrup	5.2
G2	Date syrup	7.1
H1	Agave syrup	6.7
H2	Agave syrup	10.2
I1	Stevia and erythritol	2.5
I2	Stevia and erythritol	3.9

2.11. Statistical Analysis

A multivariate analysis of variance (MANOVA) with Tukey's HSD was performed to simultaneously test the relationships between the dependent and categorical variables. The significance for all the tests was $p \leq 0.05$. All the results were analyzed using Statistica software (v. 14.1) [30]. The results of the rheological properties obtained were analyzed using Statistica 12 software. The statistical significance of the influence of the process parameters on the parameters of the descriptive statistics was determined by conducting a MANOVA. The results were considered statistically significant if $p \leq 0.05$ (95% significance level).

3. Results and Discussion

3.1. Characterization of Polyphenolic Compounds, Pigments and Antioxidant Capacity in 3D-Printed Snacks

Table 3 shows the results for the influence of four different proportions of wheat starch, namely, 0, 6, 8 and 10%, on the content of the polyphenolic compounds in the 3D samples. Considering all the phenolic compounds determined, condensed tannins were the most abundant (150.56 ± 3.31 mg 100 g^{-1}), followed by hydroxycinnamic acids (74.79 ± 2.03 mg 100 g^{-1}), flavonols (49.84 ± 1.31 mg 100 g^{-1}) and total flavonoids (9.86 ± 0.16 mg 100 g^{-1}). These results are consistent with the results of previous studies on the bioactive composition of strawberries [31].

When considering the influence of starch content, it can be seen that the content of total phenolic compounds, hydroxycinnamic acids and flavonols decreases with increasing starch content (0–10%). These results are consistent with previous reports [2], in which an increase in starch content from 4 to 8% led to a decrease in the content of total phenolic compounds. The results obtained indicate a negative effect of increasing starch content on the content of bioactive compounds, which is rather expected due to the higher content of non-phenolic compounds, i.e., starch. However, no such correlation was observed for the total flavonoids and condensed tannins. In the case of condensed tannins, it was found that there was no difference in their concentration when different proportions of starch were added. An unusual trend was observed for the content of total flavonoids, where the highest concentrations were recorded in samples without added starch (0%) and in samples with 8% starch. However, when the starch content was further increased to 10%, the TF content had the lowest value. The reason for this could be interactions within the matrix of the mixture, as well as interference in the spectrophotometric assay for the total flavonoids caused by non-flavonoid compounds. Namely, starch can react with other components of the mixture, such as some bioactive compounds, which can lead to an increase in their availability. In addition, the presence of starch can affect the solubility of bioactive compounds, which also affects their increase in concentration [32].

Table 3. Relationship between 3DP parameters and the content of bioactive compounds in the 3D-printed snacks.

Variable	n	TPC	HCAs	FLs	TFs	CTs
Starch level		$p \leq 0.01^{\dagger}$	$p \leq 0.01^{\dagger}$	$p \leq 0.01^{\dagger}$	$p \leq 0.01^{\dagger}$	$p \leq 0.01^{\dagger}$
0%	4	429.50 ± 5.87 ^{a}	84.99 ± 0.73 ^{a}	55.41 ± 0.54 ^{a}	10.66 ± 0.22 ^{a}	171.33 ± 1.66 ^{a}
6%	4	392.96 ± 5.87 ^{b}	78.16 ± 0.73 ^{b}	52.55 ± 0.54 ^{b}	9.65 ± 0.22 ^{b}	141.87 ± 1.66 ^{b}
8%	4	353.60 ± 5.87 ^{c}	71.11 ± 0.73 ^{c}	49.25 ± 0.54 ^{c}	9.90 ± 0.22 ^{a,b}	141.41 ± 1.66 ^{b}
10%	4	344.21 ± 5.87 ^{c}	64.89 ± 0.73 ^{d}	42.16 ± 0.54 ^{d}	9.22 ± 0.22 ^{b}	147.59 ± 1.66 ^{b}
3DP Program		$p = 0.43^{\ddagger}$	$p \leq 0.01^{\dagger}$	$p = 0.05^{\ddagger}$	$p = 0.17^{\ddagger}$	$p = 0.34^{\ddagger}$
Program 1	8	382.48 ± 4.15 ^{a}	76.45 ± 0.52 ^{a}	50.47 ± 0.38 ^{a}	9.69 ± 0.15 ^{a}	149.71 ± 1.17 ^{a}
Program 2	8	377.65 ± 4.15 ^{a}	73.12 ± 0.52 ^{b}	49.22 ± 0.38 ^{a}	10.02 ± 0.15 ^{a}	151.39 ± 1.17 ^{a}
Dataset average	16	380.07 ± 9.09	74.79 ± 2.03	49.84 ± 1.31	9.86 ± 0.16	150.56 ± 3.31

Results are expressed as mean ± standard error. Values represented with different letters are statistically different at $p \leq 0.05$; † significant factor in multifactor analysis; and ‡ not significant factor in multifactor analysis. TPC—total phenolic content (mg GAE 100 g^{-1}); HCAs—hydroxycinnamic acids (mg CAE 100 g^{-1}); FLs—flavonols (mg QE 100 g^{-1}); TFs—total flavonoids (mg QE 100 g^{-1}); and CTs—condensed tannins (mg CE 100 g^{-1}).

The 3D printing of the samples was carried out with two different programs (P1 and P2). These programs differed in terms of the 3DP speed, the print line thickness, the flow rate of the mixture and the nozzle height of the first layer. The only significant influence of the 3DP program was observed for the HCA content as a lower amount of HCA was found in the samples printed with program 2 compared to program 1. These two programs may have a different influence on the bioactive compound content as the 3DP process parameters, such as the nozzle movement speed, pressure and flow rate, differ between these two programs [33]. Program 2 could lead to better preservation of the bioactive compounds, as it has a higher flow rate of the mixture and a higher printing speed compared to program 1. As a result, the processing time is shorter, which means that the bioactive compounds are exposed for less time to conditions that can degrade them. The influence of printing process parameters, i.e., printing programs, on the structural quality of 3D-printed products has already been investigated, but there is not yet enough data on their influence on the stability of bioactive compounds in a 3D-printed product.

The influence of the addition of starch and the 3D programs was also observed on the influence of the stability of the pigments and the antioxidant capacity in the 3D samples (Table 4). The highest concentrations of anthocyanins and chlorophylls were determined in the samples without added starch, while the starch content of 6% had the most favorable influence on the stability of the carotenoids. Furthermore, when the starch content was increased by 6–10%, no significant differences were found in the content of the anthocyanins, chlorophylls A and chlorophylls B in the 3DP samples. A different trend was observed for the carotenoids. Increasing the starch content had a negative effect on the stability of the carotenoids. Increasing the starch content from 0 to 10% had no significant effect on the DPPH values; while higher FRAP values were obtained for the 3D samples with 0–8% added starch, the samples with 10% had the lowest values.

Table 4. Relationship between 3DP parameters and the content of pigments and antioxidant capacity in the 3D-printed snacks.

Variable	n	ANT	CAR	CHLA	CHLB	DPPH	FRAP
Starch level		$p \leq 0.01$ †	$p \leq 0.01$ †	$p \leq 0.01$ †	$p \leq 0.01$ †	$p = 0.07$ ‡	$p = 0.03$ †
0%	4	9.65 ± 0.16 [a]	0.58 ± 0.002 [b]	0.24 ± 0.01 [a]	0.40 ± 0.01 [a]	2.90 ± 0.07 [a]	290.38 ± 0.35 [a,b]
6%	4	7.75 ± 0.16 [b]	0.62 ± 0.002 [a]	0.11 ± 0.01 [b]	0.18 ± 0.01 [b]	3.00 ± 0.07 [a]	289.95 ± 0.35 [a,b]
8%	4	7.73 ± 0.16 [b]	0.52 ± 0.002 [c]	0.11 ± 0.01 [b]	0.17 ± 0.01 [b]	2.74 ± 0.07 [a]	290.82 ± 0.35 [a]
10%	4	8.09 ± 0.16 [b]	0.48 ± 0.002 [d]	0.12 ± 0.01 [b]	0.19 ± 0.01 [b]	2.73 ± 0.07 [a]	288.94 ± 0.35 [b]
3DP Program		$p = 0.64$ ‡	$p = 0.11$ ‡	$p = 0.12$ ‡	$p = 0.19$ ‡	$p = 0.15$ ‡	$p = 0.73$ ‡
Program 1	8	8.35 ± 0.11 [a]	0.55 ± 0.001 [a]	0.15 ± 0.004 [a]	0.25 ± 0.01 [a]	2.90 ± 0.05 [a]	289.96 ± 0.25 [a]
Program 2	8	8.27 ± 0.11 [a]	0.55 ± 0.001 [a]	0.14 ± 0.004 [a]	0.23 ± 0.01 [a]	2.78 ± 0.05 [a]	290.08 ± 0.25 [a]
Dataset average	16	8.31 ± 0.22	0.55 ± 0.01	0.14 ± 0.01	0.24 ± 0.03	2.84 ± 0.05	290.02 ± 0.26

Results are expressed as mean ± standard error. Values represented with different letters are statistically different at $p \leq 0.05$; † significant factor in multifactor analysis; and ‡ not significant factor in multifactor analysis. ANT—monomeric anthocyanin (mg Cy-3-Glc 100 g^{-1}); CAR—total carotenoid (mg 100 g^{-1}); CHL A—total chlorophyll A (mg 100 g^{-1}); CHL B—total chlorophyll B (mg 100 g^{-1}); DPPH assay (μmol TE 100 g^{-1}); and FRAP assay (mmol TE 100 g^{-1}).

Bebek Markovinović et al. [19] also observed a decrease in the content of carotenoids and antioxidant activity values determined by the FRAP method when the starch content in the 3DP strawberry-based products was increased to over 15%. The reason for the initial deviations from this trend in the carotenoid and FRAP values may be that this study used a strawberry and strawberry tree fruit-based blend, which has different rheological properties than a strawberry blend. Ultimately, the overall result of a decrease in the bioactive compound content was to be expected, as a printing mixture with higher starch content has a lower proportion of the fruit component, which is the source of the bioactive compounds. There is also the possibility of interactions between bioactive compounds and starch in 3DP products with higher starch content, which may result in lower levels of the bioactive compound. It is hypothesized that phenolic hydroxyl groups may interact with starch by forming non-covalent bonds, such as hydrogen bonds, and that electrostatic and ionic interactions may occur, leading to the formation of complex compounds. However, the addition of starch, which triggers such chemical reactions, can lead to an improvement in the nutritional and physico-chemical properties of the product and the digestibility of the starch [34].

The 3DP programs had no effect on the stability of the pigments and on the antioxidant capacity. The reason for this could be that the process parameters of these two programs are not sufficiently different to cause statistically significant changes in the abovementioned parameters. It can also be assumed that some other bioactive compounds, which were not considered in this study, contribute to the antioxidant capacity, and the differences in the 3DP process parameters by the two programs had no effect on them.

Table 5 shows the correlations between the investigated bioactive compounds and the antioxidant capacity (DPPH and FRAP). The TPC correlates positively with the content of all the bioactive compounds, most strongly with the HCA and least with the TF. They also correlate with the DPPH. The other bioactive compounds examined also show a positive correlation with similar chemical structures. When looking at the antioxidant capacity, the DPPH correlates significantly with the TPC, HCA, FL and CAR, while there are no significant correlations with the other bioactive compounds. On the other hand, the FRAP does not correlate significantly with any of the bioactive compounds or DPPH. This suggests that the FRAP values are related to some other bioactive compounds that were not studied. Andrés et al. [35] showed that the FRAP capacity in a smoothie of orange juice, papaya juice, melon juice, carrot puree and skim milk correlated significantly with ascorbic acid and polyphenolic compounds but not with carotenoids.

Table 5. Mutual correlations of bioactive compounds and antioxidant capacity.

	TPC	HCA	FL	TF	ANT	CT	CAR	CHL A	CHL B	DPPH	FRAP
TPC		0.92 *	0.84 *	0.56 *	0.71 *	0.69 *	0.75 *	0.77 *	0.77 *	0.51 *	0.32
HCA	0.92 *		0.91 *	0.66 *	0.63 *	0.61 *	0.78 *	0.69 *	0.69 *	0.50 *	0.41
FL	0.84 *	0.91 *		0.63 *	0.44	0.42	0.86 *	0.54 *	0.55 *	0.55 *	0.44
TF	0.56 *	0.66 *	0.63 *		0.63 *	0.64 *	0.43	0.64 *	0.65 *	0.13	0.28
ANT	0.71 *	0.63 *	0.44	0.63 *		0.90 *	0.22	0.93 *	0.92 *	0.25	0.09
CT	0.69 *	0.61 *	0.42	0.64 *	0.90 *		0.19	0.81 *	0.80 *	0.07	0.05
CAR	0.75 *	0.78 *	0.86 *	0.43	0.22	0.19		0.33	0.35	0.68 *	0.25
CHL A	0.77 *	0.69 *	0.54 *	0.64 *	0.93 *	0.81 *	0.33		1.00 *	0.26	0.20
CHL B	0.77 *	0.69 *	0.55 *	0.65 *	0.92 *	0.80 *	0.35	1.00 *		0.27	0.20
DPPH	0.51 *	0.50 *	0.55 *	0.13	0.25	0.07	0.68 *	0.26	0.27		−0.03
FRAP	0.32	0.41	0.44	0.28	0.09	0.05	0.25	0.20	0.20	−0.03	

* Correlations are significant at $p < 0.05$. TPC—total phenolic content (mg GAE 100 g^{-1}); HCA—hydroxycinnamic acid (mg CAE 100 g^{-1}); FL—flavonol (mg QE 100 g^{-1}); TF—total flavonoid (mg QE 100 g^{-1}); ANT—monomeric anthocyanin (mg Cy-3-Glc 100 g^{-1}); CT—condensed tannin (mg CE 100 g^{-1}); CAR—total carotenoid (mg 100 g^{-1}); CHL A—total chlorophyll a (mg 100 g^{-1}); CHL B—total chlorophyll b (mg 100 g^{-1}); DPPH assay (μmol TE 100 g^{-1}); and FRAP assay (mmol TE 100 g^{-1}).

3.2. Characterization of Rheological Properties in 3D-Printed Snacks

3.2.1. Texture Analysis (Forward Extrusion and Penetration Test)

The results presented in Table 6 show that by applying program 2 and increasing the starch content in the printing mixture from 6% to 8%, the extrusion force does not change significantly. However, the total flow resistance through the cylinder increases considerably during the 3DP of the mixture with a starch content of 10%. The substantially higher flow resistance observed during the extrusion of the samples with the maximum starch content results from increased sample firmness. This is consistent with the findings of the studies conducted by Feng et al. [36], Dong et al. [37] and Yang et al. [38,39], who found that the starch content in the printing mixture is proportional to the sample firmness. By applying program 1, which has a lower printing speed, the extrusion force decreases as the starch content in the mixture increases. A higher starch content results in a more consistent sample structure, increasing the flow index.

Table 6. Influence of 3DP processing parameters on textural properties of 3D-printed snacks.

Sample	3DP Program	Starch Content (%)	F (N)	W (Nmm)	F_p (N)	W_p (Nmm)
1	1	0	16.89	168.76	0.01	0.01
2	2	0	7.58	75.79	0.01	0.01
3	1	6	454.73	4545.01	0.07	0.06
4	2	6	54.08	540.43	0.05	0.05
5	1	8	112.82	1127.64	0.05	0.06
6	2	8	66.77	667.37	0.05	0.06
7	1	10	75.69	756.59	0.07	0.07
8	2	10	1003.08	10,025.82	0.06	0.04

F—mean extrusion force (firmness) (N); W—work (Nmm); F_p—maximum force (hardness); and W_p—work (Nmm).

Consequently, the sample's fluidity increases and the total resistance during extrusion through the cylinder container decreases. One possible explanation for these results, whose statistical significance is shown in Table 7, is the significant influence of the apparent viscosity value indicating the pseudoplastic character of the extruded samples. Samples 3D printed with program 1 were expected to exhibit increased resistance with increasing starch content, but the predicted effect did not materialize due to changes in the rheological parameters. Accordingly, no significant deviations from the values of the extrusion force change were observed in the extrusion work values as the printing speed and starch content in the samples changed. The aforementioned are in accordance with the findings of the research by Bebek Markovinović et al. [2], which demonstrated that the force and work required for the extrusion of the mixture significantly depend on the starch content—specifically, increasing the starch content led to increased values of work and extrusion force. The extrusion work of samples, like the extrusion force, can be considered a relevant factor in determining the textural properties of samples, with the note that, in this case, the path (the distance travelled by the piston through the cylinder) is the main factor defining the total work or the area under the curve and, consequently, the maximum extrusion force, defining its initial and final firmness.

Table 7. Influence of 3DP process parameters on texture, particle diameters and dimension in 3D-printed snacks expressed by *p*-value *.

Parameter	3DP Program	Starch Content (%)
F	0.725588	0.695556
W	0.725577	0.695550
F_p	0.178047	0.158750
W_p	0.236516	0.876531
D [3.2]	0.949818	0.003742 *
D [4.3]	0.660607	0.024213 *
d [0.1]	0.730548	0.029678 *
d [0.5]	0.616439	0.004638 *
d [0.9]	0.603835	0.037802 *
Length	0.147469	0.754530
Width	0.659869	0.795063
Height	0.122649	0.334880

* Results are statistically significant at $p \leq 0.05$.

According to the results obtained from the forward extrusion test, the penetration test method did not affect significant deviations in the characterization of the mechanical properties. Because the test is performed with smaller probe contact areas where the sensitivity of force (F_p) (hardness) sensing is more pronounced and stresses are significantly lower, differences in the sample strength are even less pronounced, considering the starch content.

3.2.2. Dimension Measurements of 3D-Printed Snacks

An increase in the starch content and the application of different printing programs did not result in statistically significant deviations in the dimensions of the printed samples. Table 8 shows the influence of the 3DP program on the dimensions of the 3DP fruit snack. The samples exhibit stability in all three observed geometry directions.

Table 8. Influence of 3DP processing parameters on the dimension of 3D-printed snacks.

Sample	3DP Program	Starch Content (%)	Length (mm)	Width (mm)	Height (mm)
1	1	0	58.88 ± 0.11	58.69 ± 0.22	9.40± 0.09
2	2	0	62.01± 0.17	59.94 ± 0.19	9.39± 0.12
3	1	6	54.85 ± 0.21	53.33 ± 0.53	12.55 ± 0.14
4	2	6	55.70 ± 0.15	53.57 ± 0.31	12.12 ± 0.15
5	1	8	55.06 ± 0.07	51.57 ± 0.23	13.47 ± 0.52
6	2	8	55.60 ± 0.32	53.99 ± 0.44	12.83 ± 0.32
7	1	10	53.73 ± 0.56	54.18 ± 0.09	13.50 ± 0.17
8	2	10	55.98 ± 0.37	53.09 ± 0.21	11.98 ± 0.26

The results are presented as an average value of triplicate measurements ± standard deviation.

3.2.3. Particle Size Distribution

An increase in the starch content has shown an influence on all the parameters of the particle size distribution. The translation of the medians toward the interval of smaller particle distribution is in correlation with a reduction in the diameter in the 90% of the total particle count (D 0.9). In accordance, the statistical significance can be attributed to the overall change in the values describing the surface area ratio of smaller diameter particles (D 3.2) compared to the volume ratio of larger diameter particles (D 4.3). From previous research by Bebek Markovinović et al. [2], it is evident that smaller diameter particles contribute to more excellent stability and durability of printed samples. Table 9 shows the influence of the printing process parameters on the particle diameters. The significance analysis confirmed that the starch content was the most important factor influencing the particle size distribution of the 3D-printed snacks.

Table 9. Influence of 3DP processing parameters on particle diameters in 3D-printed snacks.

Sample	3DP Program	Starch Content (%)	D [3.2] (µm)	D [4.3] (µm)	d [0.1] (µm)	d [0.5] (µm)	d [0.9] (µm)
1	1	0	120.24	429.11	62.18	297.58	1012.46
2	2	0	117.87	418.75	61.57	290.21	985.83
3	1	6	54.63	234.50	24.05	92.45	682.84
4	2	6	54.02	216.56	24.23	88.52	622.87
5	1	8	40.07	167.06	20.05	59.94	501.01
6	2	8	39.41	174.85	19.28	59.94	530.68
7	1	10	35.30	145.02	17.82	53.20	437.02
8	2	10	36.72	143.65	19.13	54.41	419.94

D [3.2]—surface weighted mean diameter (Sauter mean diameter); D [4.3]—volume weighted mean diameter (De Brouckere mean diameter); d [0.1]—10% of the volume distribution is below the observed diameter; d [0.5]—median diameter, 50% of the volume distribution is below and 50% is above the observed diameter; and d [0.9]—90% of the volume distribution is below the observed diameter.

Comparing the control sample with the samples containing 6% and 10% starch, an increase in the volume fraction (relative frequencies) of the particles in the distribution interval of 17.38–120.22 µm is evident (Figure 2). The addition of starch contributed to the uniformity of the particle distribution curves compared to the control sample, which is characterized by a bimodal distribution. This indicates a greater stability of the samples with added starch in all ratios.

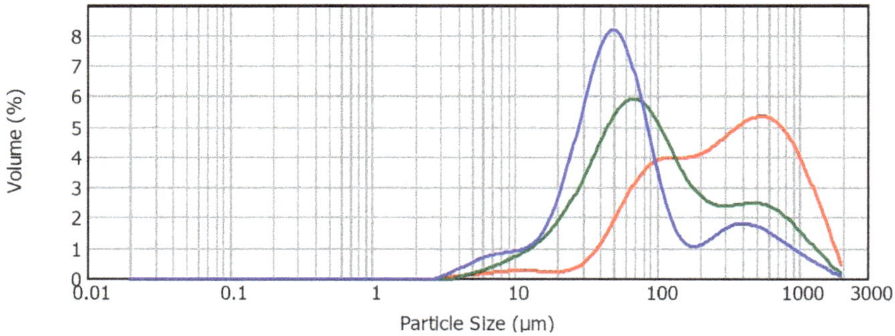

Figure 2. Volume size distribution of control sample (red), sample 4 with 6% (green) and sample 8 with 10% of starch (blue).

3.3. Characterization of Color Properties in 3D-Printed Snacks

Color plays an important role in the consumer's perception of a particular product [40]. Table 10 shows the CIELab color parameters for the 3D-printed samples as a function of the starch content (0, 6, 8 and 10%) and printing program (P1 and P2).

Table 10. Relationship between 3DP parameters and the color parameters in the 3D-printed snacks.

Variable	n	L*	a*	b*	C*	H*	ΔE
Starch level		$p \leq 0.01$ †	$p = 0.05$ ‡	$p \leq 0.01$ †	$p \leq 0.01$ †	$p \leq 0.01$ †	$p \leq 0.01$ †
0%	4	41.43 ± 0.19 [c]	22.92 ± 0.39 [a]	14.44 ± 0.30 [b]	27.09 ± 0.46 [b]	32.23 ± 0.37 [b]	0.00 ± 0.41 [c]
6%	4	46.74 ± 0.19 [a]	24.48 ± 0.39 [a]	16.98 ± 0.30 [a]	29.80 ± 0.46 [a]	34.74 ± 0.37 [a]	6.17 ± 0.41 [a]
8%	4	44.79 ± 0.19 [b]	23.45 ± 0.39 [a]	15.70 ± 0.30 [a,b]	28.22 ± 0.46 [a,b]	33.80 ± 0.37 [a,b]	3.89 ± 0.41 [b]
10%	4	44.19 ± 0.19 [b]	22.77 ± 0.39 [a]	14.82 ± 0.30 [b]	27.17 ± 0.46 [b]	33.06 ± 0.37 [a,b]	3.00 ± 0.41 [b]
3DP Program		$p = 0.29$ ‡	$p = 0.71$ ‡	$p = 0.35$ ‡	$p = 0.93$ ‡	$p = 0.11$ ‡	$p = 0.15$ ‡
Program 1	8	44.18 ± 0.13 [a]	23.48 ± 0.27 [a]	15.34 ± 0.21 [a]	28.05 ± 0.32 [a]	33.13 ± 0.26 [a]	3.59 ± 0.29 [a]
Program 2	8	44.39 ± 0.13 [a]	23.33 ± 0.27 [a]	15.64 ± 0.21 [a]	28.09 ± 0.32 [a]	33.79 ± 0.26 [a]	2.94 ± 0.29 [a]
Dataset average	16	44.29 ± 0.50	23.40 ± 0.25	15.49 ± 0.29	28.07 ± 0.35	33.46 ± 0.31	3.26 ± 0.60

Results are expressed as mean ± standard error. Values represented with different letters are statistically different at $p \leq 0.05$; † significant factor in multifactor analysis; and ‡ not significant factor in multifactor analysis. L*—lightness; a*—redness; b*—yellowness; C*—chroma; H*—hue; and ΔE—color change.

The starch content significantly influenced the lightness value (L*) of the 3D-printed samples. The samples without added starch had the lowest L* value, i.e., they were the brightest. The highest L* value was recorded for the samples with a starch content of 6%, whereupon a further increase in the starch content to 8% led to a decrease. When the starch content was further increased to 10%, no significant difference was observed in the L* parameter. Considering that there was a significant difference in the L* values between the samples without and with the addition of starch, this darkening of the samples with the addition of starch could be due to the short-term exposure to an elevated temperature by heating the fruit mixture to achieve the appropriate consistency for 3DP. During this brief heating, it is possible that a slight non-enzymatic degradation of the mixture by Maillard reactions occurred due to the elevated temperature [41].

The change in starch content (0–10%) in the 3D-printed samples had no significant effect on the color parameter a*. Because the parameter a* refers to the red color, the results obtained indicate that the addition of starch has no significant effect on the expression of the red color, which is a highly desirable sensory characteristic. In contrast, a significant influence of the starch content on the color parameter b* was found. The samples without added starch had the lowest b* value, which was close to the value of the samples with

10% added starch, i.e., the samples without starch and with 10% added starch did not differ significantly. The highest b* value was found in the samples with a starch content of 6%. Increasing the starch content from 6% to 8% decreased the b* value significantly but not less than the samples without starch and with the addition of 10% starch. In our previous studies on 3D-printed strawberry products, the a* and b* parameters ranged from 22.54 to 27.52 and 7.50 to 12.98, respectively [19], while they ranged from 19.76 to 32.78 and 18.24 to 27.90 for 3D-printed strawberry products [2]. As expected, our results of the mean a* and b* values of strawberry and strawberry tree fruit 3D-printed snacks were approximately at the mean values of the a* and b* color parameters of the strawberry and strawberry tree fruit 3D-printed snacks examined separately in the two previous studies (23.40 $\pm$ 0.25 and 15.49 $\pm$ 0.29, respectively).

The C* and H* values follow almost the same trend with the change in starch content. The addition of 6% starch led to a statistically significant increase in the C* and H* values, while a further addition of 8% starch caused a significant decrease. A further 10% increase in the starch content resulted in a significant decrease in the C* value, while there was no effect on the H* value. The C* and H values of 3DP strawberry snacks from previous studies ranged from 23.85 to 29.93 and 18.33 to 25.96, respectively [19], while the values of 3DP strawberry tree fruit snacks ranged from 27.09 to 41.89 and 35.82 to 44.63, respectively [2]. Like the a* and b* values, the mean C* and H* values of the 3DP strawberry and strawberry tree fruit snacks (28.07 $\pm$ 0.35 and 33.46 $\pm$ 0.31, respectively) were similar to the mean values of the C* and H* color parameters of the 3DP strawberry and strawberry tree fruit snacks examined separately in the two previous studies.

The color change (ΔE) in the 3D-printed products ranged from 0.00 $\pm$ 0.41 to 6.17 $\pm$ 0.41 depending on the added starch content. The 3D product with a starch content of 6% showed the highest color change, while further increasing the starch content by 8% caused a significant decrease in the ΔE. No significant difference in the ΔE value was observed when the starch content was further increased by 10%. When we compare the L* and ΔE values, we find that they follow the same trend. In both cases, the darkest samples were those with the addition of 6% starch. It is possible that here the color of the fruit pulp was degraded by the short-term heating due to Maillard reactions [41]. Except for the samples with 6% added starch, all the others had a ΔE value of less than 6, which defines these differences as appreciable differences. Only the addition of 6% starch causes ΔE values above 6, which defines them as large differences [42]. The average ΔE value was 3.26 $\pm$ 0.60, and in general, the color differences in the 3D-printed strawberry and strawberry tree fruit products were acceptable.

The 3DP programs had no statistically significant effect on any color parameter (L*, a*, b*, C*, H and ΔE). Because the choice of program had no statistically significant influence on the content of the bioactive compounds (with the exception of HCA), including the content of anthocyanins, it is to be expected that no significant changes in the color parameters were observed under the influence of the different program parameters.

3.4. Characterization of the Sensory Properties of 3D-Printed Snacks

Sample 3, which contained 6% starch and was prepared with program 1, was selected for the sensory evaluation as it proved to be the best in terms of the stability of the bioactive compounds and antioxidant capacity. The 3D-printed snacks prepared with eight different sweeteners in two different concentrations (16 samples) and the control samples (without sweeteners) were evaluated using twelve sensory descriptors (Table 11). In the first part of the table, all the samples are compared with regard to the type and amount of added sweetener, while in the second part of the table, the samples are grouped according to the type of added sweetener.

Table 11. Sensory comparison results of 3D-printed snacks with the addition of different sweeteners in two different concentrations.

Variable	n	Intensity of Orange Color	Strawberry Odor	Off-Odor	Strawberry Flavor	Strawberry Tree Fruit Flavor	Off-Flavor	Sweet Taste	Sour Taste	Harmony Taste	Off-Taste	Homogeneity	Glossy Appearance
Sample		$p = 0.81$ ‡	$p = 0.76$ ‡	$p = 0.99$ ‡	$p = 0.07$ ‡	$p = 0.99$ ‡	$p = 0.41$ ‡	$p \leq 0.01$ †	$p = 0.14$ ‡	$p \leq 0.01$ †	$p = 0.67$ ‡	$p = 0.71$ ‡	$p = 0.09$ ‡
A	13	5.85 ± 0.32 a	5.38 ± 0.37 a	1.23 ± 0.19 a	4.46 ± 0.38 a	4.15 ± 0.45 a	1.15 ± 0.18 a	2.46 ± 0.36 d	4.69 ± 0.38 a	3.69 ± 0.34 a,b	1.15 ± 0.18 a	5.15 ± 0.41 a	5.77 ± 0.35 a
B1	13	5.92 ± 0.32 a	5.08 ± 0.37 a	1.08 ± 0.19 a	4.92 ± 0.38 a	3.54 ± 0.45 a	1.08 ± 0.18 a	4.46 ± 0.36 a,b	3.23 ± 0.38 a	4.85 ± 0.34 a,b	1.08 ± 0.18 a	4.77 ± 0.41 a	5.77 ± 0.35 a
B2	13	6.08 ± 0.32 a	5.00 ± 0.37 a	1.23 ± 0.19 a	4.85 ± 0.38 a	3.46 ± 0.45 a	1.08 ± 0.18 a	5.00 ± 0.36 a,b	2.92 ± 0.38 a	4.85 ± 0.34 a,b	1.08 ± 0.18 a	5.15 ± 0.41 a	6.08 ± 0.35 a
C1	13	6.15 ± 0.32 a	4.85 ± 0.37 a	1.31 ± 0.19 a	4.31 ± 0.38 a	3.46 ± 0.45 a	1.23 ± 0.18 a	3.92 ± 0.36 a,b,c,d	3.31 ± 0.38 a	4.54 ± 0.34 a,b	1.23 ± 0.18 a	5.38 ± 0.41 a	5.92 ± 0.35 a
C2	13	5.92 ± 0.32 a	4.77 ± 0.37 a	1.23 ± 0.19 a	4.31 ± 0.38 a	3.23 ± 0.45 a	1.23 ± 0.18 a	4.77 ± 0.36 a,b	2.92 ± 0.38 a	4.85 ± 0.34 a,b	1.08 ± 0.18 a	5.31 ± 0.41 a	6.15 ± 0.35 a
D1	13	6.08 ± 0.32 a	4.69 ± 0.37 a	1.38 ± 0.19 a	3.92 ± 0.38 a	3.46 ± 0.45 a	1.38 ± 0.18 a	3.38 ± 0.36 b,c,d	3.92 ± 0.38 a	3.54 ± 0.34 b	1.38 ± 0.18 a	5.54 ± 0.41 a	5.38 ± 0.35 a
D2	13	6.23 ± 0.32 a	4.92 ± 0.37 a	1.08 ± 0.19 a	4.15 ± 0.38 a	3.46 ± 0.45 a	1.08 ± 0.18 a	4.00 ± 0.36 a,b,c,d	3.54 ± 0.38 a	4.31 ± 0.34 a,b	1.23 ± 0.18 a	5.54 ± 0.41 a	5.46 ± 0.35 a
E1	13	6.08 ± 0.32 a	4.23 ± 0.37 a	1.31 ± 0.19 a	3.54 ± 0.38 a	3.69 ± 0.45 a	1.23 ± 0.18 a	2.69 ± 0.36 c,d	4.15 ± 0.38 a	3.23 ± 0.34 b	1.46 ± 0.18 a	5.69 ± 0.41 a	4.69 ± 0.35 a
E2	13	6.15 ± 0.32 a	4.15 ± 0.37 a	1.23 ± 0.19 a	3.46 ± 0.38 a	3.54 ± 0.45 a	1.31 ± 0.18 a	3.54 ± 0.36 a,b,c,d	4.08 ± 0.38 a	3.85 ± 0.34 a,b	1.23 ± 0.18 a	5.46 ± 0.41 a	4.62 ± 0.35 a
F1	13	5.54 ± 0.32 a	5.15 ± 0.37 a	1.31 ± 0.19 a	4.62 ± 0.38 a	3.38 ± 0.45 a	1.15 ± 0.18 a	4.15 ± 0.36 a,b,c,d	3.62 ± 0.38 a	4.38 ± 0.34 a,b	1.08 ± 0.18 a	5.00 ± 0.41 a	5.77 ± 0.35 a
F2	13	5.38 ± 0.32 a	4.92 ± 0.37 a	1.15 ± 0.19 a	5.15 ± 0.38 a	3.38 ± 0.45 a	1.15 ± 0.18 a	4.62 ± 0.36 a,b	3.31 ± 0.38 a	4.69 ± 0.34 a,b	1.23 ± 0.18 a	5.38 ± 0.41 a	5.69 ± 0.35 a
G1	13	5.62 ± 0.32 a	4.69 ± 0.37 a	1.31 ± 0.19 a	4.00 ± 0.38 a	3.62 ± 0.45 a	1.54 ± 0.18 a	3.62 ± 0.36 a,b,c,d	3.92 ± 0.38 a	3.92 ± 0.34 a,b	1.62 ± 0.18 a	5.54 ± 0.41 a	5.31 ± 0.35 a
G2	13	5.62 ± 0.32 a	5.00 ± 0.37 a	1.38 ± 0.19 a	4.46 ± 0.38 a	3.38 ± 0.45 a	1.38 ± 0.18 a	4.00 ± 0.36 a,b,c,d	3.54 ± 0.38 a	4.31 ± 0.34 a,b	1.38 ± 0.18 a	5.69 ± 0.41 a	5.54 ± 0.35 a
H1	13	5.54 ± 0.32 a	4.85 ± 0.37 a	1.31 ± 0.19 a	4.62 ± 0.38 a	3.62 ± 0.45 a	1.15 ± 0.18 a	4.38 ± 0.36 a,b,c	3.62 ± 0.38 a	4.38 ± 0.34 a,b	1.38 ± 0.18 a	6.23 ± 0.41 a	5.46 ± 0.35 a
H2	13	5.54 ± 0.32 a	5.08 ± 0.37 a	1.23 ± 0.19 a	5.08 ± 0.38 a	3.69 ± 0.45 a	1.15 ± 0.18 a	5.15 ± 0.36 a	3.38 ± 0.38 a	5.23 ± 0.34 a	1.23 ± 0.18 a	5.85 ± 0.41 a	5.54 ± 0.35 a
I1	13	5.69 ± 0.32 a	5.15 ± 0.37 a	1.31 ± 0.19 a	4.38 ± 0.38 a	3.54 ± 0.45 a	1.62 ± 0.18 a	4.00 ± 0.36 a,b,c,d	3.85 ± 0.38 a	4.15 ± 0.34 a,b	1.46 ± 0.18 a	5.62 ± 0.41 a	5.08 ± 0.35 a
I2	13	5.77 ± 0.32 a	5.15 ± 0.37 a	1.15 ± 0.19 a	4.69 ± 0.38 a	3.54 ± 0.45 a	1.62 ± 0.18 a	4.53 ± 0.36 a,b	3.85 ± 0.38 a	4.38 ± 0.34 a,b	1.38 ± 0.18 a	6.00 ± 0.41 a	5.08 ± 0.35 a
Sample grouped		$p = 0.23$ ‡	$p = 0.20$ ‡	$p = 0.99$ ‡	$p \leq 0.01$ †	$p = 0.94$ ‡	$p = 0.06$ ‡	$p \leq 0.01$ †	$p = 0.01$ †	$p \leq 0.01$ †	$p = 0.27$ ‡	$p = 0.23$ ‡	$p \leq 0.01$ †
A	13	5.85 ± 0.32 a	5.38 ± 0.37 a	1.23 ± 0.19 a	4.46 ± 0.38 a,b	4.15 ± 0.45 a	1.15 ± 0.18 a	2.46 ± 0.36 b	4.69 ± 0.38 a	3.69 ± 0.34 a,b	1.15 ± 0.18 a	5.15 ± 0.41 a	5.77 ± 0.35 a,b
B	26	6.00 ± 0.23 a	5.04 ± 0.26 a	1.15 ± 0.13 a	4.88 ± 0.27 a	3.50 ± 0.32 a	1.08 ± 0.12 a	4.73 ± 0.25 a	3.08 ± 0.27 b	4.85 ± 0.24 a	1.08 ± 0.13 a	4.96 ± 0.29 a	5.92 ± 0.25 a
C	26	6.04 ± 0.23 a	4.81 ± 0.26 a	1.27 ± 0.13 a	4.31 ± 0.27 a,b	3.35 ± 0.32 a	1.23 ± 0.12 a	4.35 ± 0.25 a	3.12 ± 0.27 b	4.69 ± 0.24 a	1.15 ± 0.13 a	5.35 ± 0.29 a	6.03 ± 0.25 a
D	26	6.15 ± 0.23 a	4.81 ± 0.26 a	1.23 ± 0.13 a	4.04 ± 0.27 a,b	3.46 ± 0.32 a	1.23 ± 0.12 a	3.69 ± 0.25 a,b	3.73 ± 0.27 a,b	3.92 ± 0.24 a,b	1.31 ± 0.13 a	5.54 ± 0.29 a	5.42 ± 0.25 a,b
E	26	6.12 ± 0.23 a	4.19 ± 0.26 a	1.27 ± 0.13 a	3.50 ± 0.27 b	3.62 ± 0.32 a	1.27 ± 0.12 a	3.12 ± 0.25 b	4.12 ± 0.27 a,b	3.54 ± 0.24 b	1.35 ± 0.13 a	5.58 ± 0.29 a	4.65 ± 0.25 b
F	26	5.46 ± 0.23 a	5.04 ± 0.26 a	1.23 ± 0.13 a	4.88 ± 0.27 a	3.38 ± 0.32 a	1.15 ± 0.12 a	4.38 ± 0.25 a	3.46 ± 0.27 a,b	4.54 ± 0.24 a,b	1.15 ± 0.13 a	5.19 ± 0.29 a	5.73 ± 0.25 a,b
G	26	5.62 ± 0.23 a	4.85 ± 0.26 a	1.35 ± 0.13 a	4.23 ± 0.27 a,b	3.50 ± 0.32 a	1.46 ± 0.12 a	3.81 ± 0.25 a,b	3.73 ± 0.27 a,b	4.12 ± 0.24 a,b	1.50 ± 0.13 a	5.62 ± 0.29 a	5.42 ± 0.25 a,b
H	26	5.54 ± 0.23 a	4.96 ± 0.26 a	1.27 ± 0.13 a	4.85 ± 0.27 a	3.65 ± 0.32 a	1.15 ± 0.12 a	4.77 ± 0.25 a	3.50 ± 0.27 a,b	4.81 ± 0.24 a	1.31 ± 0.13 a	6.04 ± 0.29 a	5.50 ± 0.25 a,b
I	26	5.73 ± 0.23 a	5.15 ± 0.26 a	1.23 ± 0.13 a	4.54 ± 0.27 a,b	3.54 ± 0.32 a	1.62 ± 0.12 a	4.27 ± 0.25 a	3.85 ± 0.27 a,b	4.27 ± 0.24 a,b	1.42 ± 0.13 a	5.81 ± 0.29 a	5.08 ± 0.25 a,b
Dataset average	221	5.83 ± 0.08	4.89 ± 0.09	1.25 ± 0.04	4.41 ± 0.10	3.54 ± 0.11	1.27 ± 0.04	4.04 ± 0.10	3.64 ± 0.09	4.30 ± 0.09	1.28 ± 0.04	5.49 ± 0.10	5.49 ± 0.09

Results are expressed as mean ± standard error. Values represented with different letters are statistically different at $p \leq 0.05$; † significant factor in multifactor analysis; and ‡ not significant factor in multifactor analysis. A—control sample; 3DP fruit snacks with the addition of: B—saccharose, C—fructose, D—birch sugar (xylitol), E—erythritol, F— maple syrup, G—date syrup, H—agave syrup, I—stevia and erythritol; 1—lower level of sweeteners, 2—higher level of sweeteners.

The type of sweetener added and the level of concentration (lower vs. higher concentration) had no significant influence on the sensory properties with the exception of sweetness and harmonious taste. Therefore, no specific correlation was found between the color, taste and odor characteristics compared to the control samples. Significant differences were only found with regard to sweetness and harmonious taste. The perception of harmonious taste implies the relationship between sweetness and acidity [43] and represents a sensory descriptor that encompasses the general preference for the food product [44]. The samples C2, D2, E2, F2, H2 and I2 with a higher content of sweeteners showed a higher sweetness than their counterparts with a lower addition of sweeteners C1, D1, E1, F1, H1 and I1. As expected, the control sample had the lowest intensity of sweetness and differed significantly from the sweetened samples in almost all the comparisons. In addition, the greatest differences in harmonious taste were observed between samples D1 and D2 with the addition of birch sugar, E1 and E2 with the addition of erythritol and H1 and H2 with the addition of agave syrup. The aforementioned samples with a higher sweetener content had a higher harmonicity than their parallel samples with a lower sweetener content.

In the second part of Table 11, the samples are observed according to the type of added sweetener, without observation at the two concentration levels in which they were added.

Statistically significant differences between the samples were recorded in the sensory descriptors, such as strawberry odor, sweet taste, harmony taste and glossy appearance.

Sample E with the addition of erythritol had the least strawberry odor, while samples B, F and H had the strongest intensity of strawberry odor. The addition of saccharose, maple syrup and agave syrup in the mentioned samples had a certain positive effect on the strawberry odor.

The control sample without added sweetener had the lowest sweetness, while the samples with added saccharose (B), fructose (C), maple syrup (F), agave syrup (H) and stevia and erythritol (I) had the highest sweetness. It is not surprising that after the sample with added saccharose (B), the sample with added stevia and erythritol (I) had the highest sweetness. The sweetness of stevia comes from the steviol glycosides stevioside and rebaudioside A, which are 250 to 400 times sweeter than saccharose and do not respond to temperature and pH changes during processing [45]. The main disadvantage of using stevia in the production of functional foods is undesirable sensory properties such as a bitter and/or metallic taste [46], which was not the case here, most likely because stevia was chosen in combination with erythritol. Furthermore, the addition of sweeteners led not only to an increase in sweetness but also to an increase in the harmony of the taste, so that samples B, C and H showed the most pronounced harmonious characteristics.

The samples with the addition of saccharose and fructose, i.e., samples B and C, had the most expressive characteristic of a glossy appearance, while the other samples, with the exception of sample E, were most similar to control sample A. Ultimately, the results obtained show that functional 3D-printed snack products can be sensory-enhanced by the addition of appropriate sweeteners, which opens a perspective for exploring new formulations of 3D-printed fruit-based functional foods.

4. Conclusions

This study investigated the effects of starch content and 3D printing programs on the stability of polyphenolic compounds, pigments, antioxidant activity, color and rheological properties as well as sensory characteristics of 3D-printed snacks based on strawberries and strawberry tree fruits. The results showed a decrease in the total phenolic compounds (19.86%), hydroxycinnamic acids (23.65%) and flavonols (23.91%) with increasing starch content, while total flavonoids and condensed tannins showed different trends. The 3D printing programs showed no significant influence on most bioactive compounds, with the exception of hydroxycinnamic acids, whose content was 4.35% higher in the samples printed with program 1. This study also highlighted the influence of starch content on the pigments and antioxidant capacity, with different effects observed depending on the starch content. The color parameters were significantly affected by variations in the starch content, while the 3D printing programs had no significant effect.

This study investigated the rheological properties of 3D-printed snacks, focusing on texture, dimensions and particle size distribution. A higher starch content led to a higher extrusion force and flow resistance, enhancing sample firmness. However, the penetration tests showed only minimal effects on the mechanical properties. The dimensional measurements remained stable at different starch contents and printing programs. The particle size distribution shifted toward smaller particles with higher starch content, which improved the rheological stability of the sample.

Although process parameters such as the addition of starch and the variation in 3D programs significantly affect the CIEL*a*b* color parameters, the color change was generally satisfactory with no noticeable and/or appreciable difference. The sensory evaluation of 3D-printed snacks with different sweeteners provided interesting results. The differences in the sweetener content had no significant effect on the color, taste or odor descriptors, with the exception of sweetness and harmony. The addition of sweeteners not only increased the sweetness but also had a positive effect on the harmony of the flavor, which was particularly evident in the samples with sucrose, fructose and agave syrup. These results underline the potential of different sweeteners to improve the sensory properties of functional 3D-printed snacks.

In summary, the results obtained indicate the great potential of using fruit bases in the production of functional 3D-printed snacks, which, thanks to their good nutritional and biological potential and their rheological properties when using natural sweeteners, represent an excellent basis for the further development of functional personalized nutrition.

Author Contributions: Conceptualization, D.B.K.; methodology, D.B.K., T.B., T.J. and M.K.; software, M.K. and N.D.; validation, A.B.M., M.K., N.D. and B.P.; formal analysis, A.B.M., T.B., T.J., E.L. and K.B.; investigation, A.B.M., T.B., T.J., E.L., K.B. and B.P.; resources, D.B.K.; data curation, M.K. and N.D.; writing—original draft preparation, A.B.M., T.B. and D.B.K.; writing—review and editing, T.J., M.K., N.D., E.L., K.B. and B.P.; visualization, A.B.M., M.K., N.D. and B.P.; supervision, D.B.K.; project administration, D.B.K.; funding acquisition, D.B.K. All authors have read and agreed to the published version of the manuscript.

Funding: This research was funded by the Croatian Science Foundation through the funding of the Hurdle Technology and 3D Printing for Sustainable Fruit Juice Processing and Preservation project, IP-2019-04-2105. The work of doctoral student Anica Bebek Markovinović has been fully supported by the "Young Researchers' Career Development Project—Training of Doctoral Students" of the Croatian Science Foundation (DOK-2020–01).

Data Availability Statement: The data used to support the findings of this study can be made available by the corresponding author upon request.

Conflicts of Interest: The authors declare no conflicts of interest.

References

1. Alsubhi, M.; Blake, M.; Nguyen, T.; Majmudar, I.; Moodie, M.; Ananthapavan, J. Consumer willingness to pay for healthier food products: A systematic review. *Obes. Rev.* **2022**, *24*, e13525. [CrossRef]
2. Bebek Markovinović, A.; Brdar, D.; Putnik, P.; Bosiljkov, T.; Durgo, K.; Huđek Turković, A.; Brčić Karačonji, I.; Jurica, K.; Pavlić, B.; Granato, D.; et al. Strawberry tree fruits (*Arbutus unedo* L.): Bioactive composition, cellular antioxidant activity, and 3D printing of functional foods. *Food Chem.* **2024**, *433*, 137287. [CrossRef]
3. Zhang, Y.-J.; Gan, R.-Y.; Li, S.; Zhou, Y.; Li, A.-N.; Xu, D.-P.; Li, H.-B. Antioxidant phytochemicals for the prevention and treatment of chronic diseases. *Molecules* **2015**, *20*, 21138–21156. [CrossRef] [PubMed]
4. Sagbas, H.I.; Ilhan, G.; Zitouni, H.; Anjum, M.A.; Hanine, H.; Necas, T.; Ondrasek, I.; Ercisli, S. Morphological and biochemical characterization of diverse strawberry tree (*Arbutus unedo* L.) genotypes from northern Turkey. *Agronomy* **2020**, *10*, 1581. [CrossRef]
5. Colak, A.M. Morphological and biochemical diversity in fruits of *Arbutus unedo* L. from east aegean region in Turkey. *Erwerbs-Obstbau* **2019**, *61*, 379–383. [CrossRef]
6. Boussalah, N.; Boussalah, D.; Cebadera-Miranda, L.; Fernández-Ruiz, V.; Barros, L.; Ferreira, I.C.F.R.; Cortes Sanchez Mata, M.; Madani, K. Nutrient composition of Algerian strawberry-tree fruits (*Arbutus unedo* L.). *Fruits* **2018**, *73*, 283–297. [CrossRef]
7. Rodríguez-Delgado, M.A.; Malovaná, S.; Pérez, J.P.; Borges, T.; García Montelongo, F.J. Separation of phenolic compounds by high-performance liquid chromatography with absorbance and fluorimetric detection. *J. Chromatogr. A* **2001**, *912*, 249–257. [CrossRef] [PubMed]
8. Djekić, I.; Velebit, B.; Pavlić, B.; Putnik, P.; Šojić Merkulov, D.; Bebek Markovinović, A.; Bursać Kovačević, D. Food quality 4.0: Sustainable food manufacturing for the twenty-first century. *Food Eng. Rev.* **2023**, *15*, 577–608. [CrossRef]
9. Sundarsingh, A.; Zhang, M.; Mujumdar, A.S.; Li, J. Research progress in printing formulation for 3D printing of healthy future foods. *Food Bioprocess Technol.* **2023**. [CrossRef]
10. Ahmadzadeh, S.; Clary, T.; Rosales, A.; Ubeyitogullari, A. Upcycling imperfect broccoli and carrots into healthy snacks using an innovative 3D food printing approach. *Food Sci. Nutr.* **2023**, *12*, 84–93. [CrossRef]
11. Zhu, S.; Vazquez Ramos, P.; Heckert, O.R.; Stieger, M.; van der Goot, A.J.; Schutyser, M. Creating protein-rich snack foods using binder jet 3D printing. *J. Food Eng.* **2022**, *332*, 111124. [CrossRef]
12. Radoš, K.; Pastor, K.; Kojić, J.; Drakula, S.; Dujmić, F.; Novotni, D.; Čukelj Mustač, N. Influence of infill level and post-processing on physical parameters and betaine content of enriched 3D-printed sweet snacks. *Foods* **2023**, *12*, 4417. [CrossRef] [PubMed]
13. Molina-Montero, C.; Vicente-Jurado, D.; Igual, M.; Martínez-Monzó, J.; García-Segovia, P. Fiber enrichment of 3D printed apricot gel snacks with orange by-products. *Gels* **2023**, *9*, 569. [CrossRef] [PubMed]
14. Tan, J.D.; Lee, C.P.; Foo, S.Y.; Tan, J.C.W.; Tan, S.S.Y.; Ong, E.S.; Leo, C.H.; Hashimoto, M. 3D printability and biochemical analysis of revalorized orange peel waste. *Int. J. Bioprint.* **2023**, *9*, 510–521. [CrossRef] [PubMed]
15. Radoš, K.; Benković, M.; Čukelj Mustač, N.; Habuš, M.; Voučko, B.; Pavičić, T.V.; Ćurić, D.; Ježek, D.; Novotni, D. Powder properties, rheology and 3D printing quality of gluten-free blends. *J. Food Eng.* **2023**, *338*, 111251. [CrossRef]

16. Letras, P.; Oliveira, S.; Varela, J.; Nunes, M.C.; Raymundo, A. 3D printed gluten-free cereal snack with incorporation of Spirulina (*Arthrospira platensis*) and/or *Chlorella vulgaris*. *Algal Res.* **2022**, *68*, 102863. [CrossRef]
17. Chen, X.-h.; Zhang, M.; Teng, X.-x.; Mujumdar, A.S. Internal structure design for improved shape fidelity and crispness of 3D printed pumpkin-based snacks after freeze-drying. *Food Res. Int.* **2022**, *157*, 111220. [CrossRef] [PubMed]
18. Tomašević, I.; Putnik, P.; Valjak, F.; Pavlić, B.; Šojić, B.; Bebek Markovinović, A.; Bursać Kovačević, D. 3D printing as novel tool for fruit-based functional food production. *Curr. Opin. Food Sci.* **2021**, *41*, 138–145. [CrossRef]
19. Bebek Markovinović, A.; Putnik, P.; Bosiljkov, T.; Kostelac, D.; Frece, J.; Markov, K.; Žigolić, A.; Kaurinović, J.; Pavlić, B.; Duralija, B.; et al. 3D printing of functional strawberry snacks: Food design, texture, antioxidant bioactive compounds, and microbial stability. *Antioxidants* **2023**, *12*, 436. [CrossRef]
20. Bursać Kovačević, D.; Putnik, P.; Dragović-Uzelac, V.; Pedisić, S.; Režek Jambrak, A.; Herceg, Z. Effects of cold atmospheric gas phase plasma on anthocyanins and color in pomegranate juice. *Food Chem.* **2016**, *190*, 317–323. [CrossRef]
21. Yuan, B.; Danao, M.-G.C.; Stratton, J.E.; Weier, S.A.; Weller, C.L.; Lu, M. High pressure processing (HPP) of aronia berry purée: Effects on physicochemical properties, microbial counts, bioactive compounds, and antioxidant capacities. *Innov. Food Sci. Emerg. Technol.* **2018**, *47*, 249–255. [CrossRef]
22. Chang, C.C.; Yang, M.H.; Wen, H.M.; Chern, J.C. Estimation of total flavonoid content in propolis by two complementary colometric methods. *J. Food Drug Anal.* **2020**, *10*, 3. [CrossRef]
23. Howard, L.R.; Clark, J.R.; Brownmiller, C. Antioxidant capacity and phenolic content in blueberries as affected by genotype and growing season. *J. Sci. Food Agric.* **2003**, *83*, 1238–1247. [CrossRef]
24. Sun, B.; Ricardo-da-Silva, J.M.; Spranger, I. Critical factors of vanillin assay for catechins and proanthocyanidins. *J. Agric. Food Chem.* **1998**, *46*, 4267–4274. [CrossRef]
25. Lee, J.; Durst, R.W.; Wrolstad, R.E. Determination of total monomeric anthocyanin pigment content of fruit juices, beverages, natural colorants, and wines by the ph differential method: Collaborative study. *J. AOAC Int.* **2005**, *88*, 1269–1278. [CrossRef] [PubMed]
26. Lichtenthaler, H.K.; Buschmann, C. Chlorophylls and carotenoids: Measurement and characterization by UV-VIS spectroscopy. *Curr. Protoc. Food Anal. Chem.* **2001**, *1*, F4.3.1–F4.3.8. [CrossRef]
27. Shortle, E.; O'Grady, M.N.; Gilroy, D.; Furey, A.; Quinn, N.; Kerry, J.P. Influence of extraction technique on the anti-oxidative potential of hawthorn (*Crataegus monogyna*) extracts in bovine muscle homogenates. *Meat Sci.* **2014**, *98*, 828–834. [CrossRef] [PubMed]
28. Benzie, I.F.F. An automated, specific, spectrophotometric method for measuring ascorbic acid in plasma (EFTSA). *Clin. Biochem.* **1996**, *29*, 111–116. [CrossRef] [PubMed]
29. Vahčić, N.; Hruškar, M.; Marković, K. Quantitative descriptive analysis method in sensory evaluation of yoghurt. *Mljekarstvo* **2000**, *50*, 279–296.
30. Cloud Software Group, Inc. Data Science Workbench, Version 14.1. Available online: http://tibco.com (accessed on 5 March 2024).
31. Bebek Markovinović, A.; Putnik, P.; Duralija, B.; Krivohlavek, A.; Ivešić, M.; Mandić Andačić, I.; Palac Bešlić, I.; Pavlić, B.; Lorenzo, J.M.; Bursać Kovačević, D. Chemometric Valorization of Strawberry (*Fragaria x ananassa* Duch.) cv. 'Albion' for the Production of Functional Juice: The Impact of Physicochemical, Toxicological, Sensory, and Bioactive Value. *Foods* **2022**, *11*, 640. [CrossRef]
32. Rong, L.; Chen, X.; Shen, M.; Yang, J.; Qi, X.; Li, Y.; Xie, J. The application of 3D printing technology on starch-based product: A review. *Trends Food Sci. Technol.* **2023**, *134*, 149–161. [CrossRef]
33. Zhang, J.; Li, Y.; Cai, Y.; Ahmad, I.; Zhang, A.; Ding, Y.; Qiu, Y.; Zhang, G.; Tang, W.; Lyu, F. Hot extrusion 3D printing technologies based on starchy food: A review. *Carbohydr. Polym.* **2022**, *294*, 119763. [CrossRef] [PubMed]
34. Ngo, T.V.; Kusumawardani, S.; Kunyanee, K.; Luangsakul, N. Polyphenol-modified starches and their applications in the food industry: Recent updates and future directions. *Foods* **2022**, *11*, 3384. [CrossRef] [PubMed]
35. Andrés, V.; Villanueva, M.J.; Tenorio, M.D. The effect of high-pressure processing on colour, bioactive compounds, and antioxidant activity in smoothies during refrigerated storage. *Food Chem.* **2016**, *192*, 328–335. [CrossRef]
36. Feng, C.; Zhang, M.; Bhandari, B. Controlling the three-dimensional printing mechanical properties of *Nostoc Sphaeroides* system. *Food Biophys.* **2019**, *15*, 240–248. [CrossRef]
37. Dong, X.; Huang, Y.; Pan, Y.; Wang, K.; Prakash, S.; Zhu, B. Investigation of sweet potato starch as a structural enhancer for three-dimensional printing of *Scomberomorus niphonius* surimi. *J. Texture Stud.* **2019**, *50*, 316–324. [CrossRef] [PubMed]
38. Yang, F.; Zhang, M.; Bhandari, B.; Liu, Y. Investigation on lemon juice gel as food material for 3D printing and optimization of printing parameters. *LWT* **2018**, *87*, 67–76. [CrossRef]
39. Yang, F.; Zhang, M.; Bhandari, B. Recent development in 3D food printing. *Crit. Rev. Food Sci. Nutr.* **2015**, *57*, 3145–3153. [CrossRef] [PubMed]
40. Bhat, R.; Geppert, J.; Funken, E.; Stamminger, R. Consumers perceptions and preference for strawberries—A case study from Germany. *Int. J. Fruit Sci.* **2015**, *15*, 405–424. [CrossRef]
41. Buvé, C.; Kebede, B.T.; De Batselier, C.; Carrillo, C.; Pham, H.T.T.; Hendrickx, M.; Grauwet, T.; Van Loey, A. Kinetics of colour changes in pasteurised strawberry juice during storage. *J. Food Eng.* **2018**, *216*, 42–51. [CrossRef]
42. Chen, X.D. Food drying fundamentals. In *Drying Technologies in Food Processing*; Chen, X.D., Mujumdar, A.S., Eds.; Blackwell Publishing: Singapore, 2008; pp. 1–55.

43. Krebs, G.; Müller, M.; Becker, T.; Gastl, M. Characterization of the macromolecular and sensory profile of non-alcoholic beers produced with various methods. *Food Res. Int.* **2019**, *116*, 508–517. [CrossRef] [PubMed]
44. Muto, A.; Christofides, S.R.; Sirangelo, T.M.; Bartella, L.; Muller, C.; Di Donna, L.; Muzzalupo, I.; Bruno, L.; Ferrante, A.; Chiappetta, A.A.C.; et al. Fruitomics: The Importance of Combining Sensory and Chemical Analyses in Assessing Cold Storage Responses of Six Peach (Prunus persica L. Batsch) Cultivars. *Foods* **2022**, *11*, 2554. [CrossRef]
45. Bursać Kovačević, D.; Barba, F.J.; Granato, D.; Galanakis, C.M.; Herceg, Z.; Dragović-Uzelac, V.; Putnik, P. Pressurized hot water extraction (PHWE) for the green recovery of bioactive compounds and steviol glycosides from *Stevia rebaudiana* Bertoni leaves. *Food Chem.* **2018**, *254*, 150–157. [CrossRef] [PubMed]
46. Rocha, I.F.d.O.; Bolini, H.M.A. Passion fruit juice with different sweeteners: Sensory profile by descriptive analysis and acceptance. *Food Sci. Nutr.* **2015**, *3*, 129–139. [CrossRef] [PubMed]

foods MDPI

Article

A Characterization of Biological Activities and Bioactive Phenolics from the Non-Volatile Fraction of the Edible and Medicinal Halophyte Sea Fennel (*Crithmum maritimum* L.)

Clément Lemoine [1], Maria João Rodrigues [2], Xavier Dauvergne [1], Stéphane Cérantola [3], Luísa Custódio [2] and Christian Magné [1,*]

[1] Laboratoire Géoarchitecture_Territoires, Urbanisation, Biodiversité, Environnement, Université de Brest, CS 93837, F 29238 Brest, Cedex 3, France; lemoine1clement@gmail.com (C.L.); xavier.dauvergne@univ-brest.fr (X.D.)

[2] Centre of Marine Sciences, Faculty of Sciences and Technology, University of Algarve, Ed. 7, Campus of Gambelas, 8005-139 Faro, Portugal; mjrodrigues@ualg.pt (M.J.R.); lcustodio@ualg.pt (L.C.)

[3] Service Général des Plateformes Technologiques, Plateforme RMN-RPE, Université de Bretagne Occidentale, 6 av. le Gorgeu, CS 93837, F 29238 Brest, Cedex 3, France; stephane.cerantola@univ-brest.fr

* Correspondence: christian.magne@univ-brest.fr; Tel.: +33-298-016427

Abstract: Although the biochemical composition and biological properties of the volatile fraction of the halophyte sea fennel (*Crithmum maritimum* L.) have been largely described, little is known about its polar constituents and bioactivities. Here, a hydromethanolic extract of *Crithmum maritimum* (L.) leaves was fractionated, and the fractions were evaluated in vitro for antioxidant (using DPPH, ABTS, and FRAP bioassays), anti-inflammatory (inhibition of NO production in RAW 264.7 macrophages), antidiabetic (alpha-glucosidase inhibition), neuroprotective (inhibition of acetylcholinesterase), and skin-protective (tyrosinase and melanogenesis inhibitions) activities. Polar fractions of the extract were rich in phenolics and, correlatively, displayed a strong antioxidant power. Moreover, fractions eluted with $MeOH_{20}$ and $MeOH_{80}$ exhibited a marked inhibition of alpha-glucosidase (IC_{50} = 0.02 and 0.04 mg/mL, respectively), $MeOH_{60}$ fractions showed a strong capacity to reduce NO production in macrophages (IC_{50} = 6.4 μg/mL), and $MeOH_{80}$ and $MeOH_{100}$ fractions had strong anti-tyrosinase activities (630 mgKAE/gDW). NMR analyses revealed the predominance of chlorogenic acid in $MeOH_{20}$ fractions, 3,5-dicaffeoylquinic acid in $MeOH_{40}$ fractions, and 3-*O*-rutinoside, 3-*O*-glucoside, 3-*O*-galactoside, and 3-*O*-robinobioside derivatives of quercetin in $MeOH_{60}$ fractions. These compounds likely account for the strong antidiabetic, antioxidant, and anti-inflammatory properties of sea-fennel polar extract, respectively. Overall, our results make sea fennel a valuable source of medicinal or nutraceutical agents to prevent diabetes, inflammation processes, and oxidative damage.

Keywords: *Crithmum maritimum*; halophyte; antioxidants; anti-inflammatory agents; anti-ageing activity; antidiabetic activity; bioguiding; NMR characterization; phenolic compounds

1. Introduction

Plants have served as a valuable source of medicinal compounds for centuries, with plant-derived products being extensively used in various industries such as food, nutraceuticals, and medicine. The current shift towards natural remedies has spurred the exploration of bioactive components sourced from plants. For example, the antioxidant properties of plants have garnered significant interest in recent years, with the supplementation of exogenous antioxidants emerging as a promising strategy to combat the detrimental effects of oxidative stress [1,2].

Halophytes, in this context, offer a sustainable reservoir of bioactive substances that can be harnessed for applications in food, cosmetics, and pharmaceuticals. Despite being relatively understudied until the past decade, recent research has unveiled the rich secondary metabolite content of these plants, which underpins their diverse biological effects [3–5]. Among these compounds, polyphenols stand out for their potent antioxidant capabilities and broad spectrum of medicinal benefits [6–9].

Crithmum maritimum L. (Apiaceae), also known as sea fennel or rock samphire, is a widespread perennial halophyte which grows mostly on coastal rocks, and sometimes on sandy beaches or pebbles, along the shoreline of the Mediterranean Sea and of the Pacific and Atlantic oceans. Alternatively, it can be grown easily in open air under agronomical techniques (Figure 1), producing 1.22 TDW/ha^{-1} (C. Magné, unpublished results). This facultative halophyte is edible and has a history of being utilized in culinary dishes and traditional remedies. Its juicy leaves and fresh shoots are preserved in vinegar for flavoring purposes. Moreover, *C. maritimum* contains high amounts of minerals, vitamins (A, C, E), carotenoids, flavonoids [10,11], and essential oils [12]. These confer to the plant antiscorbutic, carminative, cytotoxic, digestive, depurative, diuretic, tonic, vermifuge, and insecticidal properties [12,13]. Although the components present in the essential oils of *C. maritimum* and their properties have been largely studied [14–17] and have found many applications, little is known about the nonvolatile fraction in this species [18]. In a recent study, mineral and phenolic compositions in sea fennel were reported, but the extraction process used (leaf decoction) was not exhaustive for phenolic compounds [19]. Lastly, the NMR characterization of phenolic compounds from *C. maritimum* leaves, and their antimicrobial and antiproliferative activities, were reported [20]. However, the work carried out on raw extracts did not allow for the characterization of flavonoid compounds. Therefore, a bioguiding study reporting the characterization of sea-fennel phenolics was exhaustively extracted from the leaves, and, relating it to some biological activities of medicinal interest, is still lacking.

Figure 1. Plants of sea fennel on a maritime rocks (**A**) and grown in open field (**B**). (Source: C. Magné).

We have previously identified free quinic acid, a carbohydrate-derived secondary metabolite previously reported in other members of Apiaceae, and chlorogenic acid as major solutes in *C. maritimum*, with possible consequences on plant extract antioxidant activity [21]. Here, a hydromethanolic extract of *C. maritimum* leaves was fractionated to characterize for the first time anti-inflammatory, neuroprotective, and anti-ageing activities, as well as the relevant bioactive compounds.

2. Materials and Methods

2.1. Chemicals, Culture Media and Supplements

Acetylcholinesterase (EC 3.1.1.7), α-amylase (EC 3.2.1.1), α-glucosidase (EC 3.2.1.20), lipase (EC 3.1.1.3), and tyrosinase (EC 1.14.18.1) enzymes, as well as murine RAW 264.7 macrophages and B16 A45 melanoma cell lines, were purchased from Sigma-Aldrich (Darmstadt, Germany). Additionally, all reagents (ATChI, ferric chloride, DTNB, Folin–Ciocalteu phenol, L-glutamine, penicillin, potassium persulfate, streptomycin, and TPTZ), standards (acarbose, galantamine, gallic and kojic acids, L-NAME, melanine, and orlistat), radicals (DPPH and ABTS), and solvents (DMSO and methanol and ethanol of analytical grade) used for chemical analyses were supplied by Sigma-Aldrich (St. Louis, MO, USA).

2.2. Plant Material and Sampling

C. maritimum plants were collected on maritime rocks along the Brittany shore, near Brest (48°21′47″ N 4°31′51″ O, France). The plant material was identified by C. Magné (Faculty of Sciences, Brest). A voucher specimen was deposited at the herbarium of Sciences and Techniques Faculty, University of Brest. The aerial parts were rinsed and the leaves were placed at −20 °C for 5 days before being subjected to freeze-drying. Dried leaves ($a_w < 0.3$) were then pulverized into a fine powder with a Dangoumeau-type grinder and passed through a fine-mesh sieve to obtain a powder with a diameter <1 mm. The dry powder was finally stored at −20 °C without light, before being used for extraction and analyses.

2.3. Extraction and Fractionation

About 200 mg of powder underwent homogenization with 5 mL of water/ethanol (1:2) under magnetic stirring at 4 °C for 20 min. Following the centrifugation of the mixture (15 min at 4 °C, 4000× *g*), the resulting pellet was subjected to extraction twice using the same procedure. The supernatants were combined, filtered through glass wool, and concentrated through rotary evaporation at 40 °C, using 50% ethanol for resuspension.

To fractionate the initial extract, solid–liquid partition chromatography was conducted on C18-bound silica gel (GRACE Davisil RP18). Solutes with decreasing polarity were eluted using methanol in increasing concentrations (successively, 0, 20, 40, 60, 80, and 100%), ultimately followed by ethyl acetate. The collected fractions were then concentrated through rotary evaporation at 40 °C and resuspended in their respective solvents prior to analysis. Then, all analyses were carried out in triplicate for three independent samples.

2.4. Estimation of Total Phenolic Content (TPC)

The TPC of the extract and fractions was assayed according to the method described by Zhang et al. [22]. Twenty µL of sample solution was mixed with 100 µL Folin–Ciocalteu reagent in a 96-well microplate. After 5 min, 80 µL of 7.5% sodium carbonate (*w*/*v* in water) was added. The microplate was gently shaken, allowed to stand at room temperature for 2 h, and its absorbance was read at 750 nm with a microplate reader (Multiskan FC, Thermo Scientific Technologies®, Waltham, MA, USA). The TPC of the samples was expressed as gallic acid equivalents (GAEs) per gram of dry weight from a calibration curve of gallic acid (0–500 mg/L).

2.5. Antioxidant Activities

2.5.1. DPPH Scavenging Activity

The scavenging activity of the stable 1,1-diphenyl-2-picrylhydrazyl (DPPH) free radical was determined by the method of Marwah et al. [23]. Briefly, the reaction medium contained 100 µL of 100 µM DPPH violet solution in ethanol and 100 µL of plant extract at different concentrations (or water for the control). The reaction mixture was incubated in the dark for 15 min and the absorbance was recorded at 517 nm. The assay was carried out in triplicate.

The decrease in absorbance upon the addition of test samples was used to calculate the inhibition percentage (%IP) of DPPH radicals, following the following equation:

$$\%IP = [(A_c - A_s)/A_c] \times 100$$

where A_c and A_s are the absorbances of the control and the test sample, respectively. From a plot of concentration against %IP, a linear-regression analysis was performed to determine the antiradical activity, as expressed by the IC_{50} (extract concentration resulting in a 50% inhibition) value for each sample.

2.5.2. Ferric Reducing Activity (FRAP)

The assay is based on the reaction of Fe^{2+} with 2,4,6-tri(pyridyl)-*s*-triazine (TPTZ) to form a violet-blue color with maximal absorbance at 593 nm [24]. The FRAP solution was prepared by mixing 10 volumes of acetate buffer (300 mM, pH 3.6) with 1 volume of TPTZ (40 mM in HCl) and 1 volume of ferric chloride (20 mM in water). The solution was prepared daily and warmed at 37 °C for 10 min before use. A 280 µL aliquot of this solution was mixed with 20 µL of samples (extract, fractions, or water for the blank) in a 96-well microplate. The mixture was incubated at 37 °C in the dark for 30 min and then read at 593 nm with a Multiskan FC microplate reader (Thermo Scientific Technologies, Shanghai, China). The increase in absorbance upon the addition of test samples was used to calculate the reducing capacity, as expressed by the following efficacy percentage (%EP):

$$\%EP = [(A_s - A_c)/A_c] \times 100$$

where A_s and A_c are the absorbances of the control and the test sample, respectively. From a plot of concentration against %EP, a linear-regression analysis was performed to determine the EC_{50} (extract concentration resulting in a 50% efficacy) value for each sample.

2.5.3. ABTS Scavenging Activity

The ABTS radical scavenging assay was based on the method described by Re et al. [25] with a slight modification. Briefly, a 7 mM ABTS stock solution was prepared by dissolving ABTS in ethanol–water (5:1 *v*/*v*). Then, an aliquot of this solution was reacted with 2.45 mM potassium persulfate in ethanol–water (1:3 *v*/*v*) and allowed to stand in the dark at room temperature for 16–20 h to prepare the ABTS radical cation ($ABTS^+$). This ABTS radical solution was diluted to an absorbance at 734 nm of 0.70 ± 0.02. Finally, the absorbance of a mixture consisting of the sample (or water for the blank) and the ABTS reagent was followed at 734 nm. The antiradical capacity of the samples was expressed as gallic acid equivalents.

2.6. *Anti-Ageing Activity*

An anti-tyrosinase assay was performed using L-DOPA as a substrate, according to the method described by Masuda et al. [26]. The samples (plant extract or fractions) were dissolved in 50% DMSO. Then, 40 µL of each sample was mixed with 80 µL of phosphate buffer (0.1 M, pH 6.8), 40 µL of tyrosinase (31 units/mL in phosphate buffer, pH 6.5), and 40 µL of 2.5 mM L-DOPA in a 96-well microplate. The absorbance of the mixture was measured at 475 nm and compared to those of a positive control containing kojic acid and a blank containing all the components except L-DOPA. The anti-tyrosinase activity was expressed as kojic acid equivalents (mg KAE/g DW).

Moreover, the anti-melanogenic activity of sea-fennel extract and fractions was assessed in vitro on B16 4A5 melanoma cells according to Bouzaiene et al. [27]. The cells were seeded at 3.5×10^4 cells/well into 12-well plates, allowed to adhere for 24 h, and then treated with sea-fennel-extract concentrations that allowed for cellular viability higher than 80% for 72 h. Thereafter, adherent cells were trypsinized and solubilized in 1 mL of sodium dodecyl sulphate (SDS; 1%, *v*/*v*). The absorbance of the samples was measured at 475 nm, and the melanin content was estimated using a standard curve of synthetic melanin (0–25 µg/mL).

2.7. Neuroprotective Activity

The neuroprotective property of sea-fennel extract was evaluated through the in vitro inhibition of acetylcholinesterase (AChE) according to Custódio et al. [28]. Samples (20 μL at concentrations of 1, 5, and 10 mg/mL) were mixed with 140 μL of sodium phosphate buffer (0.1 mM, pH 8.0) and 20 μL of AChE solution (0.28 U/mL) in a 96-well microplate. The mixture was incubated for 15 min at room temperature and the reaction was initiated by the addition of 10 μL of 4 mg/mL ATChI and 20 μL of 1.2 mg/mL DTNB. The absorbance was read at 405 nm and results were expressed as IC_{50} relative to a control containing water instead of extract. Galantamine was used as a positive control.

2.8. Antidiabetic Activity

Sea-fennel extract and fractions, at concentrations ranging from 1 to 5 mg/mL, were evaluated for their capacity to inhibit α-amylase and α-glucosidase according to Zengin [29]. Moreover, extracts and fractions were tested against porcine lipase, according to McDougall et al. [30]. Acarbose was used as a positive control for α-amylase and α-glucosidase, and orlistat was used as a positive control for lipase inhibition. The results are expressed as IC_{50} relative to a control containing DMSO.

2.9. Anti-Inflammatory Activity

Nitric oxide (NO) production by LPS-stimulated RAW 264.7 macrophages was assessed as described by Rodrigues et al. [4]. RAW 264.7 cells were cultured in RPMI 1640 culture medium, enriched with 10% heat-inactivated FBS, 1% L-glutamine (2 mM), and 1% penicillin (50 U/mL)/streptomycin (50 μg/mL), and kept at 37 °C in a 5% CO_2 humidified atmosphere. Murine cells were seeded in a 96-well plate at 2.5×10^5 cells/well and allowed to adhere overnight. Then, they were co-treated with 100 ng/mL of LPS and sea-fennel raw extract (at concentrations that allowed for cellular viability higher than 80%) for 24 h. NO production was assessed using the Griess assay. The results are expressed as a percentage of inhibition of NO production, relative to a control containing DMSO (0.5%, *v*/*v*), and compared to the positive control L-NAME.

2.10. Solute Purification

When required, the purification of solutes from a specific fraction was carried out by HPLC using the Shimadzu UFLC XR device equipped with a PDA detector (SPD-M20A, Shimadzu, Kyoto, Japan). For solute separation, a Spherisorb ODS2 column (5 μm, 250 × 4.6 mm, Waters, Milford, MA, USA) was used, and the mobile phase consisted of a mixture of 100% acetonitrile (A) and ultrapure water (B). The purification process involved the following linear gradient: t = 0 min 100% B; t = 10 min 100% A. The compounds were detected at 254 nm and collected for acid hydrolysis treatment (1 N HCl, 110 °C for 1 h) prior to structural elucidation.

2.11. NMR Analyses

To characterize bioactive compounds, a portion of crude extract and of each fraction were concentrated through rotary evaporation at 35 °C. The resulting dry residue was then dissolved in 700 μL of 99.5% deuterated water (D_2O) or methanol (MeOD) for NMR analyses. The ^{1}H NMR and ^{13}C NMR spectra were acquired using a Bruker Avance DRX-400 spectrometer (400 MHz) with a 5 mm dual ^{1}H/^{13}C probe head. The standard pulse sequences provided by the Bruker Software NMRlib 2.0 (Brüker, Wissembourg, France) were used. A typical ^{1}H NMR spectrum consisted of 32 scans, and 2,2,3,3-tetradeuterio-3-(trimethylsilyl)-propanoic acid sodium salt served as an internal standard. The identification of major solutes present in sea-fennel extracts or fractions was determined by comparing the NMR spectra with external standards. For ^{13}C (J-mod) and 2D homo- and heteronuclear NMR analyses (COSY, HMBC, HMQC, TOCSY), experiments were conducted at 298°K on a Brüker Avance III HD500 spectrometer equipped with an inverse 5 mm TCI cryoprobe (^{1}H, ^{13}C, ^{15}N) with a z gradient. The data were processed using the TopSpin v. 4.0 program (Bruker).

2.12. Statistical Analyses

All extractions and assays were conducted in triplicate. The results are expressed as mean ± standard deviation (SD), and the means were compared by using a one-way analysis of variance (ANOVA) followed by Duncan's multiple range tests performed using the "Statistica v. 5.1" software (Statsoft, 2008, Tulsa, OK, USA). The differences between individual means were deemed to be significant at $p < 0.05$. The IC_{50} values were obtained by fitting the data with a sigmoidal curve using the GraphPad Prism v. 5.0 program.

3. Results

3.1. Total Phenolic Content

The crude extract of *C. maritimum* showed a high phenolic amount, with 33.3 mg $GAE.g^{-1}DW$ (Figure 2). After the fractionation of sea-fennel leaf extract, phenolics were detected in the three fractions eluted with 20, 40, and 60% MeOH, which exhibited 3.4, 5, and 2.3 times more phenolic compounds than crude extract, respectively.

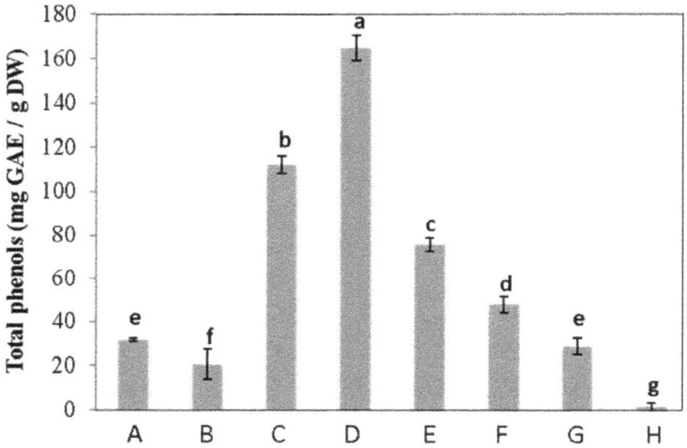

Figure 2. Total phenolic content (mg GAE/g DW) of *C. maritimum* crude extract and fractions. A: crude extract; B: water fraction; C: $MeOH_{20}$ fraction; D: $MeOH_{40}$ fraction; E: $MeOH_{60}$ fraction; F: $MeOH_{80}$ fraction; G: $MeOH_{100}$ fraction; and H: Ethyl acetate fraction. Means ± standard deviations of three replicates are represented, and different letters above the bars indicate significantly different means ($p < 0.05$).

3.2. Antioxidant Activities

The crude extract of *C. maritimum* exhibited high antioxidant activities with EC_{50} value of 0.152 $mg \cdot mL^{-1}$ for FRAP bioassay (Figure 3a). After the fractionation of sea-fennel extract, $MeOH_{20}$, $MeOH_{40}$, $MeOH_{60}$, and $MeOH_{80}$ fractions exhibited a strong reducing capacity, with FRAP EC_{50} values 5.1, 7.2, 6.3, and 2.7 times lower than that of the crude extract, respectively. The same trend was obtained with the DPPH assay, where $MeOH_{20}$, $MeOH_{40}$, $MeOH_{60}$, and $MeOH_{80}$ fractions showed IC_{50} values 3.6, 4.4, 6.7, and 3 times lower than that of the crude extract (0.136 $mg \cdot mL^{-1}$), respectively (Figure 3b). For these two bioassays, medium polar fractions (namely $MeOH_{40}$ and $MeOH_{60}$ fractions) were the most active ones. ABTS radical scavenging capacity was distributed in almost all the fractions since every fraction eluted with methanol solution exhibited a strong activity (higher than 200 mg $GAE \cdot g^{-1}$ DW) (Figure 3c). Thus, $MeOH_{20}$, $MeOH_{40}$, $MeOH_{60}$, $MeOH_{80}$, and $MeOH_{100}$ fractions showed 7, 10.8, 6.8, 7.1, and 4.6 times higher activities than the crude extract (51.7 mg $GAE \cdot g^{-1}$ DW). Conversely, the first and last fractions (namely those eluted with H_2O and ethyl acetate) showed very low or hardly detectable activities using every antioxidant bioassay.

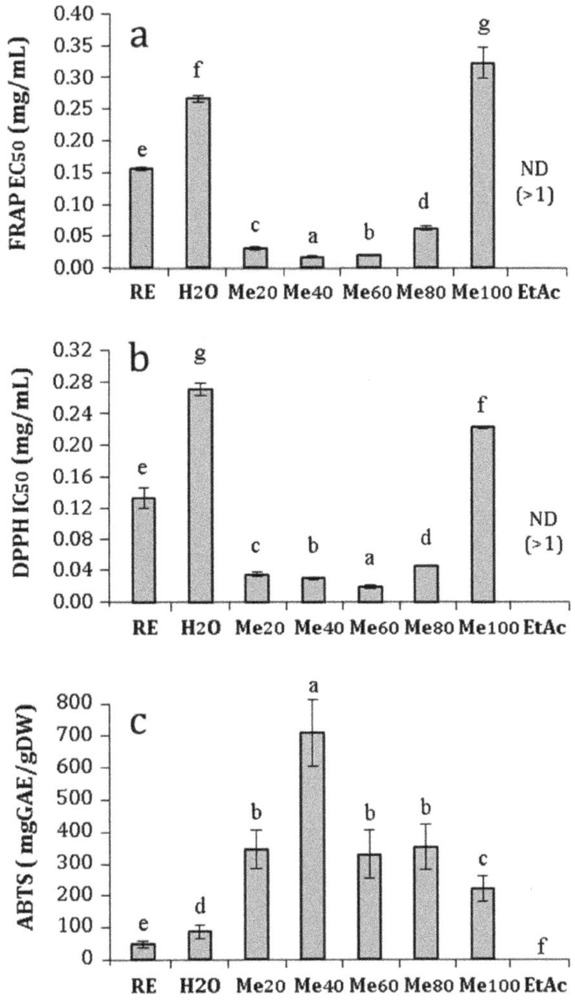

Figure 3. Antioxidant activities of *Crithmum maritimum* raw extract (RE) and fractions eluted with H_2O, 20% MeOH (Me20), 40% MeOH (Me40), 60% MeOH (Me60), 80% MeOH (Me80), 100% MeOH (Me100), and ehtylacetate (EtAc). (**a**) Ferric reducing capacity (EC50 in mg/mL); (**b**) radical scavenging activity against DPPH (IC50 in mg/mL); and (**c**) radical scavenging activity against ABTS (mg GAE/g DW). Means ± standard deviations of three replicates are represented, and different letters above the bars indicate significantly different means ($p < 0.05$).

3.3. *Other Biological Activities*

3.3.1. Anti-Ageing Activity

Sea-fennel crude extract exhibited an appreciable activity against tyrosinase, with a value of 235 mg KAE·g^{-1} DW. Of the seven fractions eluted from this extract, the last four exhibited a significantly higher activity than that of the crude extract (Table 1). Among them, the $MeOH_{100}$ and ethyl acetate fractions appeared the most active, with more than 600 mg KAE·g^{-1} DW. Moreover, neither raw extracts nor fractions of sea-fennel aerial parts showed anti-melanogenic properties on B16 4A5 melanoma cells.

Table 1. Anti-tyrosinase (diphenolase inhibition), anti-inflammatory (inhibition of NO production in RAW 264.7 macrophages), and antidiabetic (alpha-glucosidase inhibition) activities of sea-fennel polar extract and its fractions. Means ± SDs of three replicates are presented, and different letters indicate significantly different means ($p < 0.05$). ND, not detected.

	Anti-Tyrosinase (mgKAE/gDW)	NO Inhibition (IC_{50}, µg/mL)	Anti-α Glucosidase (IC_{50}, mg/mL)
Raw extract	234.29 ± 48.56 c	ND	ND
MeOH20	151.12 ± 74.89 c	ND	0.02 ± 0.01 b
MeOH40	226.37 ± 91.23 c	89.54 ± 2.16 a	ND
MeOH60	531.46 ± 68.42 b	6.41 ± 0.37 c	ND
MeOH80	563.14 ± 26.94 b	ND	0.04 ± 0.00 b
MeOH100	637.02 ± 29.23 a	ND	ND
EtAc	626.19 ± 41.13 a	ND	ND
L-NAME		27.81 ± 1.93 b	
Acarbose			3.14 ± 0.09 a

3.3.2. Anti-Inflammatory Activity

Although sea-fennel raw extract did not show any capacity to inhibit NO production by RAW 264.7 macrophages, two of its fractions were able to reduce this indicator of inflammation (Table 1). Of note, the fraction eluted with 60% MeOH showed the strongest NO inhibitory power, with an IC_{50} value (6.41 ± 0.37 µg/mL) four times lower than that of the positive control L-NAME (27.81 ± 1.93 µg/mL).

3.3.3. Neuroprotective Activity

Sea-fennel extract and fractions were evaluated for their capacity to inhibit AChE. However, neither the raw extract nor its fractions exhibited any inhibitory effect on AchE.

3.3.4. Antidiabetic and Anti-Obesity Activities

Pancreatic lipase plays a crucial role in the digestion and absorption of triglycerides. Inhibiting this enzyme is a widely used method to determine the potential efficacy of natural substances in averting obesity. Here, neither extracts nor fractions showed inhibitory activity on rat lipase enzyme in vitro. Then, their capacity to inhibit α-amylase and α-glucosidase was assessed. Though sea-fennel raw extract was inactive on these enzymes, two fractions strongly inhibited α-glucosidase (Table 1). Thus, $MeOH_{20}$ and $MeOH_{80}$ fractions exhibited a hundred-fold higher inhibition than the standard glucosidase inhibitor acarbose. However, no amylase inhibition could be detected in these fractions.

3.4. Solute Identification

Following the purification of the extract, the active fractions were analyzed using NMR spectroscopy. The ^{1}H-NMR spectrum of the $MeOH_{20}$ fraction displayed chlorogenic acid signals predominantly (Figure 4A). In the MeOH40 fraction, characteristic signals of dicaffeoyl quinic acids, such as 3,5-di-*O*-caffeoylquinic acid, were found (Figure 4B). Signals between 6 and 7.2 ppm in the ^{1}H-NMR spectrum of the $MeOH_{60}$ fraction indicated the presence of quercetins (Figure 4C), with the two doublets between 5 and 5.2 ppm suggesting glycosylated quercetins. Subsequent HPLC purification yielded four individual compounds. Following the acid hydrolysis of these compounds, the first one provided quercetin, glucose, and rhamnose, the second provided quercetin and galactose, the third provided quercetine and glucose, and the fourth provided quercetine, glucose, and galactose. Finally, ^{13}C- and 2D NMR experiments confirmed these results and led us to identify unequivocally the four glycosylated quercetins, namely quercetin 3-*O*-rutinoside, quercetin 3-*O*-galactoside, quercetin 3-*O*-glucoside, and quercetin 3-*O*-robinobioside (Supplementary Data). The $MeOH_{80}$ fraction showed small signals in the 6–8 ppm region and more pronounced ones between 1 and 1,5 ppm on its ^{1}H-NMR spectrum, corresponding to aromatic and aliphatic protons, respectively (Figure 4D). Finally, the $MeOH_{100}$ and ethyl acetate fractions contained fewer polar compounds with aliphatic protons, as indicated by the intense signals in the 1–2 ppm region (Figure 4E,F).

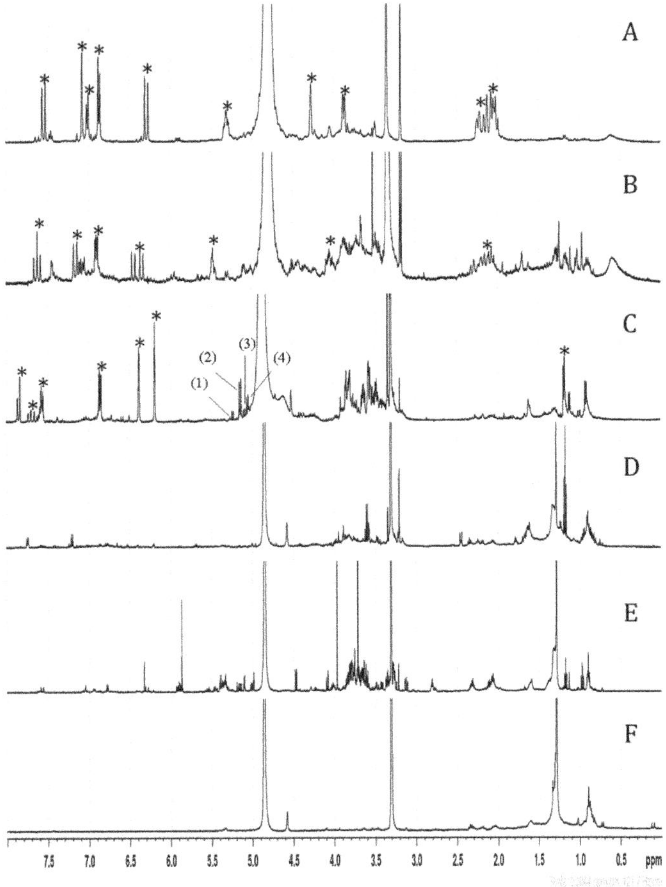

Figure 4. ^{1}H NMR spectra of *C. maritimum* bioactive fractions eluted with 20% MeOH (**A**), 40% MeOH (**B**), 60% MeOH (**C**), 80% MeOH (**D**), 100% MeOH (**E**), and ethyl acetate (**F**). Stars in spectra (**A–C**) indicate signals of chlorogenic acid, 3,5-dicaffeoylquinic acid, and quercetin glycosides, respectively. On the latter spectrum (**C**), specific signals assigned to quercetin 3-*O*-robinobioside (1), quercetin 3-*O*-glucoside (2), quercetin 3-*O*-galactoside (3), and quercetin 3-*O*-rutinoside (4) are indicated.

4. Discussion

Numerous studies have evaluated the antioxidant potential of raw extracts from *C. maritimum* [16,18,21,31]. Nevertheless, there is a lack of comprehensive investigation into the molecules accountable for these properties. Three antioxidant bioassays have been used in this study to highlight different antioxidant mechanisms: ABTS and DPPH radical scavenging, as well as ferric reducing power. Additionally, for the first time in this species, biological activities of medical interest have been assessed in a raw extract and its fractions, and major compounds in these fractions have been characterized with H1-NMR to identify the possible bioactive metabolites.

The raw extract of *C. maritimum* displayed strong antioxidant activities, supporting previous findings on other sea-fennel populations [21]. Through the bioguided fractionation of this extract, the antioxidant capacity was successfully isolated into four primary fractions:

- The fraction eluted with 40% MeOH showed the most significant antioxidant activity, as revealed by ABTS-scavenging and FRAP assays. NMR analyses of this fraction indicated the prevalence of 3,5-dicaffeoylquinic acid (syn. isochlorogenic acid). Such a compound has already been reported in *C. maritimum* [31], as well as in many other halophytic species [32]. Nevertheless, the literature on its antioxidant activity is still scarce [33], compared to that of its natural isomer chlorogenic acid. Here, we suggest that the abundance of 3,5-dicaffeoylquinic acid accounts for the very strong antioxidant capacity of the $MeOH_{40}$ fraction, as it has been reported previously in another halophytic species [34]. Of note, this fraction also inhibited NO production by LPS-stimulated RAW 264.7 cells, confirming the potent anti-inflammatory action of isochlorogenic acid [35];
- The major compounds characterized in the $MeOH_{60}$ fraction were four glycosylated quercetins: quercetin 3-*O*-glucoside, quercetin 3-*O*-robinobioside, quercetin 3-*O*-galactoside, and quercetin 3-*O*-rutinoside. Considering the strong antioxidant power of quercetin and its derivatives [36], these four flavonols are likely responsible for the marked antioxidant activity of this fraction. Quercetin 3-*O*-glucoside is widely represented in the plant kingdom, and in particular in halophytes [37]. Jallali et al. [18] have reported its presence in sea fennel, conferring acetonic extract with a strong antioxidant activity. Moreover, Kong et al. [38] reported the anti-obesity action of this flavonol isolated from the salt-marsh plant *Salicornia herbacaea*, which was not confirmed in our work with a lipase bioassay. Conversely to quercetin 3-*O*-glucoside, the 3-*O*-rutinoside derivative of quercetin (=rutin or rutoside) has been less documented in the literature, compared to that of kaempferol or luteolin. In halophytes, the presence of rutin was reported in *Crithmum maritimum* [18,39,40], as well as in *Calystegia soldanella* [41] and *Carpobrotus edulis* [42]. Quercetin 3-*O*-galactoside has been identified previously in other (non halophytic) members of the Apiaceae family [43,44], but never in sea fennel. Moreover, quercetin 3-*O*-robinobioside has only been found in three species belonging to Fabaceae, Lamiaceae, and Rosaceae [45–47], and is reported here for the first time in the Apiaceae family.

Of note, this $MeOH_{60}$ fraction exhibited a very strong NO-inhibitory activity, even more powerful than the standard anti-inflammatory L-NAME. This result is in agreement with the well-known protective effect of quercetin and its derivatives against the inflammation process [48,49], and strongly suggests that the anti-inflammatory activity of the $MeOH_{60}$ fraction of sea-fennel extract is due to these flavonols.

- Similar analyses allowed us to elucidate the antioxidant activity of the MeOH20 fraction. The major constituent detected here was chlorogenic acid, a widely recognized antioxidant compound previously reported in C. maritimum [21]. Additionally, this fraction exhibited a remarkable anti-glucosidase activity. Chlorogenic acid has been shown to stimulate glucose uptake in skeletal muscle, thus improving glucose metabolism and preventing diabetes manifestations [50]. Our results are consistent with such observations and, owing to the richness of sea fennel in chlorogenic acid, reinforce the potential antidiabetic effect of dietary sea fennel. However, interestingly, that fraction did not exhibit any anti-lipase activity though chlorogenic acid has been reported earlier to prevent obesity manifestations [51];
- The sea-fennel extract fraction eluted with 80% methanol showed a strong antioxidant activity as well as a marked inhibition of glucosidase. The constituents of this fraction have not been completely elucidated, but its NMR spectrum suggests the presence of less polar phenolics and compounds with short aliphatic chains.

The anti-tyrosinase activity was determined here for the first time in sea fennel by assessing the inhibition of L-DOPA oxidation to dopaquinone (diphenolase activity).

Interestingly, the moderate activity of crude extract was concentrated in four fractions, indicating the contribution of several constituents of differing structures in that activity. Thus, the four least polar fractions exhibited the most powerful anti-tyrosinase activity. Moreover, the inhibition level was quite close to that of the kojic acid standard, and much higher than that recently reported in other halophytic species [5]. Since tyrosinase catalyzes oxidative reactions, the inhibitory activity of $MeOH_{60}$ and $MeOH_{80}$ fractions is likely due to their major antioxidant phenolic compounds, as reported by Chang [52]. For example, one of the quercetine glycosides identified in the MeOH60 fraction, namely quercetine 3-*O*-galactoside, was reported to inhibit tyrosinase [53]. Additionally, of note was the highest anti-tyrosinase activity exhibited by the non-antioxidant fractions eluted with 100% MeOH and ethyl acetate. Their composition has not been completely elucidated yet, but preliminar NMR analyses suggest that molecules involved in anti-tyrosinase activity here are likely acyclic compounds with carbon chains such as terpenoids. Therefore, our study confirms in part that a positive correlation could be made between antioxidant and anti-tyrosinase activities, as reported previously by Choi et al. [54] and Lee et al. [55]. However, a contribution by other (non-antioxidant) substances to tyrosinase inhibition should not be excluded.

The well-known neurotransmitter acetylcholine facilitates the recovery of neuronal function after brain injury. Therefore, it is convenient to investigate the powerful neuroprotective effect of natural compounds by inhibiting AChE activity [56]. Here, sea-fennel extract and fractions did not exhibit any capacity to inhibit AChE. This result was consistent with a previous work, reporting that such activity could only be found in essential oils of Apiaceae flowers [57,58].

5. Conclusions

On the whole, we demonstrated that the antioxidant activity of *C. maritimum* hydroalcoholic extract is present in its mid-polar fractions, with chlorogenic acid, 3,5 dicaffeoyl quinic acid, and quercetin glycosides being the major contributors. Of these fractions, some also exhibited antidiabetic or anti-inflammatory properties. On the other hand, the less polar fractions are the most active against tyrosinase, and additional investigations are being conducted to ascertain the compounds (presumably terpenoids) accountable for this activity. Additionally, no anti-melanogenic nor anti-obesity properties could be found in sea-fennel hydromethanolic extract and its fractions. Overall, the findings from this study should underscore the nutraceutical value of *C. maritimum* extract or antioxidant secondary metabolites as potent agents against inflammatory, diabetes-related, and ageing processes.

Supplementary Materials: The following supporting information can be downloaded at https://www.mdpi.com/article/10.3390/foods13091294/s1. Supplementary Data: 1. NMR data (500 MHz, $CDCl_3$) of 3,5-dicaffeoylquinic acid isolated in the $MeOH_{40}$ fraction. 2. NMR data (500 MHz, $CDCl_3$) of the three quercetine derivatives isolated in the $MeOH_{60}$ fraction.

Author Contributions: Conceptualization, C.M.; methodology, C.M., X.D., M.J.R. and L.C.; software, C.L. and M.J.R.; validation, C.M., X.D., M.J.R., S.C. and L.C.; formal analysis, C.L. and M.J.R.; resources, C.M. and L.C.; writing—original draft preparation, C.L. and M.J.R.; writing—review and editing, C.M., S.C. and L.C.; supervision, C.M. and L.C.; project administration, C.M.; funding acquisition, C.M. All authors have read and agreed to the published version of the manuscript.

Funding: Agence Nationale de la Recherche (ANR) funding for SeaFennel4Med European project is also aknowledged (ANR-22-PRIM-0011)".

Institutional Review Board Statement: Not applicable.

Informed Consent Statement: Not applicable.

Data Availability Statement: The original contributions presented in the study are included in the article/Supplementary Materials, further inquiries can be directed to the corresponding author.

Conflicts of Interest: The authors declare that the research was conducted in the absence of any commercial or financial relationships that could be construed as potential conflicts of interest.

Abbreviation

AChE, acetylcholinesterase; ATChI, acetylthiocholine iodide; BHT, Butylated hydroxytoluene; DMSO, Dimethyl sulphoxide; DPPH, 1,1-diphenyl-2-picrylhydrazyl; DTNB, (5,5′-dithiobis-(2-nitro-benzoic acid); EtAc, ethyl acetate; FBS, fetal bovine serum; L-NAME, N(gamma)-nitro-L-arginine methyl ester; L-DOPA, L-3,4 dihydroxyphenylalanine; LPS, lipopolysaccharide; MeOH, methanol; PDA, photodiode array; RPMI, TPTZ, 2,4,6-Tris(2-pyridyl)-s-triazine.

References

1. Riso, P.; Klimis-Zacas, D.; Del Bo, C.; Martini, D.; Campolo, J.; Vendrame, S.; Møller, P.; Loft, S.; De Maria, R.; Porrini, M. Effect of a wild blueberry (*Vaccinium angustifolium*) drink intervention on markers of oxidative stress, inflammation and endothelial function with cardiovascular risk factors. *Eur. J. Nutr.* **2013**, *52*, 949–961. [CrossRef] [PubMed]
2. Zhang, Y.J.; Gan, R.Y.; Li, S.; Zhou, Y.; Li, A.N.; Xu, D.P.; Li, H.B. Antioxidant phytochemicals for the prevention and treatment of chronic diseases. *Molecules* **2015**, *20*, 21138–21156. [CrossRef] [PubMed]
3. Ksouri, R.; Megdiche, W.; Jallali, I.; Debez, H.; Magné, C.; Isoda, H.; Abdelly, C. Medicinal halophytes: Potent source of health promoting biomolecules with medical, nutraceutical and food applications. *Crit. Rev. Biotechnol.* **2012**, *32*, 289–326. [CrossRef] [PubMed]
4. Rodrigues, M.J.; Gangadhar, K.N.; Vizetto-Duarte, C.; Wubshet, S.G.; Nyberg, N.T.; Barreira, L.; Varela, J.; Custódio, L. Maritime halophyte species from southern Portugal as sources of bioactive molecules. *Mar. Drugs* **2014**, *12*, 2228–2244. [CrossRef]
5. Jdey, A.; Falleh, H.; Jannet, S.B.; Hammi, K.M.; Dauvergne, X.; Magné, C.; Ksouri, R. Anti-aging activities of extracts from Tunisian medicinal halophytes and their aromatic constituents. *EXCLI J.* **2017**, *16*, 755–769. [PubMed]
6. Djeridane, A.; Yousfi, M.; Brunel, J.M.; Stocker, P. Isolation and characterization of a new steroid derivative as a powerful antioxidant from *Cleome arabica* in screening the in vitro antioxidant capacity of 18 Algerian medicinal plants. *Food Chem. Toxicol.* **2010**, *48*, 2599–2606. [CrossRef]
7. Ksouri, R.; Falleh, H.; Megdiche, W.; Trabelsi, N.; Hamdi, B.; Chaieb, K.; Bakrouf, A.; Magné, C.; Abdelly, C. Antioxidant and antimicrobial activities of the edible medicinal halophyte *Tamarix gallica* L. and related polyphenolic constituents. *Food Chem. Toxicol.* **2009**, *47*, 2083–2091. [CrossRef] [PubMed]
8. Falleh, H.; Trabelsi, N.; Bonenfant-Magné, M.; Le Floch, G.; Abdelly, C.; Magné, C.; Ksouri, R. Polyphenol content and biological activities of *Mesembryanthemum edule* organs after fractionation. *Ind. Crops Prod.* **2013**, *42*, 145–152. [CrossRef]
9. Megdiche Ksouri, W.; Medini, F.; Mkadmini, K.; Legault, J.; Magné, C.; Abdelly, C.; Ksouri, R. LC–ESI-TOF–MS identification of bioactive secondary metabolites involved in the antioxidant, anti-inflammatory and anticancer activities of the edible halophyte *Zygophyllum album* Desf. *Food Chem.* **2013**, *139*, 1073–1080. [CrossRef]
10. Guil Guerrero, J.L.; Torija Isasa, M.E.; Gimenez Martinez, J.J. Composition nutricional del hinojo marino (*Crithmum maritimum* L.). *Alimentaria* **1996**, *34*, 65–72.
11. Ben Amor, N.; Ben Hamed, K.; Debez, A.; Grignon, C.; Abdelly, C. Physiological and antioxidant responses of the perennial halophyte *Crithmum maritimum* L. to salinity. *Plant Sci.* **2005**, *168*, 889–899. [CrossRef]
12. Cunsolo, F.; Ruberto, G. Bioactive metabolites from Sicilian marine fennel (*Crithmum maritimum* L.). *J. Nat. Prod.* **1993**, *56*, 1598–1600. [CrossRef] [PubMed]
13. Pateira, L.; Noguiera, T.; Antunes, A.; Venancio, F.; Taveres, R.; Capelo, J. Two chemotypes of *Crithmum maritimum* L. from Portugal. *Flavour Fragr. J.* **1999**, *14*, 333–342. [CrossRef]
14. Ruberto, G.; Baratta, M.T.; Deans, S.G.; Dorman, H.J. Antioxidant and antimicrobial activity of *Foeniculum vulgare* and *Crithmum maritimum* essential oils. *Planta Med.* **2000**, *66*, 687–693. [CrossRef] [PubMed]
15. Males, Z.; Blazevic, N.; Plazibat, M. Variations of essential oil yield and composition of *Crithmum maritimum* L. *Acta Pharm.* **2001**, *51*, 81–84.
16. Jallali, I.; Zaouali, Y.; Missaoui, I.; Smaoui, A.; Abdelly, C.; Ksouri, R. Variability of antioxidant and antibacterial effects of essential oils and acetonic extracts of two edible halophytes: *Crithmum maritimum* L. and *Inula crithmoïdes* L. *Food Chem.* **2014**, *145*, 1031–1038. [CrossRef] [PubMed]
17. Politeo, O.; Ćurlin, P.; Brzović, P.; Auzende, K.; Magné, C.; Generalić Mekinić, I. Volatiles from French and Croatian Sea Fennel Ecotypes: Chemical Profiles and the Antioxidant, Antimicrobial and Antiageing Activity of Essential Oils and Hydrolates. *Foods* **2024**, *13*, 695. [CrossRef] [PubMed]
18. Jallali, I.; Megdiche, W.; M'Hamdi, B.; Oueslati, S.; Smaoui, A.; Abdelly, C.; Ksouri, R. Changes in phenolic composition and antioxidant activities of the edible halophyte *Crithmum maritimum* L. with physiological stage and extraction method. *Acta Physiol. Plant.* **2012**, *34*, 1451–1459. [CrossRef]
19. Pereira, C.G.; Barreira, L.; da Rosa Neng, N.; Nogueira, J.M.F.; Marques, C.; Santos, T.F.; Varela, J.; Custódio, L. Searching for new sources of innovative products for the food industry within halophyte aromatic plants: In vitro antioxidant activity and phenolic and mineral contents of infusions and decoctions of *Crithmum maritimum* L. *Food Chem. Toxicol.* **2017**, *107*, 581–589. [CrossRef] [PubMed]

20. Correia, I.; Antunes, M.; Tecelão, C.; Neves, M.; Pires, C.L.; Cruz, P.F.; Rodrigues, M.; Peralta, C.C.; Pereira, C.D.; Reboredo, F.; et al. Nutritive Value and Bioactivities of a Halophyte Edible Plant: *Crithmum maritimum* L. (Sea Fennel). *Plants* **2024**, *13*, 427. [CrossRef]
21. Méot-Duros, L.; Magné, C. Antioxidant activity and phenolic content of *Crithmum maritimum* L. leaves. *Plant Physiol. Biochem.* **2009**, *47*, 37–41. [CrossRef] [PubMed]
22. Zhang, Q.; Zhang, J.; Shen, J.; Silva, A.; Dennis, D.A.; Barrow, C.J. A simple 96-well microplate method for estimation of total polyphenol content in seaweeds. *J. Appl. Phycol.* **2006**, *18*, 445–450. [CrossRef]
23. Marwah, R.G.; Fatope, M.O.; Mahrooqi, R.A.; Varma, G.B.; Abadi, H.A.; Al-Burtamani, S.K.S. Antioxidant capacity of some edible and wound healing plants in Oman. *Food Chem.* **2007**, *101*, 465–470. [CrossRef]
24. Bolanos de la Torre, A.A.S.; Henderson, T.; Nigam, P.S.; Owusu-Apenten, R.K. A universally calibrated microplate ferric reducing antioxidant power (FRAP) assay for foods and applications to Manuka honey. *Food Chem.* **2015**, *174*, 119–123. [CrossRef] [PubMed]
25. Re, R.; Pellegrini, N.; Proteggente, A.; Pannala, A.; Yang, M.; Rice-Evans, C. Antioxidant activity applying an improved ABTS radical cation decolorization assay. *Free Rad. Biol. Med.* **1999**, *26*, 1231–1237. [CrossRef] [PubMed]
26. Masuda, T.; Yamashita, D.; Takeda, Y.; Yonemori, S. Screening for tyrosinase inhibitors among extracts of seashore plants and identification of potent inhibitors from *Garcinia subelliptica*. *Biosci. Biotechnol. Biochem.* **2005**, *69*, 197–201. [CrossRef] [PubMed]
27. Bouzaiene, N.N.; Chaabane, F.; Sassi, A.; Chekir-Ghedira, L.; Ghedira, K. Effect of apigenin-7-glucoside, genkwanin and naringenin on tyrosinase activity and melanin synthesis in B16F10 melanoma cells. *Life Sci.* **2016**, *144*, 80–85. [CrossRef] [PubMed]
28. Custódio, L.; Soares, F.; Pereira, H.; Rodrigues, M.J.; Barreira, L.; Rauter, A.P.; Albericio, F.; Varela, J. *Botryococcus braunii* and *Nanochloropsis oculata* extracts inhibit cholinesterases and protect human dopaminergic SH-SY5Y cells from H_2O_2-induced cytotoxicity. *J. Appl. Phycol.* **2015**, *27*, 839–848. [CrossRef]
29. Zengin, G. A study on in vitro enzyme inhibitory properties of *Asphodeline anatolica*: New sources of natural inhibitors for public health problems. *Ind. Crops Prod.* **2016**, *83*, 39–43. [CrossRef]
30. McDougall, G.J.; Kulkarni, N.N.; Stewart, D. Berry polyphenols inhibit pancreatic lipase activity in vitro. *Food Chem.* **2009**, *115*, 193–199. [CrossRef]
31. Siracusa, L.; Kulisic-Bilusic, T.; Politeo, O.; Krause, I.; Dejanovic, B.; Ruberto, G. Phenolic composition and antioxidant activity of aqueous infusions from *Capparis spinosa* L. and *Crithmum maritimum* L. before and after submission to a two-step in vitro digestion model. *J. Agric. Food Chem.* **2011**, *59*, 12453–12459. [CrossRef] [PubMed]
32. Stanković, M.; Jakovljevićc, D.; Stojadinov, M.; Stevanović, Z.D. Halophyte species as a source of secondary metabolites with antioxidant activity. In *Ecophysiology, Abiotic Stress Responses and Utilization of Halophytes*; Hasanuzzaman, M., Nahar, K., Öztürk, M., Eds.; Springer: Singapore, 2019; pp. 289–312. [CrossRef]
33. Hong, S.; Joo, T.; Jhoo, J.W. Antioxidant and anti-inflammatory activities of 3,5-dicaffeoylquinic acid isolated from *Ligularia fischeri* leaves. *Food Sci. Biotechnol.* **2015**, *24*, 257–263. [CrossRef]
34. Kim, J.Y.; Cho, J.Y.; Ma, J.K.; Park, K.Y.; Lee, S.H.; Ham, K.S.; Lee, H.J.; Park, K.H.; Moon, J.H. Dicaffeoylquinic acid derivatives and flavonoid glucosides from glasswort (*Salicornia herbacea* L.) and their antioxidative activity. *Food Chem.* **2011**, *125*, 55–62. [CrossRef]
35. Lesjak, M.; Beara, I.; Simin, N.; Pintać, D.; Majkić, T.; Bekvalac, K.; Mimica-Dukić, N. Antioxidant and anti-inflammatory activities of quercetin and its derivatives. *J. Funct. Foods* **2018**, *40*, 68–75. [CrossRef]
36. Murota, K.; Terao, J. Antioxidative flavonoid quercetin: Implications of its intestinal absorption and metabolism. *Arch. Biochem. Biophys.* **2003**, *417*, 12–17. [CrossRef] [PubMed]
37. Stanković, M.; Jakovljević, D. Phytochemical Diversity of Halophytes. In *Handbook of Halophytes*; Grigore, M.N., Ed.; Springer Nature: Cham, Switzerland, 2021; pp. 2089–2114. [CrossRef]
38. Kong, C.-S.; Lee, J.L.; Kim, Y.A.; Kim, J.-A.; Bak, S.S.; Hong, J.W.; Park, H.Y.; Yea, S.S.; Seo, Y. Evaluation on anti-adipogenic activity of flavonoid glucopyranosides from *Salicornia herbacea*. *Process Biochem.* **2012**, *47*, 1073–1078. [CrossRef]
39. Saleh, N.; El-Negoumy, I.; El-Hadidi, N.; Hosni, M.; Hosni, H. Comparative study of the flavonoids of some local members of the Umbellifereae. *Phytochemistry* **1983**, *22*, 1417–1420. [CrossRef]
40. Bartnik, M.; Wierzchowska Renke, K.; Głowniak, P.; Głowniak, K. Phenolic acids in *Crithmum maritimum* L. (Apiaceae) after Tytanit fertilization. *Acta Soc. Bot. Pol.* **2017**, *86*, 3560. [CrossRef]
41. Murai, Y.; Setoguchi, H.; Ono, E.; Iwashina, T. Flavonoids and their qualitative variation in *Calystegia soldanella* and related species (Convolvulaceae). *Nat. Prod. Commun.* **2015**, *15*, 429–432. [CrossRef]
42. Van der Watt, E.; Pretorius, J.C. Purification and identification of active antibacterial components in *Carpobrotus edulis* L. *J. Ethnopharmacol.* **2001**, *76*, 87–91. [CrossRef]
43. Harborne, J.B.; Boardley, M. Use of high-performance liquid chromatography in the separation of flavonol glycosides and flavonol sulphates. *J. Chromatogr. A* **1984**, *299*, 377–385. [CrossRef]
44. Parejo, I.; Jauregui, O.; Sánchez-Rabaneda, F.; Viladomat, F.; Bastida, J.; Codina, C. Separation and Characterization of Phenolic Compounds in Fennel (*Foeniculum vulgare*) Using Liquid Chromatography-Negative Electrospray Ionization Tandem Mass Spectrometry. *J. Agric. Food Chem.* **2004**, *52*, 3679–3687. [CrossRef]
45. Ozga, J.A.; Saeed, A.; Wismer, W.; Reinecke, D.M. Characterization of Cyanidin- and Quercetin-Derived Flavonoids and Other Phenolics in Mature Saskatoon Fruits (*Amelanchier alnifolia* Nutt.). *J. Agric. Food Chem.* **2007**, *55*, 10414–10424. [CrossRef] [PubMed]

46. Cong, Y.; Guo, J.; Wang, T.; Li, M.; Li, K.; Wang, J.; Li, Q. Chemical constituents and antitumor activity on leukemia K562 cell of *Leonurus heterophyllus*. *Zhongguo Zhong Yao Za Zhi* **2009**, *34*, 1816–1818. [PubMed]
47. Joubert, E.; De Beer, D. Antioxidants of Rooibos beverages: Role of plant composition and processing. In *Processing and Impact on Antioxidants in Beverages*; Preedy, V.R., Ed.; Academic Press: London, UK, 2014; Chapter 14; pp. 131–144.
48. Feng, J.; Wang, Y.; Yi, X.; Yang, W.; He, X. Phenolics from Durian exert pronounced NO inhibitory and antioxidant activities. *J. Agric. Food Chem.* **2016**, *64*, 4273–4279. [CrossRef]
49. Li, Y.; Yao, J.; Han, C.; Yang, J.; Tabassum Chaudhry, M.; Wang, S.; Liu, H.; Yin, Y. Quercetin, Inflammation and Immunity. *Nutrients* **2016**, *8*, 167. [CrossRef] [PubMed]
50. Ong, K.W.; Hsu, A.; Tan, B.K. Anti-diabetic and anti-lipidemic effects of chlorogenic acid are mediated by ampk activation. *Biochem. Pharmacol.* **2013**, *85*, 1341–1351. [CrossRef]
51. Cho, A.S.; Jeon, S.M.; Kim, M.J.; Yeo, J.; Seo, K.I.; Choi, M.S.; Lee, M.K. Chlorogenic acid exhibits anti-obesity property and improves lipid metabolism in high-fat diet-induced-obese mice. *Food Chem. Toxicol.* **2010**, *48*, 937–943. [CrossRef]
52. Chang, T.S. An updated review of tyrosinase inhibitors. *Int. J. Mol. Sci.* **2009**, *10*, 2440–2475. [CrossRef]
53. Lee, S.-G.; Karadeniz, F.; Seo, Y.; Kong, C.-S. Anti-Melanogenic Effects of Flavonoid Glycosides from Limonium tetragonum (Thunb.) Bullock via Inhibition of Tyrosinase and Tyrosinase-Related Proteins. *Molecules* **2017**, *22*, 1480. [CrossRef]
54. Choi, H.K.; Lim, Y.S.; Kim, Y.S.; Park, S.Y.; Lee, C.H.; Hwang, K.W. Free-radical-scavenging and tyrosinase-inhibition activities of Cheonggukjang samples fermented for various times. *Food Chem.* **2008**, *106*, 564–568. [CrossRef]
55. Lee, K.H.; Abdel Aziz, F.H.; Syahida, A.; Abas, F.; Shaari, K.; Israf, D.A. Synthesis and biological evaluation of curcumin-like diarylpentanoid analogues for anti-inflammatory, antioxidant and anti-tyrosinase activities. *Eur. J. Med. Chem.* **2009**, *44*, 3195–3200. [CrossRef] [PubMed]
56. Singh, B.; Beg, S.; Lohan, S.; Kapil, R. Crossing blood-brain barriers using drug delivery: A successful venture using lipidic nanostructured systems. *Pharm Rev.* **2013**, *21*, 41–47.
57. Generalić Mekinić, I.; Blažević, I.; Mudnić, I.; Burčul, F.; Grga, M.; Skroza, D.; Jerčić, I.; Ljubenkov, I.; Boban, M.; Miloš, M.; et al. Sea fennel (*Crithmum maritimum* L.): Phytochemical profile, antioxidative, cholinesterase inhibitory and vasodilatory activity. *J. Food Sci. Technol.* **2016**, *53*, 3104–3112. [CrossRef]
58. Seo, S.M.; Jung, C.S.; Kang, J.; Lee, H.R.; Kim, S.W.; Hyun, J.; Park, I.K. Larvicidal and Acetylcholinesterase Inhibitory Activities of Apiaceae Plant Essential Oils and Their Constituents against *Aedes albopictus* and Formulation Development. *J. Agric. Food Chem.* **2015**, *63*, 9977–9986. [CrossRef]

foods

Article

Volatiles from French and Croatian Sea Fennel Ecotypes: Chemical Profiles and the Antioxidant, Antimicrobial and Antiageing Activity of Essential Oils and Hydrolates

Olivera Politeo [1], Pavao Ćurlin [1], Petra Brzović [2], Killian Auzende [3], Christian Magné [3,*] and Ivana Generalić Mekinić [2,*]

[1] Department of Biochemistry, Faculty of Chemistry and Technology, University of Split, R. Boškovića 35, HR-21000 Split, Croatia; olivera@ktf-split.hr (O.P.); pavao.curlin@ktf-split.hr (P.Ć.)
[2] Department of Food Technology and Biotechnology, Faculty of Chemistry and Technology, University of Split, R. Boškovića 35, HR-21000 Split, Croatia; petra.brzovic@ktf-split.hr
[3] Laboratoire Géoarchitecture_Territoires, Biodiversité, Urbanisme, Environnement, Université de Bretagne Occidentale, 6 Avenue Victor Le Gorgeu, CS 93837, CEDEX 3, 29238 Brest, France; killian.aurende@univ-brest.fr
* Correspondence: christian.magne@univ-brest.fr (C.M.); gene@ktf-split.hr (I.G.M.); Tel.: +33-298-016427 (C.M.); +385-21-329458 (I.G.M.)

Abstract: Sea fennel is a halophytic plant rich in valuable nutritional components and is characterized by pleasant organoleptic properties. While its essential oils (EOs) are well investigated, there are no reports on the volatiles from their corresponding hydrolates, which are the main by-products of EO isolation, as well as on their biological activity. Therefore, the composition and biological activities of EOs and corresponding hydrolates of sea fennel from Atlantic (French, FRA) and Mediterranean (Croatian, CRO) ecotypes were investigated and compared. The EO from the CRO sample was characterized by an abundance of sabinene and limonene, while that from the FRA ecotype was rich in dillapiol and carvacryl methyl ether. The CRO hydrolate was rich in terpinen-4-ol and 10-(acetylmethyl)-3-carene, while dillapiol, thymyl methyl ether and γ-terpinene were the main compounds in the FRA sea fennel hydrolate. The biological activities of the EOs and hydrolates were evaluated for their antioxidant (with DPPH, NO, FRAP and ORAC bioassays), antimicrobial (against some Gram+ and Gram- spoilage bacteria) and antiageing (tyrosinase, elastase and collagenase inhibition) activities. Both EOs showed low reducing powers and antiradical activities while the ability of both hydrolates to quench NO was slightly higher (35–39% if inhibition). The FRA EO showed low activity against *Staphylococcus aureus* (8 mm), while CRO moderately inhibited the growth of *P. aeruginosa* (8 mm), but strongly inhibited the other two bacterial strains. While the French EO showed no antityrosinase and anticollagenase activity, the Croatian oil significantly inhibited both enzymes (IC_{50} of 650 μg/mL and IC_{50} of 2570 μg/mL, respectively) probably due to the dominance of limonene and sabinene. Neither EO exhibited antielastase properties, while the hydrolates from both ecotypes showed no antiageing activity, regardless of the enzyme tested. The EOs from the aerial parts of sea fennel from FRA and CRO differed greatly in composition, resulting in different activities. The Croatian samples appeared to have better biological properties and are therefore good candidates for applications as preservatives or antiageing agents.

Keywords: *Crithmum maritimum*; essential oil; hydrolate; volatile organic compounds; GC-MS; antioxidants; antimicrobials; antiageing agents

1. Introduction

Essential oils (EOs) are complex mixtures of volatiles extracted from aromatic plants by distillation, in most cases by hydrodistillation or steam distillation [1–3]. Hydrodistillation is the traditional and most commonly used method for the isolation of EOs from aromatic plants, as it is simple and inexpensive, but also environmentally friendly and non-toxic [4]. In hydrodistillation, the plant material is completely immersed in water until it boils, and the EO and water vapour are simultaneously condensed and collected [5].

During EO isolation, small amounts of the EO components, volatile hydrophilic compounds that can form weak hydrogen bonds, escape into the distillation water stream and are further condensed. The resulting aqueous solutions are called hydrolate, hydrosol, hydroflorate, aromatic water, flower water, essential water, floral water or distillate water. [1–3,5,6] These are complex, highly dilute, acidic aqueous mixtures containing mainly oxygenated and hydrophilic volatile components that contribute to the aroma [1,6]. The distribution of a compound between the EO and hydrolate largely depends on its solubility [6], but the losses of oxygenated hydrophilic components of the EO that migrate into the hydrolate make the flavour of the primary EO incomplete [3], but is also responsible for the fact that the composition of hydrolates often differs from the corresponding EOs, with hydrophobic isoprenoid compounds such as hydrocarbons being absent in most cases [1]. In cases where these compounds dominate in EOs, the compositions of the hydrolates are, therefore, very different from the corresponding EOs [3]. The benefits of hydrolates are related to the presence of polar or partially miscible water volatiles that migrate from the EOs, and due to the chemical characteristics of these compounds, they are recognised for their good biological activity such as antioxidant, antimicrobial, etc. [6].

Due to the increasing consumer demand for natural and safe products, EOs have recently been widely investigated as preservatives, flavour and aromatic compounds, pharmaceutic and therapeutic agents, etc. [7]. Hydrolates are also becoming increasingly popular due to their pleasant organoleptic properties which is why they are widely used across different industries. Recent studies have reviewed the characteristics and potential applications of hydrolates in cosmetology and perfumery, aromatherapy, folk medicine and the food industry [1,6–10]. In addition, hydrolates as by-products of EO isolation, which are usually treated as waste, are less harmful to human health compared to EOs and can, therefore, be used as natural and cheap raw materials with a wide range of applications in various industries [2].

The low number of research papers on hydrolates compared to EOs is presented in Figure 1, which shows the number of publications per year (until August 2023) on EOs and hydrolates from the Scopus database. The search criteria were "article title, abstract, keyword; essential oil and hydrosol/hydrolate". However, it can be noted that the number of publications on hydrolates has increased over the last decade and continues to increase (Figure 1), although it is relatively small compared to studies on EOs.

Sea fennel is a wild perennial halophytic plant from the Apiaceae botanical family, characterised by the presence of valuable nutritional components such as vitamin C, minerals, phenolics, fatty acids, etc., as well as a high content of EOs with a pleasant flavour characterised by aromatic notes of fennel, celery and citrus peel [11–15].

Previous studies on sea fennel's EOs and its volatiles confirmed a high geographical variability in sea fennel EOs. The main difference has been found in the presence and content of dillapiole, so that, according to Pateira et al. [16], there are two different chemotypes of the plant. Moreover, according to Renna [17], there are four main chemotypes of sea fennel based on the compounds prevalent in the EOs, namely aromatic monoterpenes, monoterpene hydrocarbons, phenylpropanoids and their intermediate forms. While the EOs from sea fennel in Mediterranean populations are well studied, like those from Turkey [18–22], Italy [23–25], Tunisia [12], Spain [26], Algeria [27], Greece [28] and France [29], there are few reports on volatiles from sea fennel populations on the Atlantic coasts of Portugal [9,13,16,26,30] and France [24,29,31].

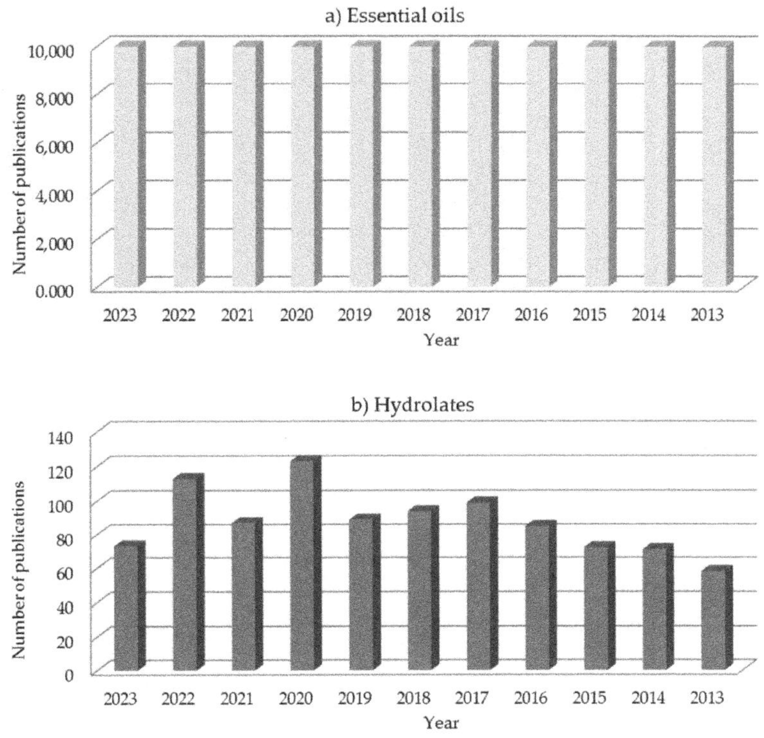

Figure 1. Number of publications per year on (**a**) essential oils and (**b**) hydrolates; obtained using search in Scopus database.

EOs from Mediterranean sea fennel are well documented, whereas there are few studies on the Atlantic sea fennel ecotype, and there are no reports on the volatiles within their corresponding hydrolates or on their biological activity. Therefore, the present study focuses on: (a) investigating and confirming the differences in the EO composition of typical Atlantic (French) and Mediterranean (Croatian) sea fennel ecotypes; (b) comparing the volatiles in the corresponding hydrolates and identifying correlations or links to the EOs; and (c) investigating and comparing the biological properties of the samples, both EOs and hydrolates, in terms of antioxidant, antimicrobial and antiageing activity.

2. Materials and Methods

2.1. Plant Material

About 2 kg of the aerial parts of the sea fennel were collected in October 2022 from two sampling sites; the Mediterranean ecotype was harvested in Dalmatia (43°39′44″ N 15°56′40″ E, Croatia), while the Atlantic ecotype was collected on the Brittany shoreline near Brest (48°21′47″ N 4°31′51″ O, France). The plant material was identified by the Tonka Ninčević Runjić (Institute for Adriatic Crops and Karst Reclamation, Split, Croatia) (CRO sample) and by Christian Magné (Université de Bretagne Occidentale, Brest, France) (FRA sample). The voucher specimens were also deposited in the herbariums of the authors' two institutions.

The samples were air-dried for 15–20 days in an aerated and shaded place before EO extraction.

2.2. Hydrodistillation

The EOs were isolated from dry plant materials by hydrodistillation (Clevenger apparatus, 3 h) following the procedure reported by Bektašević et al. [32]. The separated

EOs were dried over anhydrous Na_2SO_4. Both the EOs and the hydrolates were stored at −20 °C until analysis [33].

2.3. Chemical Analysis

The volatiles from the hydrolates were isolated by headspace solid-phase microextraction (HS-SPME) using DVB/CAR/PDMS-coated fibres (50/30 μm, Supelco, Sigma Aldrich, Bellefonte, PA, USA) according to the procedure described by Politeo et al. [33]. After thermostation (at 40 °C for 30 min), the fibre was introduced into the injection port of the gas chromatograph for thermal desorption (3 min).

EOs and hydrolates were analysed by gas chromatography–mass spectrometry (GC-MS) using a gas chromatograph (model 8890, Agilent Inc., Santa Clara, CA, USA) equipped with an automatic liquid injector (model 7693A, Agilent Inc.) and a mass spectrometer (model 7000D GC/TQ, Agilent Inc.) using the HP -5MS UI column (30 m × 0.25 mm, 0.25 μm, Agilent Inc.). The method conditions applied for the analysis of EOs and hydrolates were described in our previous study [33], and the detected compounds were identified by comparing their retention indices with a series of *n*-hydrocarbons (C8–C40, Supelco Inc., Sigma Aldrich) and their mass spectra with reference data from the Wiley 7 MS library (Wiley, NY, USA) and the NIST02 (Gaithersburg, MD, USA) database, as well as with literature reports [34]. The results are reported as mean ± standard deviation of two injections.

2.4. Biological Activity

2.4.1. Antioxidant Activity

The antioxidant activity was tested spectrometrically using a microplate reader (Synergy HTX Multi-Mode Reader, BioTek Instruments, Inc., Winooski, VT, USA) and a SPECORD 200 Plus, Edition 2010 (Analytik Jena AG, Jena, Germany).

The reducing activity of the samples was tested using the Ferric Reducing Antioxidant Power (FRAP) assay [35] and the results are expressed in μM Fe^{2+}/L.

Three different radicals were used in testing the radical scavenging activity of the samples: the synthetic stable 2,2-diphenyl-1-picrylhydrazyl radical (DPPH•) and two biologically important radicals: the nitric oxide radical (NO•) and the hydroperoxyl radical (HOO•).

DPPH inhibition activity was measured according to Katalinić et al. [36], while NO-radical inhibition was measured as described by Dastmalchi et al. [37]. The radical inhibition results are expressed as a percentage (%).

Oxygen Radical Absorbance Capacity (ORAC) assay was performed according to the method reported by Čagalj et al. [38], and the final results of inhibition of oxidation induced by peroxyl radicals are expressed in μM Trolox equivalents (μM TE).

For the FRAP, DPPH and NO assays, pure hydrolates and essential oils diluted up to a concentration of 1 mg/mL were tested, while for the ORAC measurements, EOs were tested at 10-times dilution and hydrolates at 100-times dilution. All measurements were performed with five replicates.

2.4.2. Antimicrobial Activity

The antibacterial activity of EOs and hydrolates from sea fennel against one Gram-negative (*Staphylococcus aureus* ATCC 33862) and two Gram-positive (*Escherichia coli* ATCC 1053 and *Pseudomonas aeruginosa* ATCC 27853) bacterial strains was investigated using the disk diffusion method [39].

Each bacterium was grown at 37 °C for 18 h on Tryptone Soy Broth (TSB) medium and turbidity was adjusted to 10^8 CFU/mL. Seventy microliters of the bacterial suspension was then spread onto a 9 cm Petri dish containing 15 mL of sterile Mueller Hinton Broth medium with 20 g/L agar. Paper disks (Whatman 3MM, Whatman, Inc., Clifton, NJ, USA) with a diameter of 5.5 mm were placed in the dish and soaked with 7.5 μL of EOs or hydrolates (reduced volume due to the small sample amount) at different concentrations.

Three replicates of each concentration were performed. A positive control dish contained only TSB medium inoculated with a bacterial suspension (10^8 CFU/mL), while 7.5 µL of antibiotics (5 mg/mL streptomycin) were added to the suspension of microorganisms in the negative control. The dishes were sealed with parafilm and, after incubation (24 h at 30 °C), the inhibition diameter was measured.

2.4.3. Antiageing Activity

Antityrosinase activity was evaluated using L-tyrosine as substrate and mushroom tyrosinase (SIGMA) according to the slightly modified method described by Masuda et al. [40]. The samples (plant EOs or hydrolates) were dissolved in DMSO (50%, *v*/*v*). Then, 40 µL of each sample was mixed with 80 µL phosphate buffer (0.1 M, pH 6.8), 40 µL tyrosinase (30 units/mL in phosphate buffer, pH 6.5) and 40 µL 2.5 mM L-tyrosine on a 96-well microplate. The absorbance of the sample was monitored kinetically at 475 nm for 15 min, taking a reading every 30 s, and compared to a blank containing all ingredients except EOs or hydrolates. The inhibition (%) of catalysis of L-tyrosine to L-dopa and then to dopaquinone was calculated using the following formula:

$$\text{Inhibition}(\%) = \left[1 - \left(\frac{\text{At} - \text{A0}}{\text{Ct} - \text{C0}}\right)\right] \times 100$$

where At and A0 are the absorbances of the sample at t and t = 0, respectively, and Ct and C0 are the absorbances of the control at t and t = 0, respectively. Finally, the IC_{50} (concentration of EO or hydrolate causing 50% enzyme inhibition) was determined from the regression curve of inhibition percentages.

Antielastase was assessed using elastase from porcine pancreas Type IV E0258 (Sigma) and N-Succinyl-Ala-Ala-Ala-*p*-nitroanilide (SANA) as substrate according to the method by Kalyana Sundaram et al. [41]. Twenty-five µL of the EOs or hydrolates at different concentrations in DMSO (5%, *v*/*v*) (3 replicates) were mixed with 175 µL of Tris-HCl buffer (0.2 M, pH 8) and 25 µL of enzyme (10 µg/mL) in each well of a microplate. After incubation (30 min at 37 °C), the reaction was initiated by adding 25 µL of SANA (1 mg/mL). Catalysis of SANA in *p*-nitroaniline was monitored for 15 min at 410 nm with a reading every 30 s. The negative control consisted of DMSO (5%, *v*/*v*) instead of the EOs, and the percentage of elastase inhibition (%) was calculated according to the following formula:

$$\text{Inhibition}(\%) = (1 - \frac{\text{Ss}}{\text{Sc}}) \times 100$$

where Ss and Sc are the slopes in the sample and control assays, respectively. Finally, the IC_{50} was determined using the regression curve of the inhibition percentage.

The anticollagenase activity of the EOs and hydrolates was investigated using the Enzo Life Sciences MMP-1 kit (Colorimetric Drug Discovery Kit BML-4K404, Enzo Life Sciences, Farmingdale, NY, USA). Each microplate well contained 50 µL of buffer (50 mM HEPES, 10 mM $CaCl_2$, 0.05% Brij-35, 1 mM DTNB, pH 7.5), 20 µL of the samples diluted in the buffer at various concentrations (3 replicates) and 20 µL of the enzyme at 637.5 mU/µL. The reaction was initiated by adding the chromogenic substrate Thiopeptide (Ac-PLG-[2-mercapto-4-methyl-pentanoyl]-LG-OC_2H_5) (10 µL at a concentration of 1 mM). The catalysis of thiopeptide to 2-nitro-5-thiobenzoic acid was monitored kinetically at 412 nm for 15 min, with a measurement every 30 s. The reaction rate was determined using the linear part of the kinetics (from 0 to 10 min). Controls were run with kit buffer without EO or hydrolate, and collagenase inhibition (%) was calculated using the following formula:

$$\text{Inhibition}(\%) = \left(1 - \frac{\text{Ss}}{\text{Sc}}\right) \times 100$$

where Ss and Sc are the slopes of the sample and control assays, respectively. Finally, the IC_{50} was determined using the regression curve of inhibition percentage.

2.5. Statistical Analysis

The data obtained (expressed as mean ± standard deviation) were statistically analysed using R 4.1.0 software (www.cran.r-project.org, accessed on 13 January 2024). Comparisons between data were performed using the Tukey post hoc test with statistical significance set at $p < 0.05$.

3. Results and Discussion

3.1. Chemical Composition of EOs and Hydrolates

3.1.1. Essential Oils

The composition of the sea fennel oils studied is shown in Table 1. A total of 13 compounds were detected in the French sea fennel, while 12 compounds were found in the Croatian sample. The main compounds in the French sea fennel EO were dillapiole (62.10 ± 1.83%), carvacryl methyl ether (18.00 ± 2.40%) and γ-terpinene (9.88 ± 1.10%) (89.98% of the total identified components), while the Croatian sea fennel EO was rich in sabinene (51.47 ± 3.22%) and limonene (36.28 ± 2.99%) (87.75% of the total identified components). It is clear from the results that the sea fennel samples studied are from completely different chemotypes, with the main difference being the presence or absence of dillapiole. Based on the results of previous studies, dillapiole was found to be a component of sea fennel EO in samples from Turkey [19–21], Italy [24], Tunisia [12], Greece [28], France [24,29,31] and Portugal [9,13,16], while it was never detected in Croatian samples [33,42–45].

Table 1. GC-MS chemical composition of French (FRA) and Croatian (CRO) sea fennel essential oils.

RI	Compounds	FRA (%)	CRO (%)
921	α-thujene	0.15 ± 0.02	nd
934	α-pinene	0.84 ± 0.13	tr
970	Sabinene	4.24 ± 0.10	51.47 ± 3.22
992	β-pinene	0.17 ± 0.02	0.89 ± 0.02
1012	α-terpinene	0.20 ± 0.02	0.98 ± 0.03
1020	*p*-cymene	2.97 ± 0.43	tr
1039	(*Z*)-β-ocimene	0.80 ± 0.17	nd
1032	Limonene	nd	36.28 ± 2.99
1042	(*E*)-β-ocimene	nd	tr
1056	γ-terpinene	9.88 ± 1.10	3.49 ± 0.07
1065	*cis*-sabinene hydrate	nd	0.10 ± 0.01
1086	Terpinolene	0.12 ± 0.01	0.37 ± 0.05
1118	*cis*-*p*-menth-2-en-1-ol	nd	0.10 ± 0.01
1176	terpinen-4-ol	0.25 ± 0.01	5.35 ± 0.04
1232	thymyl methyl ether	0.21 ± 0.01	nd
1242	carvacryl methyl ether	18.00 ± 2.40	nd
1620	Dillapiole	62.10 ± 1.83	nd
	TOTAL	99.93	99.03

RI = retention indices on HP-5MS UI column, tr—traces (<0.1%), nd—not detected.

However, few studies have reported that it is one of the dominant EO components. Senatore et al. [19] reported different EO compositions and chemical profiles of wild sea fennel at two sites in southern Turkey. While the presence of dillapiole could not be confirmed in the sample from Antalya, its content was significantly high (20.6%) in the sample from Mersin. Similar results were obtained by Özcan et al. [20], who also found variations in the dillapiole content in crops harvested in different years from the same site (1.9% in the first year and 21% in the second year). Jallali et al. [12] also found high dillapiole contents in samples from Tunisia (14.3 and 40.3%) by.

As is known by the authors, the presence of dillapiole has been confirmed in all sea fennel samples from the Atlantic coast, either from Portugal [13,14] or from France [24, 29,31]. Pateira et al. [16] reported variations in the composition of sea fennel EOs from plant samples at three stages of development (vegetative, flowering and fruiting), from different locations and from three years, with dillapiole content ranging from 0.3 to 46.4%. On the other hand, the samples from the French Atlantic coast had the highest dillapiole concentration, with 17.5% [31], 25% [29] and 55.7% [24].

As mentioned above, the other two dominant compounds in the studied sample from the Atlantic coast were carvacryl methyl ether (18%) and γ-terpinene ($\approx$10%). The results from Pavela et al. [24] for Atlantic sea fennel from France showed a thymyl methyl ether content of 11.8% and a γ-terpinene content of 14.0%, and Pavela et al. [31] showed a thymol methyl ether content of 2.0% and a γ-terpinene content of 33.0%.

The results obtained for Croatian sea fennel are consistent with previously published reports, in which more or less the same EO compounds are reported in different proportions. EOs from Croatian sea fennel were studied by Kulišić Bilušić et al. [42], Generalić Mekinić et al. [43], Politeo et al. [33] and Politeo et al. [44].

It is hard to draw general conclusions about the causes of the differences in the chemical composition of sea fennel EOs, as they may be due to the different abiotic and biotic factors that affect the synthesis and accumulation of plant secondary metabolites, such as the harvest location or geographical variations [24,29,31], plant vegetation/harvest period/growth cycle [16,20,28,39,45], plant part used [16,43,44], climatological factors [16,20], soil abiotic factors [26,39], etc. However, our results agreed with the conclusions of other authors [12,16,22], according to which the main reason is simply the existence of different plant chemotypes (intraspecific variability).

Kulišić Bilušić et al. [42] reported that limonene (58.4%), sabinene (26.5%), γ-terpinene (2.8%) and terpinene-4-ol (5.6%) are the main constituents of sea fennel EOs. In the research by Generalić Mekinić et al. [43] and Politeo et al. [44], the predominant compounds were the same, with slight differences in the amounts detected depending on the part of the plant analysed (leaves, flowers, stems). However, Politeo et al. [33] investigated the influence of the EO isolation method on the chemical profile of the EO and also found differences between the samples. The main EO compounds, obtained by hydrodistillation, were again limonene (51.4%) and sabinene (25.2%). These previous studies all reported a higher yield of limonene in the samples, while the sample from the present study had a higher content of sabinene. One of the reasons for this could be the when the plants were harvested, as the samples from the previous studies were collected in summer (from July to September), while the sample from this study was collected in October.

3.1.2. Hydrolates

Hydrolates are complex mixtures containing traces of EO constituents and other water-soluble compounds that give them their specific sensory properties and flavour, as well as their biological activity [3]. Hydrolates generally contain polar compounds such as alcohols (monoterpenes and sesquiterpenes), aldehydes, ketones and esters, while lipophilic hydrocarbon monoterpenes with low water solubility are usually not present [3,6]. According to Aćimović et al. [3], the ratio between hydrocarbons and oxygenated compounds in the EO affect the chemical profile of their corresponding hydrolates. In general, the hydrolate has a similar composition when the parent EO is rich in oxygenated compounds, while the composition of the hydrolate differs significantly from that of the EO when hydrocarbons are the main constituents of the EO. However, the authors also reported some studies with completely different profiles of EOs and their corresponding hydrolates.

The compounds within the volatiles from the sea fennel hydrolates are shown in Table 2 where 39 different compounds can be seen, differing in occurrence and quantity. While 32 components were detected in the French sample, only 17 volatile compounds were found in the Croatian sea fennel hydrolate. Again, the greatest differences were found in the compounds that were detected in the highest proportions. The predominant components in the French sea fennel hydrolate were dillapiole (36.66 $\pm$ 5.66%), thymyl methyl ether (26.30 $\pm$ 1.85%) and γ-terpinene (9.34 $\pm$ 0.24%), while in the Croatian sample they were terpinen-4-ol (41.93 $\pm$ 2.99%) and 10-(acetylmethyl)-3-carene (13.80 $\pm$ 2.01%) (Table 2). The results are consistent with our previous studies on sea fennel hydrolates, where terpinen-4-ol was the predominant compound with a yield of 13.86%, while 10-(acetylmethyl)-3-carene (13.45%), (*E*)-α-ionone (10.04%) and (*Z*)-β-damascenone (4.80%) were found in large amounts among the other compounds [33].

Table 2. GC-MS chemical composition of French (FRA) and Croatian (CRO) sea fennel hydrolates.

RI	Compounds	FRA (%)	CRO (%)
921	α-thujene	0.40 ± 0.02	nd
934	α-pinene	1.36 ± 0.02	nd
970	sabinene	5.64 ± 0.07	nd
986	2,3-dehydro-1,8-cineole	tr	nd
989	β-myrcene	0.61 ± 0.07	nd
998	octanal	tr	nd
1001	α-phellandrene	tr	nd
1012	α-terpinene	0.62 ± 0.04	nd
1020	*p*-cymene	6.40 ± 0.07	nd
1024	β-phellandrene	0.19 ± 0.01	nd
1039	(*Z*)-β-ocimene	2.10 ±0.01	nd
1041	benzeneacetaldehyde	0.77 ± 0.15	5.06 ± 0.03
1056	γ-terpinene	9.34 ± 0.24	nd
1065	*cis*-sabinene hydrate	tr	nd
1086	terpinolene	0.35 ± 0.02	nd
1095	*trans*-sabinene hydrate	tr	4.73 ± 0.04
1097	linalool	nd	tr
1118	*cis*-*p*-menth-2-en-1-ol	0.25 ± 0.04	4.26 ± 0.02
1138	*trans*-*p*-menth-2-en-1-ol	tr	3.37 ± 0.05
1176	terpinen-4-ol	2.12 ± 0.22	41.93 ± 2.99
1183	*p*-cymen-8-ol	tr	nd
1188	α-terpineol	0.21 ± 0.01	5.71 ± 0.51
1205	*trans*-pipertiol	nd	2.46 ± 0.01
1218	*trans*-carveol	0.20 ± 0.01	3.36 ± 0.03
1222	*cis*-carveol	nd	3.75 ± 0.02
1232	thymyl methyl ether	26.30 ± 1.85	nd
1242	carvacryl methyl ether	0.28 ± 0.02	nd
1293	thymol	0.32 ± 0.03	tr
1301	carvacrol	tr	1.07 ± 0.02
1312	*p*-vinylguaiacol	0.76 ± 0.04	nd
1327	myrtenyl acetate	nd	0.99 ± 0.01
1384	(*E*)-β-damascenone	nd	1.03 ± 0.09
1390	10-(acetylmethyl)-3-carene	nd	13.80 ± 2.01
1422	dihydrodehydro-β-ionone	nd	5.88 ± 0.09
1498	bicyclogermacrene	0.23 ± 0.01	nd
1521	myristicin	0.48 ± 0.00	nd
1557	elemicin	tr	nd
1563	germacrene B	0.39 ± 0.02	nd
1620	dillapiole	36.66 ± 5.66	1.52 ± 0.01
	TOTAL	95.98	98.92

RI = retention indices on HP-5MS UI column, tr—traces (<0.1%), nd—not detected.

Inouye et al. [46] reported that plant hydrolates can be categorised into several groups based on the functional group of the main constituent: alcohol, aldehyde, ketone, ester, phenol or phenyl methyl ether groups. Accordingly, the hydrolate from the French sea fennel belongs to the phenyl methyl ether group, while the hydrolate from the Croatian sea fennel belongs to the alcohol group.

The results in Tables 1 and 2 show differences in the chemical composition of the EOs and hydrolates of the Croatian and French samples. The French sea fennel EOs and the corresponding hydrolates were rich in dillapiole, whereas the dominant components of the Croatian EOs (limonene and sabinene) were not found in its hydrolate. This is consistent with the fact that the distribution of a component between the EO and hydrolate depends primarily on its solubility in the EO itself. The French hydrolate showed a high content of dillapiole and γ-terpinene, but also a high content of thymyl methyl ether (26.3%), which was present in low concentrations in the corresponding EO. The reverse relationship was found for carvacryl methyl ether, which was present in the EO (18.0%) and found in a much lower amount in the hydrolate (0.28%). As expected, sabinene (bicyclic unsaturated monoterpene) and limonene (aliphatic hydrocarbon, cyclic monoterpene), compounds that dominate in Croatian sea fennel EO due to their lipophilicity, were not found in the hydrolate.

3.2. Biological Activities of EOs and Hydrolates

While the biological activity of EOs has been extensively studied, especially with regard to their antimicrobial or antioxidant properties, hydrolates have only recently gained attention [3,4,6]. Here, we report the results of a comparative study on the antioxidant, antimicrobial and antiageing activities of EOs and hydrolates from Atlantic and Mediterranean ecotypes of sea fennel.

3.2.1. Antioxidant Activity

In this study, the antioxidant activity of the samples was tested using four different methods: FRAP, DPPH, NO and ORAC. While the FRAP method was used to evaluate the reducing activity of the samples, the other three methods provided information on the radical scavenging activity of the samples. The results obtained are shown in Table 3.

Table 3. Antioxidant activity of the French (FRA) and Croatian (CRO) sea fennel essential oils (EOs) and hydrolates.

Antioxidant Assay	FRA EO	CRO EO	FRA Hydrolate	CRO Hydrolate	Gallic Acid (Standard)
FRAP (μM Fe^{2+}/L)	1.34 ± 0.17 c	1.11 ± 0.08 b	0.77 ± 0.19 a	1.16 ± 0.15 b	2485.09 ± 3.73 *
DPPH (% inhibition)	1.99 ± 0.16 a	2.00 ± 0.17 a	2.39 ± 0.25 b	2.17 ± 0.13 ab	95.35 ± 0.33 **
NO (% inhibition)	10.11 ± 0.14 a	11.30 ± 0.24 b	38.94 ± 0.13 d	35.20 ± 0.19 c	11.56 ± 0.90 **
ORAC (μM Trolox equivalents/L)	115.77 ± 0.35 bc	95.48 ± 3.82 b	36.22 ± 2.86 a	138.05 ± 0.41 c	-

* at concentration 0.1 mg/mL; ** at concentration 1 mg/mL. In each line, values followed by different letter are statistically significant ($p < 0.05$, by Tukey's test).

The reducing activity of all samples is almost negligible compared to the activity of gallic acid which is recognised as one of the most potent natural phenolic antioxidants. The FRAP values of the tested samples ranged between 0.77 and 1.34 μM Fe^{2+}/L. Similar observations can be made for the DPPH inhibition results with an inhibition percentage of 2% for both EOs, and a slightly higher activity for the corresponding hydrolates. Gallic acid, which was tested at the same concentration, provided 95% of radical inhibition.

The results show that the antiradical activity against NO• differs between the samples. The activity of gallic acid was not significant in this method as it inhibited a similar share of radicals to the EO samples. Both EOs and the standard compound (at concentration of 1 mg/mL) showed more than three-times-lower activity than the corresponding hydrolates, with the CRO EO giving slightly better results. On the other hand, the hydrolates from FRA showed better activity than the sample from CRO.

The radical scavenging activity against peroxyl radicals was tested with the ORAC assay, which measures the degree of inhibition of oxidation triggered by peroxyl radicals. While the ORAC values of CRO and FRA sea fennel EOs did not differ significantly, the activity of the hydrolate CRO was 3.8 times higher (138 vs. 36 μM Trolox equivalents/L) than that of FRA. The results of previous studies are in agreement with these, showing that the hydrolates have a low or moderate antioxidant capacity [47], but their chemical composition is crucial for understanding the mechanisms of their biological activity [1,2].

The weak free radical scavenging activity of sea fennel EOs due to the high content of terpenes, which are not recognised as potent antioxidants, was also reported by Kulišić-Bilušić et al. [42] and Jallali et al. [12] Sharopov et al. [48] investigated the antioxidant activity of eighteen common essential oil constituents using the FRAP and DPPH methods and reported the high antioxidant activity of carvacrol, eugenol and thymol, while the activity of limonene was low in both methods. The chemical composition of CRO EO was characterised by the presence of non-oxygenated compounds (93.5%), while only a small amount of oxygenated compounds was detected, confirming previous reports on the low antioxidant activity of terpenes. In contrast, the FRA EO was richer in oxygenated compounds (80.6%), with dillapiol (62.1%) dominating, suggesting that its antioxidant activity is probably very low (no data were found in the literature), especially since it does not contain hydroxyl groups in its structure that could donate hydrogen to stabilise free radicals. In addition, the potentially good antioxidant activity of other compounds that were present in low amounts in the samples, such as α- and γ-terpinene and terpinolene [49], should not be neglected. On the other hand, all compounds detected in the CRO hydrolate were oxygenated compounds (oxygenated hydrocarbons), while represented 68% of the compounds in the FRA hydrolate, which probably affected the antiradical activity of the samples.

3.2.2. Antimicrobial Activity

As can be seen (Table 4), the French EO showed no antimicrobial activity against *Escherichia coli* and *Pseudomonas aeruginosa*, but a slight activity against *Staphylococcus aureus* (8 mm). Conversely, the Croatian EO moderately inhibited the growth of *P. aeruginosa* (8 mm), but strongly inhibited that of the other two bacterial strains (18 and 25 mm for *E. coli* and *S. aureus*, respectively). Interestingly, the CRO EO's inhibition of the latter bacterial strain was as strong as that of the commonly used bactericidal streptomycin. Furthermore, neither the French nor the Croatian hydrolates showed any antimicrobial effect.

Table 4. Antimicrobial activity of the French (FRA) and Croatian (CRO) sea fennel essential oils (EOs) and hydrolates, expressed as inhibition diameter (mm).

Bacterial Strains	FRA EO	CRO EO	FRA Hydrolate	CRO Hydrolate	Streptomycin
Escherichia coli	Ø	18.0 ± 1.3^{b}	Ø	Ø	28.5 ± 2.1^{a}
Staphylococcus aureus	8.0 ± 0.5^{b}	24.8 ± 3.3^{a}	Ø	Ø	22.3 ± 1.0^{a}
Pseudomonas aeruginosa	Ø	8.3 ± 0.6^{b}	Ø	Ø	26.6 ± 1.5^{a}

Data are means of inhibition diameter (mm) ± S.D. of three replicates. In each line, values followed by different letter are statistically significant ($p < 0.05$, by Tukey's test); Ø no inhibition activity.

These results can be discussed taking into account the composition of the EOs and hydrolates. As mentioned above, the CRO EO showed stronger antibacterial activity against all tested bacteria. This could be due to the predominance of monoterpene hydrocarbons in this EO [50]. In particular, the broad-spectrum bactericidal activity of limonene is well documented [51–54], which allows it to be widely used in antibacterial treatment and food preservation. In particular, the mechanism of the bactericidal effect of limonene has been elucidated [54]. A similar observation can be made for sabinene, the other main component of CRO EO [55], while terpinen-4-ol probably contributes much less to the measured antimicrobial activities [56]. Pedreiro et al. [9] also investigated the antibacterial activity of sea fennel EOs from Portugal against *E. coli* and *S. aureus*, and reported good activity, of 8.2 and 11.6 mm, respectively. According to the authors, this activity was due to the high content of γ-terpinene and sabinene in the EO studied.

In contrast, the FRA EOs showed weak antimicrobial activity and only slightly inhibited the growth of *S. aureus*. This can also be related to the composition of this EO, which differs greatly from that of the CRO ecotype. For example, the main component, dillapiole, was only found to have repulsive effects against mosquitos [57], but no antimicrobial activity. In addition, carvacryl methyl ether has no bactericidal effect despite its structure being close to carvacrol [58]. The very limited activity against *S. aureus* could be explained by the low content of sabinene in FRA EO.

Sánchez-Hernández et al. [59] reported the antibacterial activity of sea fennel EOs, rich in γ-terpinene, thymol methyl ether, and dillapiole, against *S. aureus* (13 mm) but found there was no activity against *E. coli*, which is in accordance with the results of this study. Activity against *E. coli* was also not detected in the study by Jallali et al. [12], while the diameter of inhibition against *P. aeruginosa* was 10 mm. On the other hand, the EO was more effective against the tested Gram-positive species, *B. cereus* and *S. aureus* with inhibition zones of 13 and 15 mm, respectively. Again, the tested EO was rich in γ-terpinene (19.3%), thymol methyl ether (20.6%) and dillapiole (40.2%).

Kunicka-Styczynska et al. [60] investigated the antibacterial activity of lavender hydrolates (from fresh or dried herbs or flowers) and found no activity, which could be due to the high dilution of the samples, which according to the authors, could also be the reason for the ineffectiveness of the sea fennel samples tested in this study.

3.2.3. Antiageing Activity

The results of antityrosinase activity of EOs and hydrolates from French and Croatian ecotypes are shown in Table 5. While the EO from the sea fennel from France showed no antityrosinase activity, the Croatian EO inhibited mushroom tyrosinase (IC_{50} = 649 μg/mL), although less than the arbutin standard. Since the main components of CRO EO are limonene and sabinene and these compounds have strong antityrosinase properties in other plant EOs [61,62], it is likely that the strong activity measured in CRO EO is due to the abundance of these two monoterpene hydrocarbons. Interestingly, the same observations could be made for anticollagenase activity, where only the CRO EO inhibited the enzyme (IC_{50} = 2571 μg/mL). Again, the anticollagenase activity of the two terpenoids limonene and sabinene was recently reported [63]. These values for anti-ageing activity could be considered low, but were obtained after a huge dilution (1000-fold) of the pure EOs. The EOs from the French and Croatian ecotypes of sea fennel showed no antielastase properties. Finally, the hydrolates from both ecotypes showed no antiageing activity, regardless of the enzyme used. Antimicrobial and antioxidant compounds in hydrolates are important to prevent the oxidation of ingredients in various cosmetic products and to maintain their quality, safety and shelf life. Their advantages also lie in the fact that they are of natural origin and can be used for flavouring due to the presence of compounds with pleasant aroma properties [6].

Table 5. Antiageing activities of the French (FRA) and Croatian (CRO) sea fennel essential oils (EOs) and hydrolates, as well as of positive controls. Antityrosinase, anticollagenase, and antielastase are expressed as IC_{50} in μg/mL.

Enzymes	FRA EO	CRO EO	Arbutin	EGCG	Ursolic Acid
Tyrosinase	Ø	649 ± 54	137 ± 6	/	/
Collagenase	Ø	2571 ± 334	/	51 ± 7	/
Elastase	Ø	Ø	/	/	238 ± 13

EGCG–epigallocatechin gallate. Data are means of IC50 ± S.D. of three replicates. Ø, no inhibition activity.

4. Conclusions

The investigated essential oils from the Mediterranean (chemotype II and monoterpene hydrocarbon chemotype) and Atlantic (chemotype I and phenylpropanoid chemotype) sea fennel populations differ in their chemistry and the presence/content of the major constituents, as well as the corresponding hydrolates, which mainly contain oxygenated compounds. The predominant compound in the FRA EO was dillapiole (62.10%), while sabinene (51.47%) and limonene (36.28%) were detected in the highest amounts in the Croatian sample. While dillapiole had the highest percentage in the FRA hydrolate, sabinene was the most abundant constituent of the CRO hydrolate. Although the antioxidant activity of both EOs and hydrolates was not appreciable, the antimicrobial activity of the CRO EO against *S. aureus* and *P. aeruginosa* is comparable to that of streptomycin. Probably due to the high content of limonene and sabinene, the CRO EO also showed strong antityrosinase and anticollegenase inhibitory activity, while the FRA EO showed no activity, underlining its potential for use in the cosmetics industry. Given the good chemical composition and biological activities of the Croatian sea fennel chemotype, further studies directed on its isolates and their application should be conducted. Hydrolates, as by-products of EO isolation, are usually treated as a waste and are disposed of. However, they can serve as natural, valuable and cost-effective raw materials with a wide range of applications that will also have a positive impact on the environment.

Author Contributions: Conceptualization, O.P., C.M. and I.G.M.; methodology, O.P., C.M. and I.G.M.; validation, O.P., P.Ć., P.B. and K.A.; formal analysis, O.P., P.Ć., P.B., K.A., C.M. and I.G.M.; investigation, O.P., C.M. and I.G.M.; data curation, O.P., P.B., C.M. and I.G.M.; writing—original draft preparation, I.G.M.; writing—review and editing, O.P., P.Ć., P.B., K.A., C.M. and I.G.M.; visualization, P.B. and I.G.M.; supervision, C.M. and I.G.M.; funding acquisition, I.G.M. All authors have read and agreed to the published version of the manuscript.

Funding: This work is part of the PRIMA program supported by the European Union. Project title: "Innovative sustainable organic sea fennel (*Crithmum maritimum* L.)-based cropping systems to boost agrobiodiversity, profitability, circularity, and resilience to climate changes in Mediterranean small farms" (acronym: SEAFENNEL4MED) (https://seafennel4med.com/, accessed on 25 January 2024).

Institutional Review Board Statement: Not applicable.

Informed Consent Statement: Not applicable.

Data Availability Statement: The original contributions presented in the study are included in the article, further inquiries can be directed to the corresponding author.

Acknowledgments: The authors are also thankful for the scientific-research equipment financed via the EU grant "Functional integration of the University of Split, PMFST, PFST and KTFST through the development of the scientific and research infrastructure" (KK.01.1.1.02.0018). ANR is also acknowledged for funding the French contribution to SeaFennel4Med project.

Conflicts of Interest: The authors declare no conflicts of interest.

References

1. Rajeshwara Rao, B.R. Hydrosols and water-soluble essential oils: Medicinal and biological Properties. In *Recent Progress in Medicinal Plants: Essential Oils I*, 1st ed.; Studium Press LLC: Houston, TX, USA, 2013; Chapter 6; Volume 36.
2. D'Amato, S.; Serio, A.; Chaves López, C.; Paparella, A. Hydrosols: Biological activity and potential as antimicrobials for food applications. *Food Control* **2018**, *18*, 126–137. [CrossRef]
3. Aćimović, M.; Tešević, V.; Smiljanic, K.; Cvetkovic, M.; Stankovic, J.; Kiprovski, B.; Sikora, V. Hydrolates: By-Products of Essential Oil Distillation: Chemical Composition, Biological Activity and Potential Uses. *Adv. Technol.* **2020**, *9*, 54–70. [CrossRef]
4. Filly, A.; Fabiano-Tixier, A.S.; Louis, C.; Fernandez, X.; Chemat, F. Water as a green solvent combined with different techniques for extraction of essential oil from lavender flowers. *Comptes Rendus Chim.* **2016**, *19*, 707–717. [CrossRef]
5. De Santis, D.; Turchetti, G. Hydrolates: Characteristics, Properties, and Potential Uses in the Food Industry. In *New Findings from Natural Substances*, 1st ed.; Tiezzi, A., Ovidi, E., Karpiński, T.M., Eds.; Bentham Books: Sharjah, United Arab Emirates, 2022; Volume 28, pp. 183–210.
6. Shafie, M.H.; Kamal, M.L.; Abdul Razak, N.A.; Hasan, S.; Uyup, N.H.; Rashid, N.F.A.; Zafarina, Z. Antioxidant and antimicrobial activity of plant hydrosol and Its Potential Application in Cosmeceutical Products. *Jundishapur J. Nat. Pharm. Prod.* **2022**, *17*, e124018. [CrossRef]
7. Ni, Z.J.; Wang, X.; Shen, Y.; Thakur, K.; Han, J.; Zhang, J.G.; Hu, F.; Wei, Z.J. Recent updates on the chemistry, bioactivities, mode of action, and industrial applications of plant essential oils. *Trends Food Sci Technol.* **2021**, *110*, 78–89. [CrossRef]
8. Hamedi, A.; Mahmoud Moheimani, S.; Sakhteman, S.; Etemadfard, H.; Moein, M. An overview on indications and chemical composition of aromatic waters (hydrosols) as functional beverages in Persian nutrition, culture and folk medicine for hyperlipidemia and cardiovascular conditions. *J. Evid. Based Complement. Altern. Med.* **2017**, *22*, 544–561. [CrossRef] [PubMed]
9. Pedreiro, S.; Figueirinha, A.; Cavaleiro, C.; Cardoso, O.; Donato, M.M.; Salgueiro, L.; Ramos, F. Exploiting the *Crithmum maritimum* L. Aqueous Extracts and Essential Oil as Potential Preservatives in Food, Feed, Pharmaceutical and Cosmetic Industries. *Antioxidants* **2023**, *12*, 252. [CrossRef] [PubMed]
10. Chizzola, R.; Billiani, F.; Singer, S.; Novak, J. Diversity of Essential Oils and the Respective Hydrolates Obtained from Three *Pinus cembra* Populations in the Austrian Alps. *Appl. Sci.* **2021**, *11*, 5686. [CrossRef]
11. Atia, A.; Barhoumi, Z.; Mokded, R.; Abdelly, C.; Smaoui, A. Environmental eco-physiology and economical potential of the halophyte *Crithmum maritimum* L. (Apiaceae). *J. Med. Plant Res.* **2011**, *5*, 3564–3571.

12. Jallali, I.; Zaouali, Y.; Missaoui, I.; Smeoui, M.; Abdelly, C.; Ksouri, R. Variability of antioxidant and antibacterial effects of essential oils and acetonic extracts of two edible halophytes: *Crithmum maritimum* L. and *Inula crithmoïdes* L. *Food Chem.* **2014**, *145*, 1031–1038. [CrossRef]
13. Alves-Silva, J.M.; Guerra, I.; Gonçalves, M.J.; Cavaleiro, C.; Cruz, M.T.; Figueirinha, A.; Salgueiro, L. Chemical composition of *Crithmum maritimum* L. essential oil and hydrodistillation residual water by GC-MS and HPLC-DAD-MS/MS, and their biological activities. *Ind. Crops Prod.* **2020**, *149*, 112329. [CrossRef]
14. Kraouia, M.; Nartea, A.; Maoloni, A.; Osimani, A.; Garofalo, C.; Fanesi, B.; Ismaiel, L.; Aquilanti, L.; Pacetti, D. Sea Fennel (*Crithmum maritimum* L.) as an Emerging Crop for the Manufacturing of Innovative Foods and Nutraceuticals. *Molecules* **2023**, *28*, 4741. [CrossRef] [PubMed]
15. Correia, I.; Antunes, M.; Tecelão, C.; Neves, M.; Pires, C.L.; Cruz, P.F.; Rodrigues, M.; Peralta, C.C.; Pereira, C.D.; Reboredo, F.; et al. Nutritive Value and Bioactivities of a Halophyte Edible Plant: *Crithmum maritimum* L. (Sea Fennel). *Plants* **2024**, *13*, 427. [CrossRef] [PubMed]
16. Pateira, L.; Nogueira, T.; Antunes, A.; Venâncio, F.; Tavares, R.; Capelo, J. Two chemotypes of *Crithmum maritimum* L. from Portugal. *Flavour Fragr. J.* **1999**, *14*, 333–343. [CrossRef]
17. Renna, M. Reviewing the Prospects of Sea Fennel (*Crithmum maritimum* L.) as Emerging Vegetable Crop. *Plants* **2018**, *7*, 92. [CrossRef]
18. Baser, K.H.C.; Özek, T.; Demirci, B.; Saritas, Y. Essential oil of *Crithmum maritimum* L. from Turkey. *J. Essent. Oil Res.* **2000**, *12*, 424–426. [CrossRef]
19. Senatore, F.; Napolitano, F.; Özcan, M. Composition and antibacterial activity of the essential oil from *Crithmum maritimum* L. (Apiaceae) growing wild in Turkey. *Flavour Fragr. J.* **2000**, *15*, 186–189. [CrossRef]
20. Özcan, M.; Akgül, A.; Başcr, K.H.C.; Özck, T.; Tabanca, N. Essential oil composition of sea fennel (*Crithmum maritimum*) form Turkey. *Nahrung/Food* **2001**, *45*, 353–356. [CrossRef]
21. Musa Özcan, M.; Pedro, L.G.; Cristina Figueiredo, A.; Barroso, J.G. Constituents of the essential oil of sea fennel (*Crithmum maritimum* L.) growing wild in Turkey. *J. Med. Food* **2006**, *9*, 128–130. [CrossRef]
22. Musa Özcan, M.; Uslu, N.; Figueredo, G.; Al Juhaimi, F.; Ghafoor, K.; Babiker, E.E.; Alsawmahi, O.N.; Özcan, M.M.; Ahmed, I.A.; Ahmed, M. The effect of fermentation process on bioactive properties, essential oil composition and phenolic constituents of raw fresh and fermented sea fennel (*Crithmum maritimum* L.) leaves. *Indian J. Tradit. Knowl.* **2019**, *18*, 800–804.
23. Ruberto, G.; Tiziana Baratta, M.; Deans, S.G.; Damien Dorman, H.J. Antioxidant and antimicrobial activity of *Foeniculum vulgare* and *Crithmum maritimum* essential oil. *Planta Med.* **2000**, *66*, 687–693. [CrossRef] [PubMed]
24. Pavela, R.; Maggi, F.; Lupidi, F.; Cianfaglione, K.; Dauvergne, X.; Bruno, M.; Benelli, G. Efficacy of sea fennel (*Crithmum maritimum* L., Apiaceae) essential oils against *Culex quinquefasciatus* Say and *Spodoptera littoralis* (Boisd.). *Ind. Crops Prod.* **2017**, *109*, 603–610. [CrossRef]
25. D'Agostino, G.; Giambra, B.; Palla, F.; Bruno, M.; Badalamenti, N. The Application of the Essential Oils of *Thymus vulgaris* L. and *Crithmum maritimum* L. as Biocidal on Two *Tholu Bommalu* Indian Leather Puppets. *Plants* **2021**, *10*, 1508. [CrossRef] [PubMed]
26. Moreira, X.; Pérez-Ramos, I.M.; Matías, L.; Francisco, M.; García-González, A.; Martins-Noguerol, R.; Vázquez-González, C.; Abdala-Roberts, L.; Cambrollé, J. Efects of soil abiotic factors and plant chemical defences on seed predation on sea fennel (*Crithmum maritimum*). *Plant Soil* **2021**, *465*, 289–300. [CrossRef]
27. Nabet, N.; Boudries, H.; Chougui, N.; Loupassaki, S.; Souagui, S.; Burló, F.; Hernàndez, F.; Carbonell-Barrachina, A.A.; Madani, K.; Larbat, R. Biological activities and secondary compound composition from *Crithmum maritimum* aerial parts. *Int. J. Food Prop.* **2017**, *20*, 1843–1855. [CrossRef]
28. Zafeiropoulou, V.; Tomou, E.M.; Douros, A.; Skaltsa, H. The Effect of Successive Harvesting on The Volatile Constituents of Two Essential Oils of Cultivated Populations of Sea Fennel (*Crithmum maritimum* L.) in Greece. *J. Essent. Oil-Bear. Plants* **2021**, *24*, 1–11. [CrossRef]
29. Coiffard, L.; Piron-Frenet, M.; Amicel, L. Geographical variations of the constituents of the essential oil of *Crithmum maritimum* L., Apiaceae. *Int. J. Cosmet. Sci.* **1993**, *15*, 15–21. [CrossRef]
30. Pereira, C.G.; Moraes, C.B.; Franco, C.H.; Feltrin, C.; Grougnet, R.; Barbosa, E.G.; Panciera, M.; Correia, C.R.D.; Rodrigues, M.J.; Custódio, L. In Vitro Anti-*Trypanosoma cruzi* Activity of Halophytes from Southern Portugal Reloaded: A Special Focus on Sea Fennel (*Crithmum maritimum* L.). *Plants* **2021**, *10*, 2235. [CrossRef]
31. Pavela, R.; Benelli, G.; Pavoni, L.; Bonacucina, G.; Cespi, M.; Cianfaglione, K.; Bajalan, I.; Morshedloo, M.R.; Lupidi, G.; Romano, D.; et al. Microemulsions for delivery of Apiaceae essential oils—Towards highly effective and eco-friendly mosquito larvicides? *Ind. Crops Prod* **2019**, *129*, 631–640. [CrossRef]
32. Bektasevic, M.; Jurin, M.; Roje, M.; Politeo, O. Phytochemical profile, antioxidant activity and cholinesterase inhibition potential of essential oil and extracts of *Teucrium montanum* from Bosnia and Herzegovina. *Separations* **2023**, *10*, 421. [CrossRef]
33. Politeo, O.; Popović, M.; Veršić Bratinčević, M.; Koceić, P.; Ninčević Runjić, T.; Generalić Mekinić, I. Conventional vs. microwave-assisted hydrodistillation: Influence on the chemistry of sea fennel essential oil and its by-products. *Plants* **2023**, *12*, 1466. [CrossRef] [PubMed]
34. Adams, R.P. *Identification of Essential Oil Components by Gas Chromatography/Mass Spectroscopy*, 4th ed.; Allured Publishing Corporation: Carol Stream, IL, USA, 2017.

35. Benzie, I.F.F.; Strain, J.J. Ferric reducing/antioxidant power assay: Direct measure of total antioxidant activity of biological fuids and modifed version for simultaneous measurement of total antioxidant power and ascorbic acid concentration. *Methods Enzymol.* **1999**, *299*, 15–27.
36. Katalinić, V.; Smole Možina, S.; Skroza, D.; Generalić, I.; Abramovič, H.; Miloš, M.; Ljubenkov, I.; Piskernik, S.; Pezo, I.; Terpinc, P.; et al. Polyphenolic profile, antioxidant properties and antimicrobial activity of grape skin extracts of 14 *Vitis vinifera* varieties grown in Dalmatia (Croatia). *Food Chem.* **2010**, *119*, 715–723. [CrossRef]
37. Dastmalchi, K.; Dorman, H.J.D.; Oinonen, P.P.; Darwis, Y.; Laakso, I.; Hiltunen, R. Chemical composition and in vitro antioxidative activity of a lemon balm (*Melissa officinalis* L.) extract. *Lebensmittel-Wissenschaft Technol.* **2008**, *41*, 391–400. [CrossRef]
38. Čagalj, M.; Skroza, D.; Razola-Díaz, M.D.C.; Verardo, V.; Bassi, D.; Frleta, R.; Generalić Mekinić, I.; Tabanelli, G.; Šimat, V. Variations in the Composition, Antioxidant and Antimicrobial Activities of *Cystoseira compressa* during Seasonal Growth. *Mar. Drugs* **2022**, *20*, 64. [CrossRef] [PubMed]
39. Meot-Duros, L.; Le Floch, G.; Magné, C. Radical scavenging, antioxidant and antimicrobial activities of halophytic species. *J. Ethnopharmac.* **2008**, *116*, 258–262. [CrossRef] [PubMed]
40. Masuda, T.; Yamashita, D.; Takeda, Y.; Yonemori, S. Screening for Tyrosinase Inhibitors among extracts of seashore plants and identification of potent inhibitors from *Garcinia subelliptica*. *Biosci. Biotechnol. Biochem.* **2005**, *69*, 197–201. [CrossRef] [PubMed]
41. Kalyana Sundaram, I.; Deeptirekha Sarangi, D.; Sundararajan, V.; George, S.; Sheik Mohideen, S. Poly herbal formulation with anti-elastase and anti-oxidant properties for skin anti-aging. *BMC Complement. Altern. Med.* **2018**, *18*, 33. [CrossRef]
42. Kulišić-Bilušić, T.; Blažević, I.; Dejanović, B.; Miloš, M.; Pifat, G. Evaluation of the antioxidant activity of essential oils from caper (*Capparis Spinosa*) and sea fennel (*Crithmum Maritimum*) by different methods. *J. Food Biochem.* **2010**, *34*, 286–302. [CrossRef]
43. Generalić Meknić, I.; Blažević, I.; Mudnić, I.; Burčul, F.; Grga, M.; Skroza, D.; Jerčić, I.; Ljubenkov, I.; Boban, M.; Miloš, M.; et al. Sea fennel (*Crithmum maritimum* L.): Phytochemical profile, antioxidative, cholinesterase inhibitory and vasodilatory activity. *J. Food Sci. Technol.* **2016**, *53*, 3104–3112. [CrossRef]
44. Politeo, O.; Popović, M.; Veršić Bratinčević, M.; Kovačević, K.; Urlić, B.; Generalić Mekinić, I. Chemical profiling of sea fennel (Crithmum maritimum L., Apiaceae) essential oils and their isolation residual waste-waters. *Plants* **2023**, *12*, 214. [CrossRef] [PubMed]
45. Generalić Meknić, I.; Šimat, V.; Ljubenkov, I.; Burčul, F.; Grga, M.; Mihajlovski, M.; Lončar, R.; Katalinić, V.; Skroza, D. Influence of the vegetation period on sea fennel, *Crithmum maritimum* L. (Apiaceae), phenolic composition, antioxidant and anticholinesterase activities. *Ind. Crop. Prod.* **2018**, *124*, 947–953. [CrossRef]
46. Inouye, S.; Takahashi, M.; Abe, S. A comparative study on the composition of fourty four hydrosols and their essential oils. *Int. J. Essent. Oil Ther.* **2008**, *2*, 89–104.
47. Jakubczyk, K.; Tuchowska, A.; Janda-Milczarek, K. Plant hydrolates—Antioxidant properties, chemical composition and potential applications. *Biomed. Pharmacother.* **2021**, *142*, 112033. [CrossRef]
48. Sharopov, F.S.; Wink, M.; Setzer, W.N. Radical scavenging and antioxidant activities of essential oil components—an experimental and computational investigation. *Nat Prod Commun.* **2015**, *10*, 153–156. [CrossRef] [PubMed]
49. Grassmann, J. Terpenoids as plant antioxidants. *Vitam Horm.* **2005**, *72*, 505–535.
50. Chouhan, S.; Sharma, K.; Guleria, S. Antimicrobial activity of some essential oils-present status and future perspectives. *Medicines* **2017**, *4*, 58. [CrossRef]
51. Di Pasqua, R.; Hoskins, N.; Betts, G.; Mauriello, G. Changes in Membrane Fatty Acids Composition of microbial cells induced by addiction of thymol, carvacrol, limonene, cinnamaldehyde, and eugenol in the growing media. *J. Agric. Food Chem.* **2006**, *54*, 2745–2749. [CrossRef]
52. Lee, S.B.; Cha, K.H.; Kim, S.N.; Altantsetseg, S.; Shatar, S.; Sarangerel, O.; Nho, C.W. The antimicrobial activity of essential oil from *Dracocephalum foetidum* against pathogenic microorganisms. *J. Microbiol.* **2007**, *45*, 53–57.
53. Celaya, L.S.; Alabrudzińska, M.H.; Molina, A.C.; Viturro, C.I.; Silvia, M. The Inhibition of Methicillin-Resistant *Staphylococcus aureus* by essential oils isolated from leaves and fruits of *Schinus areira* depending on their chemical compositions. *Acta Biochim. Pol.* **2014**, *61*, 41–46. [CrossRef]
54. Han, Y.; Chen, W.; Sun, Z. Antimicrobial activity and mechanism of limonene against *Staphylococcus aureus*. *Food Saf.* **2021**, *41*, e12918. [CrossRef]
55. Zhou, S.; Wei, C.; Zhang, C.; Han, C.; Kuchkarova, N.; Shao, H. Chemical composition, phytotoxic, antimicrobial and insecticidal activity of the essential oils of *Dracocephalum integrifolium*. *Toxins* **2019**, *11*, 598. [CrossRef]
56. Aluyor, E.O.; Oboh, I.O. Peservatives | Traditional Preservatives—Vegetable Oils. In *Encyclopedia of Food Microbiology*, 2nd ed.; Batt, C.A., Tortorello, M.L., Eds.; Elsevier Inc.: Amsterdam, The Netherlands, 2014; pp. 137–140.
57. De Almeida, R.R.P.; Souto, R.N.P.; Bastos, C.N.; da Silva, M.H.L.; Maia, J.G.S. Chemical Variation in *Piper aduncum* and biological properties of its dillapiole-rich essential oil. *Chem. Biodivers.* **2009**, *6*, 1427–1434. [CrossRef]
58. Ben Arfa, A.; Combes, S.; Preziosi-Belloy, L.; Gontard, N.; Chalier, P. Antimicrobial activity of carvacrol related to its chemical structure. *Lett. Appl. Microbiol.* **2006**, *43*, 149–154. [CrossRef] [PubMed]
59. Sánchez-Hernández, E.; Buzón-Durán, L.; Andrés-Juan, C.; Lorenzo-Vidal, B.; Martín-Gil, J.; Martín-Ramos, P. Physicochemical characterization of *Crithmum maritimum* L. and daucus carota subsp. gummifer (syme) hook.fil. and their antimicrobial activity against apple tree and grapevine phytopathogens. *Agronomy* **2021**, *11*, 886. [CrossRef]

60. Kunicka-Styczynska, A.; Smigielski, K.; Prusinowska, R.; Rajkowska, K.; Kusmider, B.; Sikora, M. Preservative activity of lavender hydrosols in moisturizing body gels. *Lett. Appl. Microbiol.* **2014**, *60*, 27–32. [CrossRef] [PubMed]
61. Aumeeruddy-Elalfi, Z.; Gurib-Fakim, A.; Mahomoodally, M.F. Kinetic studies of tyrosinase inhibitory activity of 19 essential oils extracted from endemic and exotic medicinal plants. *S. Afr. J. Bot.* **2016**, *103*, 89–94. [CrossRef]
62. Cheraif, K.; Bakchiche, B.; Gherib, A.; Bardaweel, S.K.; Çol Ayvaz, M.; Flamini, G.; Ascrizzi, R.; Ghareeb, M.A. Chemical composition, antioxidant, anti-tyrosinase, anti-cholinesterase and cytotoxic activities of essential oils of six algerian plants. *Molecules* **2020**, *25*, 1710. [CrossRef]
63. Jugreet, B.S.; Lall, N.; Anina Lambrechts, I.; Reid, A.M.; Maphutha, J.; Nel, M.; Hassan, A.H.; Khalid, A.; Abdalla, A.N.; Van, B.L.; et al. In vitro and in silico pharmacological and cosmeceutical potential of ten essential oils from aromatic medicinal plants from the Mascarene Islands. *Molecules* **2022**, *27*, 8705. [CrossRef] [PubMed]

foods

Article

Exploitation of Black Olive (*Olea europaea* L. cv. Piantone di Mogliano) Pomace for the Production of High-Value Bread

Federica Cardinali [1], Luca Belleggia [1], Anna Reale [2], Martina Cirlini [3], Floriana Boscaino [2], Tiziana Di Renzo [2], Lorenzo Del Vecchio [3], Natascia Cavalca [3], Vesna Milanović [1], Cristiana Garofalo [1], Cristiana Cesaro [1], Giorgia Rampanti [1], Andrea Osimani [1,*] and Lucia Aquilanti [1]

[1] Dipartimento di Scienze Agrarie, Alimentari ed Ambientali, Università Politecnica delle Marche, Via Brecce Bianche, 60131 Ancona, Italy; f.cardinali@univpm.it (F.C.); l.belleggia@pm.univpm.it (L.B.); v.milanovic@univpm.it (V.M.); c.garofalo@univpm.it (C.G.); c.cesaro@pm.univpm.it (C.C.); g.rampanti@pm.univpm.it (G.R.); l.aquilanti@univpm.it (L.A.)

[2] Istituto di Scienze dell'Alimentazione, Consiglio Nazionale delle Ricerche, Via Roma 64, 83100 Avellino, Italy; anna.reale@isa.cnr.it (A.R.); floriana.boscaino@isa.cnr.it (F.B.); tiziana.direnzo@isa.cnr.it (T.D.R.)

[3] Dipartimento di Scienze degli Alimenti e del Farmaco, Università di Parma, Viale Parco Area delle Scienze 27/A, 43124 Parma, Italy; martina.cirlini@unipr.it (M.C.); lorenzo.delvecchio@unipr.it (L.D.V.); natascia.cavalca@unipr.it (N.C.)

* Correspondence: a.osimani@univpm.it

Abstract: In this study, the morpho-textural features, total phenolic content (TPC), and antioxidant capacity (AOC) of bread fortified with olive (*Olea europaea* L.) pomace were evaluated. Fresh olive pomace was subjected to microbiological and chemical (TPC, AOC, and fiber) analyses; then, the same olive pomace was analyzed during 1 to 6 months of storage at 4 °C or −20 °C. All olive pomace samples were used in 10%, 15%, or 20% amounts to produce type 0 soft wheat (*Triticum aestivum*) and whole wheat bread samples. The volatile organic compounds (VOCs) in the bread samples were also analyzed to assess the effect of the addition of the olive pomace on the flavor profile of the baked products. The TPC and AOC evaluation of olive pomace showed no differences among the analyzed samples (fresh, refrigerated, or frozen). Regarding the bread containing olive pomace, the specific volume was not affected by the amount or the storage methods of the added pomace. Bread samples produced with soft wheat flour showed the lowest hardness values relative to those produced with whole wheat flour, irrespective of the amount or storage method of the olive pomace. Regarding color, the crust and crumb of the bread samples containing 20% olive pomace were significantly darker. The bread samples containing 20% olive pomace had the highest TPC. The bread samples with fresh olive pomace were characterized by terpenoids, ketones, and aldehydes, whereas the bread samples containing refrigerated olive pomace were characterized by alcohols (mainly ethanol), acids, esters, and acetate. Finally, the bread samples with frozen olive pomace showed a volatile profile similar to that of bread produced with fresh olive pomace. Olive pomace was shown to be a suitable ingredient for producing bread with high nutritional value.

Keywords: antioxidant capacity; olive oil by-product; shelf life; bread-making; wholemeal bread; volatilome

1. Introduction

In the Mediterranean basin, the cultivation of *Olea europaea* L. provides virgin olive oil that, due to its nutritional benefits in the human diet, has been recognized as one of the best vegetable oils since ancient times [1,2]. However, the production of olive oil also engenders large amounts of by-products and wastes, including so-called olive pomace, olive mill waste waters, olive leaves, olive stones, and seeds [1]. Furthermore, the disposal of these by-products and wastes represents a threat to the environment and a further cost for the olive oil food industry [3]. As an example, between 2020 and 2021, more than

8 million tons of olive mill waste waters were produced worldwide [4], with a considerable impact on the environment in terms of phytotoxicity, pollution of natural waters, threat to aquatic life, and offensive odors [5]. Based on the global society's growing awareness of sustainable food production, the European Union (EU) issued the Circular Economy Action Plan within the European Green Deal. This action, which is a prerequisite for achieving the EU's 2050 climate neutrality target and halting the loss of biodiversity, is aimed at reducing pressure on natural resources and creating sustainable growth and job opportunities.

In such a context, the olive oil industry has developed new strategies to convert its by-products and wastes into commercially viable products, such as fuels, fertilizers, feed ingredients, compost, cement and bricks, and phytochemicals [4,6–8]. As reviewed by Roselló-Soto et al. [9], by-products and wastes from olive oil production include compounds of considerable nutritional value such as fatty acids, pigments (carotenoids and chlorophylls), tocopherols, phytosterols, squalene, volatile and aromatic compounds, and polyphenols. Therefore, academic researchers and the food industry have started to jointly design and promote new high-value food products that incorporate variable amounts of by-products from the olive oil industry to (i) foster circular economy processes, (ii) encourage sustainable consumption, (iii) ensure waste prevention, and (iv) exploit the nutritional value of food by-products.

Among the by-products obtained during olive oil production, olive pomace has already proved to be a very versatile ingredient for enriching conventional food products. Indeed, Ying et al. [10] showed the feasibility of preparing fiber- and polyphenol-enriched extruded food products based on mixed rice–oat flour and maize–oat flour containing olive pomace. Moreover, Balli et al. [11] have successfully used olive pomace to enrich tagliatelle pasta with phenolic compounds and fiber, whereas Simonato et al. [12] observed a positive impact on the glycemic response of pasta containing olive pomace. Interestingly, Ribeiro et al. [13] produced yogurt enriched in fiber and hydroxytyrosol by incorporating olive pomace into milk. Finally, a few authors have exploited olive pomace to produce bakery products including biscuits, taralli, and bread with enhanced nutritional value [14–17].

Bread represents a staple food in the typical Western diet. In bread-making, refined wheat flour (commonly classified as type 00 or 0) represents the most used raw material; however, bread lacks in essential amino acids and fiber, the latter of which is abundant in bran and germ [18,19]. Conversely, bread produced with whole wheat flour is rich in fiber, thereby conferring health benefits to the consumer including the reduction of glycemic index and blood cholesterol [19]. Notwithstanding, whole wheat flour has a low gluten content, thus producing a bread dough with low technological features (reduced stability, resistance, and extensibility) [19]. To produce bread, the leavening of the dough can be obtained through (i) the direct method, consisting of the addition of baker's yeast to a mixture of water and flour, or (ii) through indirect methods, which involve the use of either "sourdough" or "sponge" (with the latter also referred to as *biga*) as the leavening agent [20,21]. The *biga* method consists of two steps. The first step produces a light "sponge" by mixing flour, water, and baker's yeast, followed by ~16 h of fermentation; the second step leads to a "dough" by mixing the remaining ingredients and the *biga*, followed by a final fermentation (leavening) before baking [20].

To date, many studies have focused on improving the nutritional and rheological features of bread produced with new functional ingredients [18]. To the best of the authors' knowledge, for bread, the added amount of olive pomace usually ranges between 4% [16] and 10% [15], and knowledge of the use of this ingredient in bread-making is still limited. Hence, bread represents an optimal low-cost candidate to valorize olive pomace and obtain a healthy and value-added food.

This study aimed to evaluate the chemical, nutritional, and volatile characteristics of bread fortified with olive pomace. For this purpose, fresh olive pomace obtained from the milling of black olives (*Olea europaea* L. cv. Piantone di Mogliano) was first subjected to microbiological and chemical analyses. The same olive pomace was then stored under different refrigeration conditions and analyzed for the same parameters after 1, 2, 3, 4,

5, and 6 months of storage. Then, fresh olive pomace and that maintained under the abovementioned storage conditions were used at different concentrations to produce experimental bread samples. The doughs were subjected to microbiological analyses, whereas the baked bread samples were subjected to morpho-textural and chemical analyses. As is widely acknowledged, the flavor of bread is influenced by enzymatic reactions during dough fermentation and by thermal reactions occurring during the baking process, such as the non-enzymatic Maillard reactions and the caramelization of sugars. Moreover, the presence and quantity of new ingredients in the bread recipe can influence the aroma of the end product. Hence, to evaluate the influence of the added olive pomace in the flavor profiles of the experimental bread samples, an analysis of volatile organic compounds (VOCs) using solid-phase microextraction (SPME) coupled with gas chromatography–mass spectrometry (GC–MS) was also performed.

2. Materials and Methods

2.1. Raw Materials

Fresh olive pomace from black olives (*Olea europaea* L. cv. Piantone di Mogliano) was obtained by a three-phase olive oil production process at the Corradini oil mill (Mogliano, Italy). Olive pomace was transported to the laboratory under refrigerated conditions (4 °C) and analyzed immediately for chemical and microbiological characteristics. Then, the fresh olive pomace was vacuum-packed in sterile food-grade plastic bags and stored for 6 months at −20 °C or 4 °C. The stored samples were analyzed after 1, 2, 3, 4, 5, and 6 months to assess their shelf life. No treatment was applied to the raw olive pomace.

Thereafter, the olive pomace stored for 6 months under the two different conditions was used as an ingredient in the bread-making process, as detailed in Section 2.3. Type 0 soft wheat flour was supplied by Molino Bianchi (Osimo, Italy), whereas whole wheat flour was supplied by Molino Fratini (Pollenza, Italy). The experimental design is shown in Figure 1.

2.2. Analyses of Olive Pomace during the 6-Month Storage Period

2.2.1. Microbiological Analyses

For the microbiological analyses, 10 g of fresh or stored olive pomace was added to 90 mL of sterile peptone water (1 g L^{-1}) and homogenized in a 400 Circulator Stomacher apparatus (International PBI, Milan, Italy) at 260 rpm for 2 min. Aliquots of 1 mL were serially ten-fold diluted in sterile peptone water for the enumeration of specific microbial groups. In more detail, mesophilic aerobic bacteria were counted in Plate Count Agar (PCA) (VWR International Srl, Milan, Italy) incubated at 30 °C for 48 h; presumptive mesophilic lactobacilli were counted on De Man, Rogosa, and Sharpe (MRS) agar (VWR) supplemented with cycloheximide (250 mg L^{-1}) (VWR) incubated for 48–72 h at 37 °C; eumycetes (yeasts and molds) were counted on Rose Bengal Chloramphenicol Agar (RBA) (VWR) incubated for 4 days at 25 °C; and Enterobacteriaceae were counted in Violet Red Bile Glucose Agar (VRBGA) (VWR) incubated for 24 h at 37 °C.

The results were expressed as the mean Log cfu (colony forming units) g^{-1} of two independent analyses ± standard deviation.

2.2.2. Evaluation of Phenolic Fraction

Antioxidant compounds were extracted from the olive pomace samples following the protocol reported by Dall'Asta et al. [22] with some modifications. Briefly, 0.5 g of sample was weighted and added to 5 mL of methanol/water solution (80/20 v/v) and then extracted on an HS 501 digital shaker (IKA-Werke, Staufen, Germany) at 200 strokes min^{-1} for 30 min at room temperature. Then, the extracts were centrifuged at 5000 rpm for 10 min at room temperature using a Centrifugette 4206 centrifuge (Alc International, Pévy, France), and the supernatants were recovered and kept in the dark at −20 °C until analysis. Each sample was extracted in duplicate.

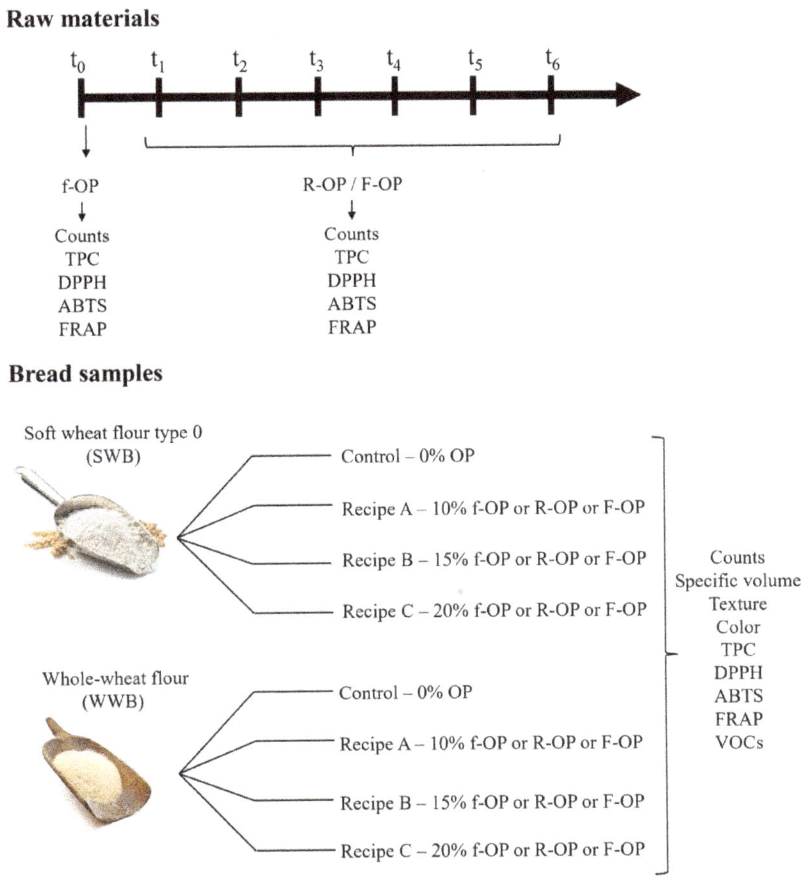

Figure 1. Experimental design. f-OP: fresh olive pomace; R-OP: refrigerated olive pomace; F-OP: frozen olive pomace.

Total phenolic content (TPC) was evaluated based on the Folin–Ciocalteau method in accordance with the protocol already described by Martin-Diana et al. [23] with a few modifications. In more detail, 250 μL of sample extract was mixed with 1 mL of Folin–Ciocalteau reagent previously diluted in bi-distilled water (1/10 *v*/*v*) and 2 mL of sodium carbonate aqueous solution (10% *w*/*v*) and then kept in the dark for 30 min at room temperature. Then, the absorbance of the solution was measured in triplicate at 760 nm using a V-530 spectrophotometer (JASCO, Easton, MD, USA). To calculate TPC, a calibration curve was constructed by analyzing—under the same conditions—five gallic acid standard solutions (0.01^{-1} mg GAE g^{-1}); a blank sample was also analyzed.

2.2.3. Determination of the Antioxidant Capacity (AOC)

DPPH Radical-Scavenging Activity Assay

To determine the radical-scavenging capacity of the extracts, the DPPH (2,2-Diphenyl-1-picrylhydrazyl) radical-scavenging assay was applied according to Dall'Asta et al. [22]. In more detail, 100 μL aliquots of each sample extract were mixed with 2.9 mL of a methanolic DPPH solution (0.05 mM) and kept in the dark at room temperature for 30 min. The absorbance of the samples was then recorded in triplicate at 517 nm using a JASCO V-530 spectrophotometer. A blank sample was prepared using 100 μL of extraction solution and then measured after incubation with the DPPH reagent solution. To quantify the antioxidant capacity of the extracts, a calibration curve was prepared using Trolox as the reference standard in the concentration range of 0.025–0.25 mg g^{-1} (five points). The radical-scavenging capacity (I%) was calculated considering the percentage of radical inhibition. The following mathematical formula was applied:

$$I\% = [(Abs_{blank} - Abs_{sample})/Abs_{blank}] \times 100$$

where Abs_{blank} is the absorbance of the blank sample and Abs_{sample} is the absorbance of the standard solution or sample. The results were expressed as mg TEAC (Trolox Equivalent Antioxidant Capacity) g^{-1}.

ABTS Radical-Scavenging Activity Assay

ABTS [2,2-azinobis (3-ethylbenzothiazoline-6-sulfonic acid)] assay was carried out in accordance with the method reported by Jemai et al. [24] with some modifications. For the generation of the ABTS radical cation (ABTS+), an aqueous solution containing ABTS (7 mM) and potassium persulfate (2.45 mM) was prepared and kept in the dark for ~16 h. Then, the solution was fully diluted with ethanol (1.70^{-1} v/v), and its absorbance at 734 nm was measured using a JASCO V-530 spectrophotometer, checking that the value did not exceed 0.70 ± 0.2. For the photometric assay, 20 µL aliquots of sample extract (either blank or standard solution) were treated with 1.98 mL of the ABTS+-diluted solution. The reaction was performed in the dark at room temperature and, then, the absorbance of all the samples was measured in triplicate at 734 nm. The quantification was performed based on Trolox, as already described in the DPPH determination procedure.

Determination of Ferric-Ion-Reducing Power (FRAP)

The antioxidant capacity of the sample extracts was also evaluated using FRAP assay. For this analysis, the FRAP reagent solution was prepared by mixing 2.5 mL of an aqueous solution of ferric chloride hexahydrate (20 mM) with 2.5 mL of an aqueous solution of 2,4,6-tripyridyl-s-triazine (TPTZ) (10 mM) acidified with hydrochloric acid (40 mM) and 25 mL of 300 mM acetate buffer prepared from sodium acetate and acetic acid (pH 3.6) [25]. The solution was heated to 37 °C prior to use. Sample extracts, blank samples, and Trolox standard solutions (150 µL) were submitted to the reaction with the FRAP solution (2.85 mL) in the dark at room temperature for 30 min; then, the absorbance was measured in triplicate at 593 nm using a JASCO V-530 spectrophotometer. The ferric-ion-reducing activity in the samples was estimated based on Trolox, by constructing a calibration curve in the same concentration range used in the previously described tests.

2.3. Bread-Making Trials

Bread-making trials were carried out in duplicate using fresh, frozen-stored (−20 °C), or refrigerated-stored (4 °C) olive pomace. For each trial, bread loaves were produced according to the recipes reported in Table 1.

Table 1. Recipes for the bread-making process for bread containing olive pomace.

Raw Material	Control		Recipe A		Recipe B		Recipe C	
	SWB	WWB	SWB	WWB	SWB	WWB	SWB	WWB
Olive pomace	0%		10.0%		15.0%		20.0%	
Whole wheat flour		38.2%		31.7%		28.4%		25.2%
Type 0 soft wheat flour	60.1%	20.6%	50.1%	17.1%	45.1%	15.4%	40.1%	13.6%
Biga	6.0%	5.9%	6.0%	5.9%	6.0%	5.9%	6.0%	5.9%
Water	33.5%	35.1%	33.5%	35.1%	33.5%	35.1%	33.5%	35.1%
Baker's yeast	0.4%	0.2%	0.4%	0.2%	0.4%	0.2%	0.4%	0.2%

SWB: bread produced with type 0 soft wheat flour; WWB: bread produced with whole wheat flour.

First, a *biga* was prepared by mixing 67.50% (*w*/*w*) type 0 soft wheat flour, 31.5% (*v*/*w*) water, and 1% (*w*/*w*) fresh baker's yeast, which was left to ferment at 18 °C for 18 h.

For each trial, all the ingredients and the *biga* were mixed for 15 min in a Model K12-2V spiral mixer (Kosmitech Srl, Morrovalle, Italy) and the doughs were divided into portions of 300 g. After 1.5 h of proofing at 30 °C and 70% relative humidity (R.H.), the doughs were baked in a Model FOSTR6 electric oven (Fimar S.p.a., Villa Verucchio, Italy) at 200 °C for 1 h. Bread loaves without olive pomace were used as controls.

2.3.1. Analyses of Dough and Bread Samples

Microbiological Analyses

For the microbiological analyses, 10 g dough samples collected before and after leavening were added to 90 mL of sterile peptone water and homogenized in a Stomacher apparatus (International PBI) for 2 min at 260 rpm. Aliquots of 1 mL were serially ten-fold diluted in sterile peptone water for the enumeration of presumptive lactobacilli on MRS agar (VWR) supplemented with cycloheximide (250 mg L^{-1}) (VWR) and incubated at 30 °C for 48–72 h and eumycetes on RBA (VWR) incubated at 25 °C for 4 days. For the viable counts of the bread samples, 10 g aliquots were used for the enumeration of bacterial spores in PCA medium (VWR). The results were expressed as the mean of the Log cfu g^{-1} of two independent measurements ± standard deviation.

Determination of Bread Specific Volume, Hardness, and Color

Bread volume was determined using the AACC 10-05.01 rapeseed displacement method; the specific volume, expressed as $cm^3\ g^{-1}$, was calculated as the ratio between the loaf bread volume and loaf weight. All determinations were performed in triplicate and the results were expressed as the mean value ± standard deviation.

The hardness of the bread was measured with a Texture Analyzer (model CT3-4500, Brookfield Engineering Laboratories Inc., Middleboro, MA, USA) using a 36 mm diameter cylindrical probe (mod. TA-AACC36). A 4500 g load cell was used. The probe compressed the crumb to a 40% compression limit (10 mm compression depth) at a speed of 100 mm min^{-1}. The measurements were performed at room temperature in the middle of bread slices (20 mm thickness). For each sample, three independent measurements were carried out.

The color measurements were performed using a Chroma Meter CR-200 (Minolta, Osaka, Japan) with a D65 illuminant. Color parameters were determined only for the bread samples produced with fresh olive pomace, which were used as a reference for future product development. In more detail, the color of the crust and crumb was determined according to the CIE $L^*a^*b^*$ system (L^*, brightness; a^*, redness/greenness; b^*, blueness/yellowness) [26]. Bread slices were analyzed by cutting them longitudinally (10 mm thickness) and reproducing their cross sections using a scanner (ENVY 6200 Series, HP, Palo Alto, CA, USA) [27].

Evaluation of Bread Phenolic Fraction

Antioxidant compounds were extracted from the bread samples following the protocol reported by Dall'Asta et al. [22] with the same modifications already described in Section 2.2.2.

The TPC determination of the bread samples was carried out as already described in Section 2.2.2.

The determinations of the AOC, DPPH radical-scavenging activity assay, the ABTS radical-scavenging activity assay, and the FRAP of bread samples were carried out as already described in Section 2.2.3.

Evaluation of Bread Dietary Fiber

Dietary fiber content was determined using the AOAC 991.43 enzymatic–gravimetric method.

SPME–GC/MS Analysis of Bread Volatile Components

The volatile fractions of the breads were analyzed using headspace sampling based on the solid-phase microextraction technique (SPME) according to Cardinali et al. [28]. In detail, for each SPME analysis, 2 g of sample was placed into a 20 mL headspace vial, and 5 μL of 4-methyl-2 pentanol (internal standard, 100 mg L^{-1} standard solution) was added. Each vial was placed in a thermostatic block (40 °C) on a stirrer, and the fiber was inserted and maintained in the sample headspace for 30 min. Then, the fiber was removed and immediately inserted into the GC/MS injector for the desorption of compounds. The extraction was automatically performed by the multipurpose sampler of the GC/MS system.

For the analyses, a silica fiber coated with 75 μm of Carboxen/Polydimethylsiloxane (CAR/PDMS) was used (Supelco, Bellefonte, PA, USA). The SPME–GC/MS analysis was performed using an Agilent GC 7890A/MSD 5975 system with an automatic sampler Gerstel MPS2 (Agilent Technologies, Santa Clara, CA, USA). The experimental setup and operating conditions were as follows: an HP-Innowax capillary column was used (Agilent Technologies, 30 m × 0.25 mm ID, film thickness 0.25 μm), the gas carrier was helium (flow rate = 1.5 mL min^{-1}), and SPME injections were splitless (straight glass line, 0.75 mm I.D.) at 240 °C for 20 min, during which time thermal desorption of the analytes from the fiber occurred. The oven parameters were as follows: the initial temperature was 40 °C, held for 3 min, followed by an increase to 240 °C at a rate of 5 °C min^{-1}, and then held for 10 min. The injector temperature was 240 °C. The mass spectrometer was operated in scan mode over a mass range of 33 to 300 amu (2 s $scan^{-1}$) at an ionization potential of 70 eV. VOC identification was achieved by comparing the mass spectra with the Nist library (NIST 20) and by matching the retention indices (RI) calculated according to the equation of Van Den Dool and Kratz [29] and based on a series of alkanes. Data were expressed as the relative peak area with respect to the internal standard. Blank experiments were conducted in two different modalities—blank fiber and blank empty vial. These controls were carried out after every 10 analyses. The analyses were performed in duplicate.

2.4. Statistical Analysis

A one-way analysis of variance (ANOVA) was applied to compare the data (IBM SPSS Statistics 26.0 software, Chicago, IL, USA). In more detail, for the olive pomace samples, the one-way ANOVA test was used to compare the results obtained among the refrigerated sample or frozen sample groups, whereas, for the bread, all the samples were included in the same dataset. Moreover, a t-test for independent samples was used to compare the data obtained from the refrigerated and frozen olive pomace samples, considering a p-value of 0.05 as statistically significant. Finally, the correlations among the results obtained from the different assays were also tested using a two-tailed Pearson correlation analysis.

To evaluate how the different breads were distributed according to the detected chemical groups of volatile compounds, a Principal Component Analysis (PCA) was performed using Tanagra 1.4 software.

3. Results

3.1. Microbiological Analyses of Olive Pomace

The results of the viable counts of the olive pomace samples are reported in Table 2.

Table 2. Microbial viable counts of olive pomace samples.

Panel a					
		Total Mesophilic Aerobes	**Presumptive Lactobacilli**	**Enterobacteriaceae**	**Eumycetes**
f-OP	t_0	5.00 ± 0.12 [a]	2.96 ± 0.01 [a]	<1.00	4.81 ± 0.15 [d]
R-OP	t_1	5.02 ± 0.04 [a]	1.87 ± 0.04 [b]	<1.00	5.69 ± 0.02 [c]
	t_2	4.07 ± 0.16 [b]	1.65 ± 0.07 [c]	<1.00	6.04 ± 0.07 [b]
	t_3	1.95 ± 0.07 [c]	1.54 ± 0.09 [c]	<1.00	6.24 ± 0.02 [b]
	t_4	1.83 ± 0.18 [c]	1.86 ± 0.03 [b]	<1.00	6.64 ± 0.00 [a]
	t_5	1.69 ± 0.12 [c]	<1.00	<1.00	6.54 ± 0.01 [a]
	t_6	<1.00	<1.00	<1.00	6.54 ± 0.00 [a]
Panel b					
		Total Mesophilic Aerobes	**Presumptive Lactobacilli**	**Enterobacteriaceae**	**Eumycetes**
f-OP	t_0	5.00 ± 0.12 [a]	2.96 ± 0.01 [a]	<1.00	4.81 ± 0.15 [a]
F-OP	t_1	3.59 ± 0.03 [b]	2.13 ± 0.07 [b]	<1.00	3.33 ± 0.02 [b]
	t_2	3.51 ± 0.00 [b]	2.06 ± 0.03 [b]	<1.00	3.33 ± 0.01 [b]
	t_3	3.09 ± 0.01 [c]	1.99 ± 0.02 [b]	<1.00	3.27 ± 0.01 [b]
	t_4	2.74 ± 0.03 [d]	1.98 ± 0.04 [b]	<1.00	3.05 ± 0.05 [bc]
	t_5	2.87 ± 0.04 [d]	1.81 ± 0.05 [c]	<1.00	2.75 ± 0.21 [c]
	t_6	2.89 ± 0.06 [cd]	2.08 ± 0.05 [b]	<1.00	3.08 ± 0.01 [bc]

f-OP: fresh olive pomace; R-OP: refrigerated olive pomace; F-OP: frozen olive pomace. t_0: fresh olive pomace; t_1, t_2, t_3, t_4, t_5, and t_6 represent olive pomace samples after 1, 2, 3, 4, 5, and 6 months of storage, respectively. Results are expressed as mean Log cfu g^{-1} ± standard deviation. For each panel, the different letters in the same column indicate significant differences according to the Tukey–Kramer (HSD) test ($p < 0.05$).

In more detail, total mesophilic aerobes were counted at 5.00 ± 0.12 Log cfu g^{-1} in the fresh olive pomace. During storage, the counts of the refrigerated olive pomace varied between 5.02 ± 0.04 (t_1) and 1.69 ± 0.12 Log cfu g^{-1} (t_5), whereas the counts of the frozen olive pomace varied between 3.59 ± 0.03 (t_1) and 2.89 ± 0.06 Log cfu g^{-1} (t_6), with a progressive and significant decrease from t_1 to t_6.

Regarding lactic acid bacteria, for the fresh olive pomace, counts were 2.96 ± 0.01 Log cfu g^{-1}. During storage, viable counts of the refrigerated olive pomace varied between 1.87 ± 0.04 (t_1) and 1.86 ± 0.03 Log cfu g^{-1} (t_4), whereas for frozen olive pomace, the viable counts varied between 2.13 ± 0.07 (t_1) and 2.08 ± 0.05 Log cfu g^{-1} (t_6), with no significant differences among the samples.

In the fresh olive pomace, eumycetes were counted at 4.81 ± 0.15 Log cfu g^{-1}. In the refrigerated olive pomace, the counts of eumycetes ranged between 5.69 ± 0.02 (t_1) and 6.54 ± 0.00 Log cfu g^{-1} (t_6), whereas the counts in frozen olive pomace ranged between 3.33 ± 0.02 (t_1) and 3.08 ± 0.01 Log cfu g^{-1} (t_6), with no significant differences among the samples.

Finally, all samples of olive pomace showed viable counts of Enterobacteriaceae of <1 Log cfu g^{-1}.

3.2. Total Phenolic Content (TPC) and Antioxidant Capacity of Olive Pomace

The results of the total phenolic content and antioxidant capacity analyses of the olive pomace samples are reported in Table 3.

The TPC of the fresh olive pomace was 7.18 ± 0.12 mg GAE g^{-1}; no significant differences were observed among the values of the fresh olive pomace and refrigerated or frozen olive pomace during storage. In more detail, the TPC of the refrigerated olive pomace ranged between 7.35 ± 0.16 (t_1) and 7.39 ± 0.28 mg GAE g^{-1} (t_6), whereas the TPC values of the frozen olive pomace ranged between 8.40 ± 0.59 (t_1) and 9.05 ± 0.29 mg GAE g^{-1} (t_6).

Table 3. Total phenolic content determination and evaluation of the antioxidant capacity of olive pomace.

Panel a					
		TPC mg GAE g^{-1}	**DPPH mg TEAC g^{-1}**	**ABTS mg TEAC g^{-1}**	**FRAP mg TEAC g^{-1}**
f-OP	t_0	7.18 ± 0.12 ^{a}	16.21 ± 0.41 ^{a}	15.36 ± 0.51 ^{a}	15.95 ± 0.19 ^{a}
R-OP	t_1	7.35 ± 0.16 ^{a}	16.15 ± 0.11 ^{a}	14.69 ± 0.18 ^{a}	16.48 ± 0.24 ^{a}
	t_2	6.80 ± 0.19 ^{a}	15.47 ± 0.08 ^{a}	14.78 ± 0.02 ^{a}	14.67 ± 0.66 ^{ab}
	t_3	7.01 ± 0.12 ^{a}	15.41 ± 0.32 ^{a}	14.57 ± 0.25 ^{a}	14.65 ± 0.35 ^{ab}
	t_4	7.40 ± 0.09 ^{a}	16.47 ± 0.67 ^{a}	16.49 ± 0.32 ^{a}	15.71 ± 0.60 ^{a}
	t_5	6.67 ± 0.15 ^{b}	14.87 ± 0.28 ^{a}	14.42 ± 0.32 ^{a}	14.34 ± 0.28 ^{b}
	t_6	7.39 ± 0.28 ^{a}	16.23 ± 0.52 ^{a}	15.86 ± 0.28 ^{a}	15.36 ± 0.02 ^{a}
Panel b					
		TPC mg GAE g^{-1}	**DPPH mg TEAC g^{-1}**	**ABTS mg TEAC g^{-1}**	**FRAP mg TEAC g^{-1}**
f-OP	t_0	7.18 ± 0.12 ^{a}	16.21 ± 0.41 ^{a}	15.36 ± 0.51 ^{a}	15.95 ± 0.19 ^{a}
F-OP	t_1	8.40 ± 0.59 ^{a}	17.25 ± 0.44 ^{a}	16.42 ± 0.27 ^{ab}	18.16 ± 0.46 ^{ab}
	t_2	7.92 ± 0.40 ^{a}	17.03 ± 0.45 ^{a}	16.94 ± 0.04 ^{ab}	18.56 ± 0.62 ^{b}
	t_3	7.38 ± 0.82 ^{a}	16.27 ± 0.04 ^{a}	17.13 ± 0.50 ^{ab}	17.26 ± 0.31 ^{a}
	t_4	8.24 ± 0.31 ^{a}	16.77 ± 0.22 ^{a}	17.31 ± 0.58 ^{ab}	18.88 ± 0.31 ^{b}
	t_5	7.98 ± 1.11 ^{a,*}	16.80 ± 0.06 ^{a,*}	18.33 ± 0.82 ^{b,*}	18.99 ± 0.70 ^{b,*}
	t_6	9.05 ± 0.29 ^{a,*}	17.41 ± 0.55 ^{a,*}	15.80 ± 1.03 ^{a}	20.20 ± 0.47 ^{b,*}

TPC: Total phenolic content, DPPH radical-scavenging activity test; ABTS radical-scavenging activity assay; FRAP: ferric-ion-reducing power. f-OP: fresh olive pomace; R-OP: refrigerated olive pomace; F-OP: frozen olive pomace. t_0: fresh olive pomace; t_1, t_2, t_3, t_4, t_5, and t_6 represent olive pomace samples after 1, 2, 3, 4, 5, and 6 months of storage, respectively. Results are expressed as mg GAE g^{-1} and mg TEAC g^{-1} on a wet basis; different letters indicate significant differences among samples from the two groups (refrigerated and frozen samples; one-way ANOVA test); stars indicate statistical differences among samples in comparison to the corresponding samples of the other sample group (comparison between refrigerated versus frozen samples analyzed at the same storing time; *t*-test).

Regarding antioxidant capacity, the fresh olive pomace samples yielded values of 16.21 ± 0.41 mg TEAC g^{-1} when the DPPH radical was used (corresponding to 64.68 ± 1.62 mmol Trolox kg^{-1}), 15.95 ± 0.19 mg TEAC g^{-1} when applying the FRAP method, and 15.36 ± 0.51 mg TEAC g^{-1} using the ABTS assay. The antioxidant capacities measured using the three different applied assays showed a significant, positive Pearson correlation with TPC (DPPH $p = 0.924$; ABTS $p = 0.561$; FRAP $p = 0.933$). The results obtained for the refrigerated and frozen samples were compared to the values obtained for the fresh olive pomace, and, at the same time, a comparison among the two storage temperatures (+4 °C and −20 °C) was also evaluated. Based on the one-way ANOVA, the refrigerated and frozen samples maintained the initial characteristics in terms of TPC and antioxidant capacity over time, especially when comparing t_0 with the prolonged storage time. In addition, a *t*-test for independent samples was used to determine differences among the two storage methods (refrigerated or frozen). By comparing the results obtained from the refrigerated and stored samples for the four applied tests, statistically higher TPC and AOC values were observed for the frozen olive pomace, especially at the end of the storage period.

3.3. Microbiological Analyses of Dough and Bread

The results of the viable counts of lactobacilli and eumycetes in the dough samples, before and after leavening, produced with the addition of different amounts (10%, 15%, or 20%) of fresh, refrigerated, or frozen olive pomace are reported in Table 4.

For the dough made with fresh olive pomace produced with type 0 soft wheat flour, the viable counts of lactobacilli before leavening ranged between 1.83 ± 0.18 and 2.69 ± 0.04 Log cfu g^{-1} with a 0% or 20% ratio, respectively, with statistically significant differences among the samples. After leavening, the lactobacilli count ranged from 2.02 ± 0.09 to 3.04 ± 0.33 Log cfu g^{-1} in the doughs made with fresh olive pomace with a 0% or 20% ratio, respectively, with no statistically significant differences among the samples. Regarding the dough produced with the fresh olive pomace and whole wheat flour, the viable counts of lactobacilli before leavening ranged between 1.72 ± 0.17 and 3.04 ± 0.10 Log cfu g^{-1} with a 0% or 20% pomace ratio, respectively, with statistically significant differences among the samples. After leavening, the lactobacilli counts ranged from 1.39 ± 0.12 to 3.02 ± 0.13 Log cfu g^{-1} with a 0% or 20% pomace ratio, respectively, with statistically significant differences among the samples. No differences were observed among the lactobacilli counts of the analyzed samples before or after leavening, irrespective of the added amount of fresh olive pomace or flour used.

Table 4. Viable counts of experimental bread doughs.

	% OP	Presumptive Lactobacilli				Eumycetes			
		SWD		WWD		SWD		WWD	
		BL	AL	BL	AL	BL	AL	BL	AL
t_0-f	0%	1.83 ± 0.18 ^{b,A}	2.02 ± 0.09 ^{a,A}	1.72 ± 0.17 ^{b,A}	1.39 ± 0.12 ^{b,A}	6.40 ± 0.00 ^{b,A}	6.30 ± 0.06 ^{ab,A}	6.50 ± 0.01 ^{a,A}	6.30 ± 0.00 ^{b,B}
	10%	2.60 ± 0.18 ^{a,A}	1.98 ± 0.71 ^{a,A}	2.40 ± 0.02 ^{ab,A}	2.55 ± 0.04 ^{a,A}	6.59 ± 0.02 ^{a,A}	6.37 ± 0.04 ^{a,B}	6.45 ± 0.03 ^{a,A}	6.24 ± 0.05 ^{b,B}
	15%	2.76 ± 0.11 ^{a,A}	2.94 ± 0.12 ^{a,A}	2.69 ± 0.51 ^{ab,A}	2.88 ± 0.18 ^{a,A}	6.57 ± 0.00 ^{ab,A}	6.20 ± 0.01 ^{b,B}	6.54 ± 0.10 ^{a,A}	6.78 ± 0.03 ^{a,A}
	20%	2.69 ± 0.04 ^{a,A}	3.04 ± 0.33 ^{a,A}	3.04 ± 0.10 ^{a,A}	3.02 ± 0.13 ^{a,A}	6.55 ± 0.09 ^{ab,A}	6.21 ± 0.04 ^{ab,B}	6.59 ± 0.01 ^{a,A}	5.96 ± 0.01 ^{c,B}
t_6-R	0%	1.48 ± 0.40 ^{b,A}	1.50 ± 0.24 ^{c,A}	1.86 ± 0.45 ^{b,A}	2.32 ± 0.22 ^{a,A}	6.92 ± 0.10 ^{b,A}	7.05 ± 0.08 ^{a,A}	6.26 ± 0.10 ^{b,A}	6.72 ± 0.03 ^{c,A}
	10%	3.21 ± 0.24 ^{a,A}	3.06 ± 0.33 ^{b,A}	2.96 ± 0.36 ^{ab,A}	2.73 ± 0.84 ^{a,A}	7.28 ± 0.13 ^{a,A}	7.08 ± 0.10 ^{a,A}	7.14 ± 0.04 ^{a,A}	7.00 ± 0.01 ^{a,B}
	15%	3.31 ± 0.25 ^{a,A}	3.32 ± 0.32 ^{ab,A}	3.48 ± 0.92 ^{a,A}	3.33 ± 1.11 ^{a,A}	7.27 ± 0.17 ^{a,A}	7.10 ± 0.12 ^{a,A}	7.03 ± 0.02 ^{a,A}	7.05 ± 0.01 ^{a,A}
	20%	3.48 ± 0.13 ^{a,A}	3.66 ± 0.20 ^{a,A}	3.69 ± 0.66 ^{a,A}	3.34 ± 0.77 ^{a,A}	7.35 ± 0.11 ^{a,A}	7.17 ± 0.11 ^{a,A}	7.09 ± 0.01 ^{a,A}	6.81 ± 0.06 ^{b,B}
t_6-F	0%	1.48 ± 0.40 ^{b,A}	1.50 ± 0.24 ^{a,A}	1.86 ± 0.45 ^{a,A}	2.32 ± 0.22 ^{a,A}	6.92 ± 0.10 ^{b,A}	7.05 ± 0.08 ^{ab,A}	6.26 ± 0.10 ^{d,B}	6.72 ± 0.03 ^{c,A}
	10%	1.86 ± 0.28 ^{ab,A}	1.56 ± 0.40 ^{a,A}	1.76 ± 0.15 ^{a,A}	1.84 ± 0.17 ^{a,A}	7.29 ± 0.14 ^{a,A}	7.29 ± 0.12 ^{a,A}	6.38 ± 0.02 ^{c,A}	6.23 ± 0.06 ^{d,B}
	15%	2.31 ± 0.39 ^{a,A}	1.51 ± 0.37 ^{a,B}	2.08 ± 0.27 ^{a,A}	1.75 ± 0.50 ^{a,A}	7.41 ± 0.14 ^{a,A}	7.15 ± 0.12 ^{ab,B}	7.10 ± 0.01 ^{a,A}	6.91 ± 0.04 ^{b,B}
	20%	1.15 ± 0.30 ^{b,A}	1.80 ± 0.55 ^{a,A}	2.60 ± 0.96 ^{a,A}	1.83 ± 1.59 ^{a,A}	7.37 ± 0.14 ^{a,A}	7.23 ± 0.07 ^{ab,A}	6.83 ± 0.01 ^{b,B}	7.07 ± 0.01 ^{a,A}

% OP: percentage of olive pomace added to dough; SWD: dough produced with soft wheat flour type 0; WWD: dough produced with whole wheat flour; BL: dough before leavening; AL: dough after leavening. t_0-f: dough made using freshly sampled OP; t_6-R: dough made using olive pomace after 6 months of refrigeration; t_6-F: dough made using olive pomace after 6 months of freezing. Results are expressed as mean Log cfu g^{-1} ± standard deviation. For each parameter, within each type of olive pomace storage technique, and considering the flour type used, means followed by different letters within each column (lowercase) and each row (capital letters) indicate significant differences ($p < 0.05$).

As for the dough with refrigerated olive pomace produced using type 0 soft wheat flour, the viable counts of lactobacilli before leavening ranged between 1.48 ± 0.40 and 3.48 ± 0.13 Log cfu g^{-1} with a 0% or 20% pomace ratio, respectively, with statistically significant differences among the samples. After leavening, the lactobacilli count ranged from 1.50 ± 0.24 to 3.66 ± 0.20 Log cfu g^{-1} in the dough made with fresh olive pomace at a 0% or 20% ratio, respectively, with statistically significant differences among the samples. Regarding the dough produced with refrigerated olive pomace and whole wheat flour, the viable counts of lactobacilli before leavening ranged between 1.86 ± 0.45 and 3.69 ± 0.66 Log cfu g^{-1} with a 0% or 20% pomace ratio, respectively, with statistically significant differences among the samples. After leavening, the lactobacilli counts ranged from 2.32 ± 0.22 to 3.34 ± 0.77 Log cfu g^{-1} with a 0% or 20% pomace ratio, respectively, with no statistically significant differences among the samples. No differences were observed among the lactobacilli counts of the analyzed samples before or after leavening, irrespective of the added amount of fresh olive pomace or flour used.

As for dough with the frozen olive pomace produced with type 0 soft wheat flour, the viable counts of lactobacilli before leavening ranged between 1.48 ± 0.40 and 1.15 ± 0.30 Log cfu g^{-1} with a 0% or 20% pomace ratio, respectively, with no statistically significant differences among these samples. After leavening, the lactobacilli counts ranged from 1.50 ± 0.24 to 1.80 ± 0.55 Log cfu g^{-1} in the dough made with fresh olive pomace with a 0% or 20% pomace ratio, respectively, with no statistically significant differences among these samples. Regarding the dough produced with refrigerated olive pomace and whole wheat flour, the viable counts of lactobacilli before leavening ranged between 1.86 ± 0.45 and 2.60 ± 0.96 Log cfu g^{-1} with a 0% or 20% pomace ratio, respectively, with no statistically significant differences among the samples. After leavening, the lactobacilli count ranged from 2.32 ± 0.22 to 1.83 ± 1.59 Log cfu g^{-1} with a 0% or 20% pomace ratio, respectively, with no statistically significant differences among the samples. No differences were observed among the lactobacilli counts of the analyzed samples before or after leavening, irrespective of the added amount of olive pomace or flour used.

Eumycetes were counted to assess the viability of the baker's yeast used to allow leavening, and data are reported here in brief. For this microbial group, counts between 6.20 ± 0.01 and 7.41 ± 0.14 Log cfu g^{-1} were obtained, with a few samples showing statistically significant differences.

Finally, no bacterial spores were detected in any of the experimental bread samples that were analyzed soon after baking.

3.4. Specific Volume, Hardness, and Color of the Experimental Breads

The results of the specific volume assessments of experimental bread loaves produced with different amounts of olive pomace are reported in Table 5.

Table 5. Specific volume and hardness of experimental bread samples.

	%OP	Specific Volume (cm^3 g^{-1})		Hardness (N)	
		SWB	WWB	SWB	WWB
t_0-f	0%	22.14 ± 0.01 ^{a,A}	20.81 ± 0.05 ^{a,B}	22.75 ± 3.17 ^{a,A}	25.86 ± 1.73 ^{a,A}
	10%	21.23 ± 1.06 ^{a,A}	20.76 ± 0.03 ^{a,A}	10.83 ± 1.20 ^{c,B}	21.57 ± 2.31 ^{c,A}
	15%	20.91 ± 1.30 ^{a,A}	21.93 ± 0.72 ^{a,A}	11.74 ± 1.22 ^{c,B}	26.29 ± 1.73 ^{b,A}
	20%	22.56 ± 0.20 ^{a,A}	21.95 ± 0.07 ^{a,A}	18.75 ± 0.47 ^{b,B}	23.00 ± 2.01 ^{b,A}
t_6-R	0%	22.65 ± 0.82 ^{a,A}	19.21 ± 1.20 ^{a,B}	22.40 ± 2.22 ^{a,B}	26.90 ± 0.95 ^{a,A}
	10%	20.40 ± 1.04 ^{a,A}	20.11 ± 0.77 ^{a,A}	8.09 ± 0.94 ^{b,B}	16.66 ± 1.95 ^{b,A}
	15%	20.31 ± 1.35 ^{a,A}	21.48 ± 1.01 ^{a,A}	9.73 ± 1.33 ^{b,B}	24.30 ± 1.87 ^{a,A}
	20%	21.68 ± 0.76 ^{a,A}	21.15 ± 2.59 ^{a,A}	13.34 ± 1.52 ^{b,B}	24.01 ± 1.51 ^{ab,A}
t_6-F	0%	22.65 ± 0.82 ^{a,A}	19.21 ± 1.20 ^{a,B}	22.40 ± 2.22 ^{b,B}	26.90 ± 0.94 ^{b,A}
	10%	21.07 ± 0.62 ^{a,A}	19.41 ± 2.34 ^{a,A}	19.26 ± 2.74 ^{b,B}	25.59 ± 4.91 ^{b,A}
	15%	21.07 ± 1.09 ^{a,A}	21.67 ± 1.48 ^{a,A}	28.21 ± 1.18 ^{a,B}	31.26 ± 2.47 ^{a,A}
	20%	22.56 ± 0.95 ^{a,A}	20.80 ± 0.43 ^{a,B}	29.12 ± 3.73 ^{a,B}	33.72 ± 1.27 ^{a,A}

% OP: percentage of added olive pomace. SWB: bread produced with soft wheat flour type 0; WWB: bread produced with whole wheat flour. t_0-f: bread made using freshly sampled olive pomace; t_6-R: bread made using olive pomace after 6 months of refrigeration; t_6-F: bread made using olive pomace after 6 months of freezing. For each parameter, for each type of olive pomace storage technique, and considering the flour type used, means followed by different letters within each column (lowercase) and each row (capital letters) indicate significant differences ($p < 0.05$).

In more detail, for the bread produced with type 0 soft wheat flour, no statistically significant differences were observed among the samples, irrespective of the olive pomace inclusion amounts or storage methods. Similarly, for the bread produced with whole wheat flour, no statistically significant differences were observed among the samples, irrespective of the olive pomace inclusion amounts or storage methods.

The bread samples produced with soft wheat flour generally showed no statistically significant differences in specific volume compared to those produced with whole wheat flour without, except those without olive pomace. In these latter samples, the bread produced with soft wheat flour showed the highest average specific volume values.

The results of the hardness measurements of the experimental bread loaves are reported in Table 5.

In more detail, the samples produced with soft wheat flour and 0% fresh olive pomace showed the highest values; the same trend was observed for the bread samples produced with wholemeal wheat flour and 0% olive pomace. As for the samples produced with soft wheat flour and refrigerated olive pomace, the highest hardness value was observed in the bread with 0% olive pomace, whereas, for the bread made with whole wheat flour, significant differences were observed among the samples, with the sample containing 10% olive pomace showing the lowest average value.

Regarding the bread produced with soft wheat flour and frozen olive pomace, the highest hardness values were observed in samples containing 15% and 20% olive pomace; the same trend was observed for the bread samples produced with whole wheat flour.

The bread samples produced with soft wheat flour showed the lowest hardness values compared to those produced with whole wheat flour, irrespective of the amount or storage method of the olive pomace.

Finally, the color parameters evaluated for bread loaves produced with fresh olive pomace are reported in Figure 2.

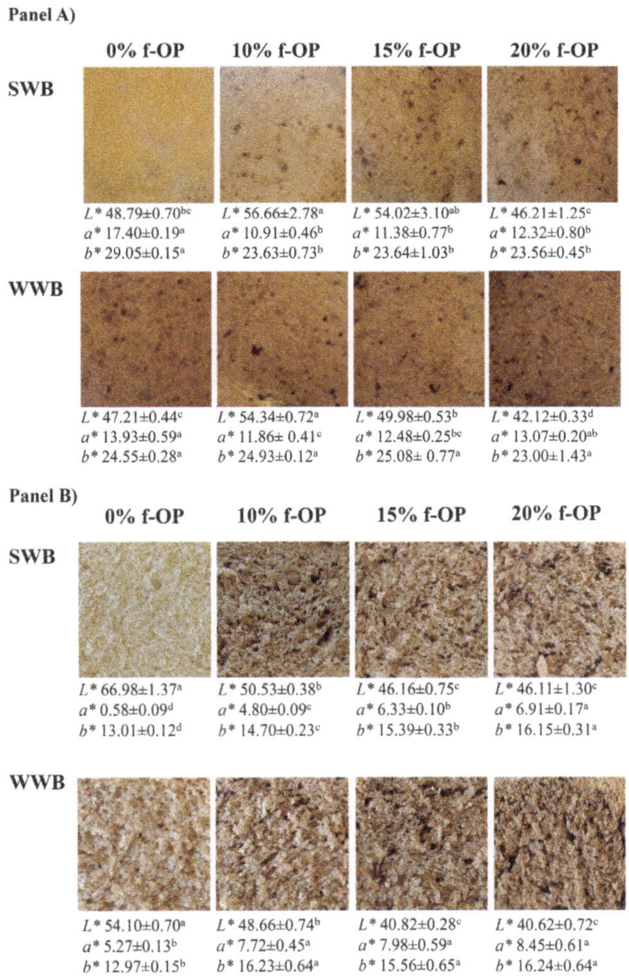

Figure 2. Images of the crust (panel **A**) and crumb (panel **B**) of bread samples containing fresh olive pomace (f-OP) produced with soft wheat flour type 0 (SWB) or whole wheat flour (WWB). Means ± standard deviations of triplicate independent measurements are shown. Within each panel and each type of bread (SWB or WWB), for the same color parameter, means followed by different letters are significantly different ($p < 0.05$). L^* values describe the lightness; a^* values describe the redness/greenness; b^* values describe the blueness/yellowness.

For the crust of the bread loaves produced with soft wheat flour, the average value of the L^* parameter was the lowest in the samples containing 20% olive pomace. The average values of the a^* and b^* parameters were the highest in the samples produced with no olive pomace. The same differences were observed for the crust of the bread loaves made with whole wheat flour.

As for the crumb of the bread loaves made with soft wheat flour, the average value of the L^* parameter was the highest in the samples made with no olive pomace. The average values of the a^* and b^* parameters were the highest in the samples produced with 20% olive pomace. The same differences were observed for the crumb of the bread loaves made with whole wheat flour.

3.5. *Total Phenolic Content, Antioxidant Capacity, and Dietary Fiber of the Breads*

The total phenolic content, antioxidant activity, and dietary fiber values of the bread samples studied in this research are reported in Table 6.

As expected, the bread produced with type 0 soft wheat flour resulted in samples with the lowest TPC; the whole wheat bread showed slightly higher TPC values than those of the bread produced with soft wheat flour, although this difference was not significant. The addition of higher percentages of olive pomace in the bread dough led to an augmentation of the TPC for both the bread types, and the highest values were measured in the bread samples containing 20% olive pomace.

The antioxidant activity of the tested samples reflected the concentrations of polyphenols found in the same products. This was proved by calculating the Pearson correlation coefficient among the four datasets (TPC, DPPH, ABTS, and FRAP values) and obtaining a high value, ranging between 0.845 and 0.969. A positive relationship between the phenolic amount and the antioxidant properties of the samples was observed, with a strong relationship among the different assays applied in the AOC determination. The bread samples without olive pomace had significantly lower AOC, whereas those containing 20% olive pomace showed the highest AOC.

As for dietary fiber, the bread samples produced with type 0 soft wheat flour and without olive pomace showed statistically significant lower values than those containing different percentages of olive pomace. In more detail, the bread made with soft wheat flour and 20% olive pomace showed the highest values, irrespective of the pomace storage conditions. In the bread samples produced with whole wheat flour, no significant differences were observed, irrespective of the pomace storage conditions or the percentage of olive pomace added. Finally, for the samples without olive pomace (control), those produced with soft wheat flour showed the lowest dietary fiber content compared to those containing whole wheat flour. No significant differences emerged among the samples containing 10% to 20% olive pomace, irrespective of the flour type used.

3.6. *SPME–GC/MS Analysis of the Volatile Components of the Breads*

The SPME–GC/MS analysis allowed the volatile profiles of the bread samples to be obtained for those made with type 0 soft wheat (Table 7) and whole wheat (Table 8) flours and with the addition of different amounts of olive pomace stored under different conditions.

Table 6. Total phenolic content, antioxidant capacity (AOC) as determined using the FOLIN, DPPH, ABTS, and FRAP assays, and dietary fiber content of bread samples.

	%OP	TPC (mg GAE g^{-1})		DPPH (mg TEAC g^{-1})		ABTS (mg TEAC g^{-1})		FRAP (mg TEAC g^{-1})		Dietary Fiber (g 100 g^{-1})	
		SWB	WWB	SWB	WWB	SWB	WWB	SWB	WWB	SWB	WWB
t_0-f	0%	0.27 ± 0.01 ^{a,A}	0.29± 0.061 ^{a,A}	0.19 ± 0.03 ^{a,A}	0.28 ± 0.10 ^{a,A}	0.32 ± 0.01 ^{a,A}	0.56 ± 0.13 ^{a,A}	0.14 ± 0.01 ^{a,A}	0.25 ± 0.47 ^{a,A}	3.27 ± 0.08 ^{c,B}	6.09 ± 0.83 ^{a,A}
	10%	0.70 ± 0.08 ^{b,A}	0.68 ± 0.05 ^{b,A}	1.26 ± 0.17 ^{bc,A}	1.29 ± 0.00 ^{b,A}	1.23 ± 0.22 ^{b,A}	1.41 ± 0.04 ^{b,A}	1.15 ± 0.10 ^{b,A}	1.23 ± 0.04 ^{b,A}	4.43 ± 0.52 ^{b,A}	6.83 ± 0.59 ^{a,A}
	15%	0.84 ± 0.09 ^{bc,A}	0.90 ± 0.10 ^{bc,A}	1.67 ± 0.09 ^{bc,A}	1.78 ± 0.05 ^{bc,A}	1.55 ± 0.01 ^{bc,A}	1.73 ± 0.10 ^{b,A}	1.63 ± 0.07 ^{b,A}	1.65 ± 0.05 ^{b,A}	5.49 ± 0.09 ^{ab,A}	7.00 ± 0.69 ^{a,A}
	20%	0.91 ± 0.05 ^{c,A}	0.75 ± 0.11 ^{bc,A}	1.90 ± 0.18 ^{c,A}	1.12 ± 0.00 ^{b,A}	1.76 ± 0.05 ^{c,A}	1.24 ± 0.04 ^{b,B}	1.76 ± 0.16 ^{b,A}	1.11 ± 0.01 ^{b,B}	6.57 ± 0.09 ^{a,A}	8.71 ± 0.83 ^{a,A}
t_6-R	0%	0.19 ± 0.01 ^{a,A}	0.37 ± 0.04 ^{a,B}	0.25 ± 0.04 ^{a,A}	0.33 ± 0.03 ^{a,A}	0.77 ± 0.11 ^{a,A}	0.97 ± 0.05 ^{a,A}	0.21 ± 0.00 ^{a,A}	0.31 ± 0.01 ^{a,A}	3.36 ± 0.20 ^{b,B}	5.20 ± 0.44 ^{a,A}
	10%	1.08 ± 0.28 ^{b,A}	0.99 ± 0.05 ^{b,A}	0.89 ± 0.21 ^{b,A}	0.82 ± 0.19 ^{b,A}	1.93 ± 0.49 ^{b,A}	1.82 ± 0.04 ^{b,A}	1.38 ± 0.49 ^{b,A}	1.21 ±0.30 ^{b,A}	4.83 ± 0.60 ^{ab,A}	6.58 ± 1.27 ^{a,A}
	15%	1.46 ± 0.33 ^{b,A}	1.20 ± 0.13 ^{b,A}	1.21 ± 0.21 ^{b,A}	0.97 ± 0.01 ^{b,A}	2.57 ± 0.50 ^{c,A}	2.15 ± 0.04 ^{bc,A}	2.02 ± 0.59 ^{bc,A}	1.45 ±0.03 ^{b,B}	4.87 ± 1.03 ^{ab,A}	6.65 ± 0.88 ^{a,A}
	20%	1.70 ± 0.20 ^{b,A}	1.61 ± 0.52 ^{b,A}	1.36 ± 0.27 ^{b,A}	1.34 ± 0.33 ^{b,A}	2.81 ± 0.45 ^{c,A}	2.64 ± 0.58 ^{c,A}	2.35 ± 0.51 ^{c,A}	2.14 ±0.81 ^{c,A}	6.44 ± 0.59 ^{a,A}	7.34 ± 0.44 ^{a,A}
t_6-F	0%	0.19 ± 0.01 ^{a,A}	0.37 ± 0.04 ^{a,B}	0.25 ± 0.04 ^{a,A}	0.33 ± 0.03 ^{a,A}	0.77 ± 0.11 ^{a,A}	0.97 ± 0.05 ^{a,A}	0.21 ± 0.00 ^{a,A}	0.31 ± 0.01 ^{a,A}	3.36 ± 0.20 ^{b,B}	5.20 ± 0.44 ^{a,A}
	10%	0.91 ± 0.18 ^{b,A}	0.95 ± 0.05 ^{b,A}	0.62 ± 0.08 ^{b,A}	0.66 ± 0.16 ^{b,A}	1.56 ± 0.18 ^{b,A}	1.64 ± 0.71 ^{b,A}	1.01 ± 0.22 ^{b,A}	0.99 ± 0.21 ^{b,A}	4.32 ± 0.72 ^{ab,A}	5.32 ± 0.71 ^{a,A}
	15%	1.01 ± 0.33 ^{b,A}	0.72 ± 0.14 ^{b,A}	1.01 ± 0.38 ^{b,A}	0.82 ± 0.03 ^{b,A}	2.11 ± 0.72 ^{c,A}	1.58 ± 0.13 ^{b,A}	1.53 ± 0.73 ^{b,A}	1.13 ± 0.10 ^{b,A}	5.65 ± 0.83 ^{a,A}	5.97 ± 1.35 ^{a,A}
	20%	1.39 ± 0.60 ^{b,A}	1.22 ± 0.13 ^{b,A}	1.07 ± 0.39 ^{b,A}	1.27 ± 0.26 ^{b,A}	2.36 ± 0.40 ^{c,A}	2.30 ± 0.01 ^{c,A}	1.85 ± 0.64 ^{c,A}	1.69 ± 0.24 ^{c,A}	6.21 ± 0.67 ^{a,A}	6.75 ± 0.50 ^{a,A}

% OP: percentage of olive pomace added to bread; TPC: total phenolic content; DPPH radical-scavenging activity test; ABTS radical-scavenging activity assay; FRAP: ferric-ion-reducing power. SWB: bread produced with soft wheat flour type 0; WWB: bread produced with whole wheat flour. t_0-f: bread made using freshly sampled olive pomace; t_6-R: bread made using olive pomace after 6 months of refrigeration; t_6-F: bread made olive pomace after 6 months of freezing. Results are expressed as mg GAE g^{-1} and mM TEAC g^{-1} on a wet basis. For each parameter, within each type of olive pomace storage technique and considering the flour type used, means followed by different letters within each column (lowercase) and each row (capital letters) indicate significant differences ($p < 0.05$).

Table 7. Volatile organic compounds (VOCs) identified in bread samples made with soft wheat flour (SWB) and with different concentration of olive pomace (0%, 10%, 15%, or 20%) and stored under different conditions (f, fresh; F, frozen; R, refrigerated).

RI	Compounds	t_0-f 0%	t_0-f 10%	t_0-f 15%	t_0-f 20%	t_6-(R or F) 0%	t_6-R 10%	t_6-R 15%	t_6-R 20%	t_6-F 10%	t_6-F 15%	t_6-F 20%
	Terpenoids											
1084	beta pinene	1.90 ± 0.03 ^{a}	3.00 ± 0.05 ^{c}	4.10 ± 0.05 ^{d}	5.46 ± 0.51 ^{e}	1.95 ± 0.04 ^{a}	1.43 ± 0.26 ^{b}	3.19 ± 0.53 ^{c}	3.17 ± 0.12 ^{c}	2.98 ± 0.20 ^{c}	3.48 ± 0.17 ^{c}	4.44 ± 0.23 ^{d}
1191	limonene	0.20 ± 0.01 ^{b}	0.30 ± 0.02 ^{c}	0.35 ± 0.03 ^{c}	0.50 ± 0.02 ^{d}	0.26 ± 0.01 ^{a}	0.24 ± 0.05 ^{a}	0.28 ± 0.01 ^{a}	0.34 ± 0.05 ^{c}	0.30 ± 0.08 ^{c}	0.32 ± 0.05 ^{c}	0.49 ± 0.08 ^{d}
1480	Copaene	Nd	2.14 ± 0.00 ^{d}	2.50 ± 0.02 ^{e}	2.84 ± 0.01 ^{f}	Nd	0.87 ± 0.02 ^{a}	1.04 ± 0.06 ^{b}	1.33 ± 0.19 ^{c}	1.20 ± 0.04 ^{c}	1.26 ± 0.29 ^{c}	1.40 ± 0.05 ^{c}
	Ketones											
812	2-Butanone	Nd	0.20 ± 0.00 ^{d}	0.21 ± 0.01 ^{d}	0.37 ± 0.01 ^{b}	Nd	0.60 ± 0.12 ^{a}	0.49 ± 0.03 ^{a}	0.37 ± 0.02 ^{b}	Nd	Nd	0.16 ± 0.01 ^{c}
938	2-pentanone	Nd	5.61 ± 0.14 ^{a}	6.56 ± 0.52 ^{a}	6.51 ± 0.42 ^{a}	Nd	Nd	6.01 ± 1.06 ^{a}	5.73 ± 0.34 ^{a}	Nd	Nd	Nd
1065	2,3-pentanedione	0.22 ± 0.01 ^{b}	2.18 ± 0.09 ^{f}	5.88 ± 0.51 ^{g}	7.48 ± 0.51 ^{h}	0.26 ± 0.01 ^{a}	0.92 ± 0.17 ^{c}	0.79 ± 0.21 ^{c}	0.73 ± 0.12 ^{c}	0.54 ± 0.02 ^{d}	0.22 ± 0.01 ^{a}	0.15 ± 0.03 ^{e}
1198	2-heptanone	0.48 ± 0.15 ^{a}	0.48 ± 0.04 ^{a}	1.10 ± 0.09 ^{b}	1.85 ± 0.04 ^{c}	0.52 ± 0.06 ^{a}	Nd	Nd	Nd	Nd	Nd	Nd
1283	2-octanone	0.16 ± 0.01 ^{b}	Nd	Nd	Nd	0.11 ± 0.01 ^{a}	0.35 ± 0.04 ^{c}	0.47 ± 0.08 ^{d}	0.78 ± 0.03 ^{e}	Nd	0.22 ± 0.02 ^{f}	0.33 ± 0.01 ^{c}
1281	Acetoin	0.77 ± 0.03 ^{b}	2.18 ± 0.01 ^{d}	2.57 ± 0.04 ^{d}	2.75 ± 0.01 ^{d}	3.16 ± 0.59 ^{a}	4.01 ± 0.84 ^{c}	2.21 ± 0.49 ^{d}	1.85 ± 0.47 ^{ed}	1.34 ± 0.08 ^{e}	2.04 ± 0.22 ^{d}	2.10 ± 0.35 ^{d}
1303	Acetol	0.50 ± 0.01 ^{b}	1.40 ± 0.00 ^{c}	0.99 ± 0.00 ^{c}	0.12 ± 0.03 ^{e}	0.66 ± 0.03 ^{a}	1.38 ± 0.19 ^{c}	1.08 ± 0.24 ^{c}	0.90 ± 0.09 ^{c}	0.57 ± 0.04 ^{d}	0.42 ± 0.09 ^{b}	0.40 ± 0.08 ^{b}
	Aldehydes											
820	2-methylbutanal	11.96 ± 0.63 ^{b}	16.28 ± 0.49 ^{d}	10.22 ± 0.22 ^{a}	5.52 ± 0.12 ^{c}	10.18 ± 0.06 ^{a}	10.38 ± 0.78 ^{a}	5.71 ± 0.42 ^{c}	5.11 ± 1.10 ^{c}	10.89 ± 0.12 ^{b}	9.68 ± 0.14 ^{a}	5.28 ± 0.13 ^{c}

Table 7. *Cont.*

RI	Compounds	t_0-f 0%	t_0-f 10%	t_0-f 15%	t_0-f 20%	t_6-(R or F) 0%	t_6-R 10%	t_6-R 15%	t_6-R 20%	t_6-F 10%	t_6-F 15%	t_6-F 20%
825	3-methylbutanal	14.07 ± 0.59 ^{b}	30.84 ± 0.46 ^{g}	21.18 ± 0.32 ^{h}	13.62 ± 0.17 ^{b}	10.63 ± 0.08 ^{a}	16.75 ± 1.01 ^{c}	10.54 ± 0.43 ^{a}	10.25 ± 1.22 ^{d}	10.99 ± 0.18 ^{ad}	9.21 ± 0.15 ^{e}	7.25 ± 0.24 ^{f}
1078	Hexanal	7.78 ± 0.50 ^{b}	6.46 ± 0.19 ^{f}	8.99 ± 0.47 ^{g}	9.56 ± 0.20 ^{h}	5.62 ± 0.43 ^{a}	3.81 ± 0.46 ^{c}	2.55 ± 0.19 ^{d}	2.46 ± 0.13 ^{d}	1.17 ± 0.22 ^{e}	1.58 ± 0.05 ^{e}	1.68 ± 0.32 ^{e}
1325	2-heptenal	0.10 ± 0.00 ^{a}	1.24 ± 0.08 ^{e}	1.07 ± 0.05 ^{b}	0.92 ± 0.02 ^{b}	Nd	1.07 ± 0.13 ^{b}	0.90 ± 0.05 ^{c}	1.19 ± 0.20 ^{b}	0.54 ± 0.14 ^{d}	0.45 ± 0.09 ^{d}	0.63 ± 0.07 ^{d}
1522	Benzaldehyde	1.16 ± 0.06 ^{b}	9.76 ± 0.19 ^{e}	12.98 ± 0.75 ^{f}	15.90 ± 0.10 ^{g}	0.97 ± 0.04 ^{a}	3.91 ± 0.88 ^{c}	4.46 ± 0.33 ^{c}	4.73 ± 0.92 ^{c}	4.28 ± 0.72 ^{c}	4.97 ± 0.23 ^{c}	7.72 ± 1.24 ^{d}
1640	Benzeneacetaldehyde	Nd	0.87 ± 0.09 ^{c}	1.18 ± 0.03 ^{d}	1.23 ± 0.03 ^{d}	Nd	0.23 ± 0.05 ^{a}	0.52 ± 0.11 ^{b}	0.60 ± 0.10 ^{b}	Nd	0.84 ± 0.10 ^{c}	0.78 ± 0.12 ^{c}
	Esters and acetates											
1433	Ethyl octanoate	0.30 ± 0.01 ^{a}	0.30 ± 0.00 ^{c}	0.32 ± 0.00 ^{d}	0.86 ± 0.04 ^{e}	Nd	0.25 ± 0.05 ^{a}	0.30 ± 0.02 ^{a}	0.74 ± 0.17 ^{b}	0.50 ± 0.11 ^{b}	0.61 ± 0.05 ^{b}	0.64 ± 0.10 ^{b}
1660	Ethyl benzoate	Nd	0.30 ± 0.01 ^{a}	0.31 ± 0.02 ^{a}	0.28 ± 0.01 ^{a}	Nd	0.33 ± 0.09 ^{a}	0.87 ± 0.10 ^{b}	0.88 ± 0.03 ^{b}	0.57 ± 0.03 ^{c}	0.55 ± 0.04 ^{c}	0.47 ± 0.07 ^{c}
801	Ethyl acetate	1.96 ± 0.01 ^{a}	2.60 ± 0.00 ^{e}	3.75 ± 0.04 ^{b}	4.01 ± 0.05 ^{d}	2.18 ± 0.35 ^{a}	1.86 ± 0.13 ^{a}	3.64 ± 0.67 ^{b}	5.75 ± 0.48 ^{c}	3.40 ± 0.31 ^{b}	3.88 ± 0.19 ^{b}	4.18 ± 0.24 ^{d}
	Alcohols											
888	Ethanol	134.31 ± 8.04 ^{b}	86.71 ± 7.33 ^{f}	128.4 ± 3.27 ^{g}	120.2 ± 0.67 ^{h}	105.61 ± 5.58 ^{a}	101.1 ± 20.77 ^{a}	183.8 ± 37.12 ^{c}	174.2 ± 9.15 ^{c}	254.3 ± 12.73 ^{d}	225.7 ± 36.31 ^{d}	212.3 ± 25.10 ^{e}
1211	Isoamylalcohol	17.53 ± 0.12 ^{a}	26.90 ± 3.63 ^{a}	30.08 ± 0.28 ^{b}	42.26 ± 0.17 ^{b}	16.57 ± 3.19 ^{a}	17.95 ± 4.31 ^{a}	20.86 ± 0.86 ^{a}	20.15 ± 3.22 ^{a}	21.12 ± 4.88 ^{a}	19.09 ± 4.92 ^{a}	19.33 ± 2.98 ^{a}
1108	Isobutanol	2.29 ± 0.07 ^{b}	2.50 ± 0.01 ^{b}	3.40 ± 0.02 ^{a}	9.46 ± 0.44 ^{c}	3.50 ± 0.47 ^{a}	2.85 ± 0.51 ^{b}	2.47 ± 0.25 ^{b}	2.38 ± 0.13 ^{b}	3.18 ± 0.31 ^{a}	3.58 ± 0.31 ^{a}	3.94 ± 1.18 ^{a}
1355	1-Hexanol	1.72 ± 0.12 ^{a}	7.37 ± 0.04 ^{d}	8.10 ± 0.26 ^{e}	15.03 ± 0.90 ^{f}	1.57 ± 0.43 ^{a}	2.83 ± 0.53 ^{b}	2.99 ± 0.80 ^{b}	2.50 ± 0.20 ^{b}	3.94 ± 0.85 ^{c}	4.38 ± 0.28 ^{c}	5.21 ± 0.78 ^{c}
1383	3-hexen-1-ol	Nd	Nd	1.10 ± 0.10 ^{c}	1.78 ± 0.02 ^{d}	Nd	Nd	Nd	0.39 ± 0.02 ^{a}	Nd	0.82 ± 0.02 ^{b}	0.90 ± 0.10 ^{b}
1905	Benzeneethanol	1.74 ± 0.01 ^{b}	1.21 ± 0.01 ^{d}	1.31 ± 0.03 ^{e}	1.51 ± 0.00 ^{f}	2.14 ± 0.61 ^{a}	2.31 ± 0.27 ^{a}	2.49 ± 0.18 ^{a}	3.40 ± 0.2 ^{c}	1.69 ± 0.26 ^{b}	1.83 ± 0.26 ^{b}	2.09 ± 0.04 ^{a}
	Pyrazines											
1268	Methylpyrazine	5.00 ± 0.11 ^{b}	5.50 ± 0.01 ^{b}	7.47 ± 0.42 ^{c}	9.86 ± 0.01 ^{f}	4.50 ± 0.18 ^{a}	5.85 ± 1.03 ^{b}	8.72 ± 1.05 ^{c}	10.27 ± 1.35 ^{d}	3.26 ± 0.19 ^{e}	4.81 ± 0.55 ^{a}	7.61 ± 1.18 ^{c}
1325	2,5-dimethyl pyrazine	0.71 ± 0.02 ^{a}	0.75 ± 0.01 ^{a}	0.71 ± 0.04 ^{a}	1.50 ± 0.01 ^{e}	0.65 ± 0.11 ^{a}	0.93 ± 0.12 ^{b}	1.21 ± 0.09 ^{c}	2.96 ± 0.57 ^{d}	0.86 ± 0.16 ^{b}	0.81 ± 0.04 ^{b}	1.55 ± 0.07 ^{e}
1349	2,3-dimethylpyrazine	1.11 ± 0.14 ^{a}	Nd	Nd	Nd	1.00 ± 0.12 ^{a}	1.45 ± 0.15 ^{b}	0.45 ± 0.02 ^{c}	0.29 ± 0.03 ^{d}	0.71 ± 0.04 ^{e}	0.19 ± 0.01 ^{f}	0.51 ± 0.02 ^{g}
1386	2-ethyl-6-methylpyrazine	0.78 ± 0.12 ^{b}	1.98 ± 0.06 ^{g}	2.15 ± 0.03 ^{f}	2.35 ± 0.01 ^{h}	0.97 ± 0.07 ^{a}	0.98 ± 0.13 ^{a}	1.38 ± 0.17 ^{c}	2.60 ± 0.13 ^{d}	1.11 ± 0.01 ^{e}	1.52 ± 0.02 ^{c}	2.12 ± 0.03 ^{f}
1391	2-ethyl-5-methylpyrazine	1.55 ± 0.09 ^{a}	1.77 ± 0.04 ^{e}	2.01 ± 0.02 ^{f}	2.65 ± 0.02 ^{g}	1.65 ± 0.08 ^{a}	1.17 ± 0.12 ^{b}	1.58 ± 0.10 ^{a}	3.18 ± 0.25 ^{c}	0.91 ± 0.16 ^{b}	1.20 ± 0.04 ^{b}	1.35 ± 0.11 ^{d}
	Acids											
1447	Acetic acid	11.83 ± 0.47 ^{b}	16.41 ± 0.25 ^{a}	19.83 ± 0.43 ^{c}	20.02 ± 0.53 ^{c}	13.77 ± 0.64 ^{a}	21.44 ± 1.76 ^{c}	23.16 ± 2.46 ^{c}	27.76 ± 0.45	15.40 ± 2.06 ^{a}	20.39 ± 2.73 ^{c}	22.57 ± 2.89 ^{c}
1628	Butanoic acid	0.54 ± 0.00 ^{b}	0.85 ± 0.02 ^{c}	1.03 ± 0.03 ^{d}	1.15 ± 0.02 ^{e}	0.62 ± 0.04 ^{a}	0.76 ± 0.06 ^{c}	0.94 ± 0.13 ^{d}	0.98 ± 0.11 ^{d}	0.91 ± 0.22 ^{d}	1.23 ± 0.15 ^{e}	1.23 ± 0.18 ^{e}
1840	Hexanoic acid	Nd	0.45 ± 0.01 ^{c}	0.54 ± 0.01 ^{d}	0.59 ± 0.00 ^{d}	0.21 ± 0.01 ^{a}	0.35 ± 0.04 ^{b}	0.48 ± 0.08 ^{c}	0.58 ± 0.08 ^{d}	0.41 ± 0.03 ^{c}	0.40 ± 0.01 ^{c}	0.50 ± 0.09 ^{c}
2058	Octanoic acid	0.25 ± 0.01 ^{b}	0.22 ± 0.00 ^{a}	0.26 ± 0.06 ^{b}	0.52 ± 0.10 ^{f}	0.21 ± 0.04 ^{a}	0.37 ± 0.06 ^{c}	0.20 ± 0.00 ^{a}	0.14 ± 0.01 ^{d}	0.09 ± 0.00 ^{e}	0.14 ± 0.01 ^{d}	0.16 ± 0.01 ^{d}
2281	Decanoic acid	0.30 ± 0.01 ^{b}	0.34 ± 0.03 ^{a}	0.46 ± 0.04 ^{f}	0.70 ± 0.04 ^{g}	0.35 ± 0.02 ^{a}	0.10 ± 0.01 ^{c}	0.32 ± 0.02 ^{ab}	0.94 ± 0.17 ^{d}	0.23 ± 0.03 ^{e}	0.25 ± 0.04 ^{e}	0.35 ± 0.04 ^{a}
	Furans and pyrans											
1223	2-pentylfuran	4.02 ± 0.03 ^{b}	3.78 ± 0.10 ^{f}	1.97 ± 0.04 ^{g}	1.91 ± 0.01 ^{h}	3.50 ± 0.02 ^{a}	Nd	0.49 ± 0.05 ^{c}	0.81 ± 0.03 ^{d}	0.49 ± 0.06 ^{c}	0.45 ± 0.05 ^{c}	0.40 ± 0.01 ^{e}
1462	2-furfural	2.10 ± 0.02 ^{b}	5.08 ± 0.06 ^{g}	5.24 ± 0.34 ^{g}	5.50 ± 0.31 ^{d}	2.64 ± 0.31 ^{a}	9.65 ± 1.87 ^{c}	7.70 ± 1.49 ^{d}	6.35 ± 0.65 ^{d}	3.37 ± 0.49 ^{e}	4.05 ± 0.21 ^{f}	2.04 ± 0.91 ^{b}
1658	2-Furanmethanol	1.61 ± 0.06 ^{a}	8.43 ± 0.47 ^{c}	7.59 ± 0.24 ^{g}	7.38 ± 0.21 ^{g}	2.10 ± 0.62 ^{a}	12.11 ± 2.41 ^{b}	8.61 ± 0.57 ^{c}	3.88 ± 0.25 ^{d}	5.89 ± 0.92 ^{e}	3.36 ± 0.17 ^{f}	2.27 ± 0.58 ^{a}
1503	2-acetylfuran	0.28 ± 0.00 ^{a}	1.07 ± 0.01 ^{c}	0.79 ± 0.05 ^{f}	0.73 ± 0.03 ^{f}	0.27 ± 0.06 ^{a}	1.27 ± 0.17 ^{b}	1.14 ± 0.21 ^{c}	0.49 ± 0.06 ^{d}	0.64 ± 0.06 ^{e}	0.59 ± 0.08 ^{de}	0.55 ± 0.11 ^{d}
2501	5-(hydroxymethyl)-2-furfural (HMF)	0.31 ± 0.00 ^{b}	0.46 ± 0.02 ^{a}	0.51 ± 0.03 ^{a}	0.62 ± 0.03 ^{d}	0.45 ± 0.12 ^{a}	0.84 ± 0.15 ^{c}	0.52 ± 0.07 ^{a}	0.41 ± 0.06 ^{a}	0.45 ± 0.03 ^{a}	0.52 ± 0.01 ^{a}	0.44 ± 0.06 ^{a}
	Phenols											
2203	Vinylguaiacol	Nd	0.30 ± 0.05 ^{c}	0.35 ± 0.04 ^{c}	0.42 ± 0.02 ^{d}	Nd	0.19 ± 0.01 ^{a}	0.18 ± 0.04 ^{ab}	0.13 ± 0.03 ^{b}	0.24 ± 0.04 ^{c}	0.24 ± 0.05 ^{c}	0.31 ± 0.05 ^{c}
1862	Guaiacol	Nd	0.48 ± 0.02 ^{c}	0.51 ± 0.01 ^{c}	0.60 ± 0.02 ^{d}	Nd	0.32 ± 0.07 ^{a}	0.36 ± 0.01 ^{a}	0.42 ± 0.04 ^{b}	0.32 ± 0.07 ^{a}	0.40 ± 0.03 ^{b}	0.41 ± 0.04 ^{b}

RI: retention index. RIs were calculated using the van Den Dool and Kratz formula. Calculated RIs were compared using the online NIST database (http://webbook.nist.gov/chemistry/; accessed on 1 June 2023) for a high polar column for InnoWAX or similar stationary phases. All the compounds were identified by the matching RI and MS. The results are expressed as RAP ± SD; relative peak area (area peak compound/area peak internal standard) × 100 ± standard deviation. Values labeled with different lowercase letters in the same row are significantly different ($p < 0.05$). Nd: not detected.

Table 8. Volatile organic compounds (VOCs) identified in breads made with whole wheat flour (WWB) and with different concentrations of olive pomace (0%, 10%, 15%, or 20%) and stored under different conditions (t, fresh; F, frozen; R, refrigerated).

RI	Compounds	t_0-f 0%	t_0-f 10%	t_0-f 15%	t_0-f 20%	t_6-(R or F) 0%	t_6-R 10%	t_6-R 15%	t_6-R 20%	t_6-F 10%	t_6-F 15%	t_6-F 20%
	Terpenoids											
1084	Beta pinene	3.23 ± 0.65 ^{a}	4.60 ± 0.50 ^{a}	4.80 ± 0.60 ^{a}	6.10 ± 0.50 ^{b}	3.50 ± 1.0 ^{a}	4.10 ± 0.36 ^{a}	4.50 ± 0.73 ^{a}	4.60 ± 0.47 ^{a}	4.48 ± 0.58 ^{a}	4.85 ± 0.98 ^{c}	2.98 ± 0.15 ^{a}
1191	Limonene	0.55 ± 0.04 ^{a}	0.57 ± 0.05 ^{a}	0.95 ± 0.01 ^{d}	2.44 ± 0.15 ^{e}	0.60 ± 0.05 ^{a}	0.47 ± 0.01 ^{b}	0.47 ± 0.02 ^{b}	0.60 ± 0.02 ^{a}	0.60 ± 0.03 ^{a}	0.74 ± 0.10 ^{a}	1.12 ± 0.01 ^{c}
1480	Copaene	Nd	1.57 ± 0.03 ^{c}	2.94 ± 0.06 ^{d}	3.18 ± 0.02 ^{e}	Nd	0.95 ± 0.05 ^{a}	1.14 ± 0.28 ^{a,b}	1.23 ± 0.21 ^{b}	0.85 ± 0.01 ^{c,b}	0.87 ± 0.08 ^{c,b}	1.04 ± 0.04 ^{b}
	Ketones											
812	2-Butanone	0.33 ± 0.00 ^{a}	0.31 ± 0.00 ^{a}	0.36 ± 0.03 ^{a}	0.85 ± 0.03 ^{e}	0.34 ± 0.02 ^{a}	0.21 ± 0.04 ^{b}	0.39 ± 0.02 ^{c}	0.57 ± 0.09 ^{d}	0.19 ± 0.04 ^{b}	0.25 ± 0.03 ^{b}	0.36 ± 0.03 ^{a}
938	2-pentanone	3.84 ± 0.36 ^{b}	4.78 ± 0.15 ^{e}	4.91 ± 0.05 ^{e}	7.20 ± 0.40 ^{f}	3.10 ± 0.05 ^{a}	2.04 ± 0.20 ^{c}	2.47 ± 0.20 ^{d}	2.54 ± 0.05 ^{d}	2.32 ± 0.23 ^{c,d}	2.80 ± 0.30 ^{d}	2.96 ± 0.59 ^{d}
1065	2,3-pentanedione	0.40 ± 0.08 ^{a}	1.10 ± 0.03 ^{d}	2.08 ± 0.03 ^{e}	2.20 ± 0.05 ^{f}	0.38 ± 0.02 ^{a}	0.31 ± 0.03 ^{a}	0.47 ± 0.05 ^{a}	0.72 ± 0.04 ^{b}	0.23 ± 0.02 ^{c}	0.32 ± 0.02 ^{a}	0.67 ± 0.02 ^{b}
1198	2-heptanone	0.70 ± 0.01 ^{a}	Nd	Nd	Nd	0.75 ± 0.07 ^{a}	Nd	Nd	Nd	Nd	Nd	Nd
1283	2-octanone	1.10 ± 0.05 ^{a}	0.95 ± 0.05 ^{c}	1.10 ± 0.06 ^{a}	1.45 ± 0.02 ^{e}	1.18 ± 0.27 ^{a}	0.62 ± 0.12 ^{b}	0.85 ± 0.05 ^{c}	0.90 ± 0.05 ^{c}	0.87 ± 0.03 ^{c}	1.10 ± 0.05 ^{a}	1.29 ± 0.07 ^{d}
1281	Acetoin	1.17 ± 0.08 ^{b}	2.29 ± 0.04 ^{c}	2.34 ± 0.03 ^{c}	2.85 ± 0.04 ^{c}	2.34 ± 0.26 ^{a}	2.81 ± 0.13 ^{c}	3.18 ± 0.14 ^{d}	3.44 ± 0.54 ^{d,a}	2.60 ± 0.53 ^{c}	2.68 ± 0.34 ^{c}	3.06 ± 0.07 ^{d}
1303	Acetol	0.85 ± 0.03 ^{a}	0.20 ± 0.03 ^{e}	0.34 ± 0.01 ^{f}	0.74 ± 0.00 ^{g}	0.90 ± 0.04 ^{a}	0.99 ± 0.04 ^{b}	1.09 ± 0.07 ^{b}	1.04 ± 0.13 ^{b}	0.47 ± 0.03 ^{c}	0.57 ± 0.03 ^{d}	1.37 ± 0.02 ^{d}
	Aldehydes											
820	2-methylbutanal	9.56 ± 0.21 ^{a}	10.69 ± 0.43 ^{g}	11.49 ± 0.35 ^{g}	8.84 ± 0.35 ^{a}	8.79 ± 0.67 ^{a}	3.25 ± 0.46 ^{c}	3.67 ± 0.11 ^{d}	7.13 ± 0.50 ^{a}	2.87 ± 0.34 ^{c}	7.30 ± 0.05 ^{e}	7.75 ± 0.05 ^{f}
825	3-methylbutanal	13.31 ± 2.41 ^{a}	21.62 ± 0.01 ^{c}	20.66 ± 0.19 ^{c}	18.67 ± 0.76 ^{d}	12.97 ± 2.17 ^{a}	15.36 ± 0.48 ^{a}	9.40 ± 0.24 ^{b}	8.91 ± 0.67 ^{b}	14.98 ± 2.11 ^{a,c}	15.06 ± 2.11 ^{a,c}	15.81 ± 0.19 ^{a}
1078	Hexanal	6.91 ± 0.84 ^{a}	12.90 ± 0.39 ^{d}	9.83 ± 0.68 ^{e}	8.40 ± 0.30 ^{f}	6.41 ± 1.09 ^{a}	4.72 ± 0.41 ^{b}	3.81 ± 0.58 ^{b}	2.84 ± 0.12 ^{c}	4.88 ± 0.99 ^{b}	3.85 ± 0.31 ^{b}	3.08 ± 0.18 ^{b}
1325	2-heptenal	0.19 ± 0.01 ^{a}	0.57 ± 0.03 ^{c}	0.73 ± 0.05 ^{e}	0.95 ± 0.02 ^{f}	0.20 ± 0.01 ^{a}	1.14 ± 0.04 ^{b}	1.11 ± 0.05 ^{b}	1.12 ± 0.21 ^{b}	0.56 ± 0.09 ^{c}	0.85 ± 0.02 ^{d}	1.12 ± 0.09 ^{b}
1522	Benzaldehyde	1.92 ± 0.02 ^{a}	5.03 ± 0.06 ^{b}	7.15 ± 0.42 ^{d}	7.65 ± 0.41 ^{d}	1.72 ± 0.19 ^{a}	4.07 ± 0.93 ^{b}	4.90 ± 0.86 ^{b}	5.87 ± 0.44 ^{b}	3.75 ± 0.23 ^{c}	4.34 ± 0.30 ^{b}	5.20 ± 0.75 ^{b}
1640	Benzeneacetaldehyde	Nd	2.10 ± 0.05 ^{b}	2.40 ± 0.06 ^{b}	3.12 ± 0.07 ^{d}	Nd	1.76 ± 0.19 ^{a}	2.32 ± 0.19 ^{b}	2.45 ± 0.09 ^{b}	1.78 ± 0.04 ^{a}	1.85 ± 0.01 ^{a}	6.65 ± 0.05 ^{c}
	Esters and acetates											
1020	Ethyl butanoate	Nd	0.30 ± 0.01 ^{d}	0.33 ± 0.02 ^{d}	0.45 ± 0.03 ^{c}	Nd	0.15 ± 0.01 ^{a}	0.26 ± 0.02 ^{b}	0.26 ± 0.01 ^{b}	0.23 ± 0.02 ^{b}	0.26 ± 0.01 ^{b}	0.42 ± 0.04 ^{c}
1229	Ethyl hexanoate	Nd	0.71 ± 0.03 ^{c}	1.20 ± 0.03 ^{d}	1.39 ± 0.05 ^{b}	Nd	2.24 ± 0.13 ^{a}	2.20 ± 0.14 ^{a}	1.53 ± 0.24 ^{b}	0.80 ± 0.14 ^{c}	1.11 ± 0.20 ^{d}	1.36 ± 0.01 ^{b}
1433	Ethyl octanoate	0.14 ± 0.01 ^{a}	0.72 ± 0.03 ^{e}	0.78 ± 0.00 ^{f}	0.82 ± 0.00 ^{b}	Nd	0.87 ± 0.10 ^{b}	0.82 ± 0.10 ^{b}	0.92 ± 0.06 ^{b}	0.67 ± 0.01 ^{c}	0.60 ± 0.03 ^{d}	0.68 ± 0.11 ^{c}
801	Ethyl acetate	1.94 ± 0.04 ^{b}	2.71 ± 0.03 ^{f}	3.20 ± 0.05 ^{h}	3.77 ± 0.09 ^{i}	1.14 ± 0.01 ^{a}	1.43 ± 0.09 ^{c}	2.30 ± 0.15 ^{d}	2.02 ± 0.31 ^{d}	1.74 ± 0.06 ^{a}	2.55 ± 0.15 ^{d,f}	5.71 ± 1.23 ^{g}
	Alcohols											
888	Ethanol	93.06 ± 0.10 ^{b}	97.78 ± 4.10 ^{a,b}	103.00 ± 0.69 ^{f}	110.00 ± 1.66 ^{c}	95.53 ± 0.23 ^{a}	119.94 ± 6.55 ^{c}	186.97 ± 2.70 ^{d}	215.73 ± 4.86 ^{a}	92.78 ± 4.10 ^{a,b}	104.77 ± 0.69 ^{f}	105.93 ± 1.66 ^{f}
1211	Isoamylalcohol	26.73 ± 0.42 ^{a}	24.17 ± 0.07 ^{f}	27.96 ± 0.58 ^{g}	34.36 ± 0.14 ^{h}	28.97 ± 2.15 ^{a}	17.48 ± 0.50 ^{b}	20.23 ± 1.20 ^{c}	32.55 ± 0.51 ^{d}	12.60 ± 0.60 ^{e}	16.90 ± 0.15 ^{b}	24.46 ± 0.50 ^{f}
1108	Isobutanol	3.25 ± 0.15 ^{a}	3.60 ± 0.50 ^{a}	2.64 ± 0.75 ^{b}	1.76 ± 0.04 ^{c}	3.30 ± 0.48 ^{a}	3.31 ± 0.99 ^{a}	2.51 ± 0.16 ^{b}	2.33 ± 0.29 ^{b}	3.33 ± 0.60 ^{a}	2.51 ± 0.06 ^{b}	1.81 ± 0.03 ^{c}
1355	1-Hexanol	3.97 ± 0.14 ^{a}	5.10 ± 0.55 ^{a}	8.64 ± 0.56 ^{e}	12.13 ± 0.69 ^{f}	4.32 ± 0.30 ^{a}	6.70 ± 0.36 ^{b}	4.64 ± 0.42 ^{a}	3.61 ± 0.41 ^{c}	4.74 ± 0.15 ^{d}	8.60 ± 0.21 ^{e}	9.43 ± 0.40 ^{f}
1905	Benzeneethanol	0.92 ± 0.02 ^{a}	1.23 ± 0.01 ^{c}	1.14 ± 0.03 ^{c}	1.02 ± 0.06 ^{a}	0.92 ± 0.05 ^{a}	3.35 ± 0.51 ^{b}	3.12 ± 0.65 ^{b}	2.34 ± 0.58 ^{b}	1.30 ± 0.09 ^{c}	3.27 ± 0.19 ^{b}	1.30 ± 0.21 ^{c}
	Pyrazines											
1268	Methylpyrazine	2.60 ± 0.74 ^{a}	5.39 ± 0.03 ^{b}	5.48 ± 0.07 ^{b}	6.10 ± 0.07 ^{b}	2.80 ± 0.50 ^{a}	5.86 ± 0.52 ^{b}	7.79 ± 1.55 ^{c}	12.89 ± 1.51 ^{d}	5.26 ± 0.35 ^{b}	5.87 ± 0.35 ^{b}	9.63 ± 0.60 ^{c}
1325	2,5-dimethyl pyrazine	0.75 ± 0.11 ^{a}	1.42 ± 0.04 ^{d}	1.50 ± 0.01 ^{d}	1.67 ± 0.00 ^{e}	0.80 ± 0.11 ^{a}	1.19 ± 0.10 ^{b}	1.83 ± 0.47 ^{c}	1.91 ± 0.12 ^{c}	1.10 ± 0.03 ^{b}	1.25 ± 0.02 ^{b}	1.47 ± 0.29 ^{d}
1349	2.3-dimethylpyrazine	Nd	0.80 ± 0.04 ^{c}	1.20 ± 0.06 ^{f}	1.98 ± 0.05 ^{g}	Nd	0.35 ± 0.01 ^{a}	0.42 ± 0.03 ^{b}	0.86 ± 0.05 ^{c}	0.74 ± 0.01 ^{d}	0.85 ± 0.05 ^{c}	1.74 ± 0.01 ^{e}
1386	2-ethyl-6-methylpyrazine	1.12 ± 0.10 ^{b}	0.90 ± 0.02 ^{g}	1.20 ± 0.04 ^{b}	1.84 ± 0.00 ^{d}	0.10 ± 0.01 ^{a}	1.15 ± 0.04 ^{b}	1.38 ± 0.04 ^{c}	1.83 ± 0.04 ^{d}	0.97 ± 0.02 ^{e}	1.40 ± 0.04 ^{c}	1.60 ± 0.02 ^{f}
1391	2-ethyl-5-methylpyrazine	0.65 ± 0.02 ^{b}	1.30 ± 0.05 ^{f}	1.34 ± 0.10 ^{g}	1.85 ± 0.20 ^{e}	0.79 ± 0.02 ^{a}	1.14 ± 0.07 ^{c}	1.64 ± 0.12 ^{d}	1.87 ± 0.13 ^{e}	1.25 ± 0.01 ^{f}	1.45 ± 0.03 ^{g}	1.87 ± 0.02 ^{e}

Table 8. *Cont.*

RI	Compounds	t_0-f 0%	t_0-f 10%	t_0-f 15%	t_0-f 20%	t_6-(R or F) 0%	t_6-R 10%	t_6-R 15%	t_6-R 20%	t_6-F 10%	t_6-F 15%	t_6-F 20%
	Acids											
1447	Acetic acid	12.19 ± 0.30 ^{b}	20.50 ± 0.40 ^{c}	22.35 ± 0.55 ^{c}	27.20 ± 0.43 ^{e}	13.76 ± 0.79 ^{a}	20.73 ± 0.72 ^{c}	21.13 ± 0.75 ^{c}	22.71 ± 2.00 ^{c,d}	20.17 ± 3.66 ^{c}	22.42 ± 1.01 ^{c}	25.02 ± 1.53 ^{d}
1628	Butanoic acid	0.64 ± 0.01 ^{a}	0.43 ± 0.01 ^{c}	0.55 ± 0.12 ^{a}	0.67 ± 0.08 ^{a}	0.59 ± 0.07 ^{a}	1.03 ± 0.11 ^{b}	0.98 ± 0.05 ^{b}	1.03 ± 0.10 ^{b}	0.44 ± 0.09 ^{a,c}	0.45 ± 0.02 ^{c}	0.70 ± 0.07 ^{a}
1840	Hexanoic acid	0.20 ± 0.01 ^{b}	1.12 ± 0.03 ^{e}	1.23 ± 0.01 ^{e}	1.40 ± 0.05 ^{f}	0.27 ± 0.02 ^{a}	0.70 ± 0.11 ^{c}	0.90 ± 0.04 ^{d}	1.19 ± 0.04 ^{e}	1.07 ± 0.02 ^{f}	1.08 ± 0.02 ^{f}	1.17 ± 0.09 ^{e}
2058	Octanoic acid	0.18 ± 0.00 ^{a}	0.12 ± 0.00 ^{c}	0.16 ± 0.06 ^{c}	0.18 ± 0.06 ^{c,a}	0.22 ± 0.04 ^{a}	0.35 ± 0.02 ^{b}	0.23 ± 0.02 ^{a}	0.13 ± 0.01 ^{c}	0.11 ± 0.02 ^{c}	0.12 ± 0.00 ^{c}	0.15 ± 0.01 ^{c}
2281	Decanoic acid	0.30 ± 0.04 ^{a}	0.34 ± 0.06 ^{a}	0.40 ± 0.05 ^{a}	0.48 ± 0.01 ^{e}	0.35 ± 0.04 ^{a}	0.25 ± 0.06 ^{b}	0.56 ± 0.02 ^{c}	0.81 ± 0.01 ^{d}	0.30 ± 0.01 ^{a}	0.25 ± 0.01 ^{b}	0.28 ± 0.03 ^{b}
	Furans and pyrans											
1223	2-pentylfuran	3.43 ± 0.26 ^{a}	3.51 ± 0.05 ^{g}	3.83 ± 0.05 ^{d}	4.85 ± 0.68 ^{e}	3.37 ± 0.35 ^{a}	1.25 ± 0.04 ^{b}	1.38 ± 0.03 ^{c}	3.80 ± 0.09 ^{d}	1.82 ± 0.02 ^{e}	2.65 ± 0.10 ^{f}	3.88 ± 0.6 ^{d}
1462	2-furfural	4.22 ± 0.07 ^{b}	6.10 ± 0.01 ^{g}	5.48 ± 0.04 ^{e}	4.25 ± 0.03 ^{b}	3.80 ± 0.17 ^{a}	10.37 ± 0.22 ^{c}	8.61 ± 0.79 ^{d}	4.25 ± 0.85 ^{b}	5.42 ± 0.12 ^{e}	3.95 ± 0.14 ^{a,b}	3.05 ± 0.01 ^{f}
1658	2-Furanmethanol	3.48 ± 0.39 ^{a}	6.71 ± 0.06 ^{e}	2.06 ± 0.08 ^{f}	1.44 ± 0.08 ^{g}	3.81 ± 0.55 ^{a}	9.15 ± 0.07 ^{c}	7.13 ± 0.54 ^{d}	4.1 ± 0.56 ^{a}	6.64 ± 0.11 ^{d,e}	2.11 ± 0.28 ^{f}	2.17 ± 0.25 ^{f}
1503	2-acetylfuran	0.65 ± 0.05 ^{b}	0.88 ± 0.02 ^{i}	0.81 ± 0.00 ^{a}	0.73 ± 0.02 ^{a}	0.77 ± 0.06 ^{a}	0.80 ± 0.03 ^{a}	1.17 ± 0.22 ^{d}	1.77 ± 0.10 ^{e}	1.31 ± 0.02 ^{f}	0.47 ± 0.01 ^{g}	0.27 ± 0.01 ^{h}
2501	5-(hydroxymethyl)-2-furfural (HMF)	0.35 ± 0.01 ^{a}	0.30 ± 0.01 ^{c}	0.35 ± 0.09 ^{a}	0.46 ± 0.02 ^{b}	0.31 ± 0.07 ^{a}	0.46 ± 0.04 ^{b}	0.48 ± 0.01 ^{b}	0.39 ± 0.02 ^{a}	0.27 ± 0.04 ^{a}	0.24 ± 0.01 ^{a}	0.28 ± 0.02 ^{a}
	Phenols											
2203	Vinylguaiacol	Nd	0.71 ± 0.00 ^{b}	0.76 ± 0.01 ^{c}	0.81 ± 0.00 ^{d}	Nd	0.61 ± 0.02 ^{a}	0.63 ± 0.05 ^{a}	0.69 ± 0.01 ^{a,b}	0.70 ± 0.01 ^{b}	0.75 ± 0.01 ^{c}	0.79 ± 0.03 ^{d}
1862	Guaiacol	Nd	0.60 ± 0.01 ^{c}	0.75 ± 0.02 ^{d}	0.88 ± 0.01 ^{b}	Nd	0.46 ± 0.01 ^{a}	0.87 ± 0.07 ^{b}	0.99 ± 0.08 ^{b}	0.52 ± 0.05 ^{a}	0.63 ± 0.05 ^{c}	0.72 ± 0.03 ^{d}

RI: retention index. RIs were calculated using the van Den Dool and Kratz formula. Calculated RIs were compared using the online NIST database (http://webbook.nist.gov/chemistry/; accessed on 1 June 2023) for a high polar column for InnoWAX or similar stationary phases. All the compounds were identified by the matching RI and MS. The results are expressed as RAP ± SD; relative peak area (area peak compound/area peak internal standard) × 100 ± standard deviation. Values labeled with different lowercase letters in the same row are significantly different ($p < 0.05$). Nd: not detected.

The detected compounds belonged to nine classes— ketones (7), aldehydes (6), alcohols (5), acids (5), furans and pyrans (5), pyrazines (5), esters and acetates (4), terpenoids (3), and phenols (2).

Among the carboxylic acids, acetic acid was the most represented type in all the samples. Beta pinene, limonene, and copaene were the detected terpenoids.

Among the phenols, vinylguaiacol and guaiacol were found in trace amounts.

Among the furans, 2-pentylfuran, 2-furfural, and 2-furanmethanol were detected in almost all of the samples and at the highest amounts.

Regarding the alcohols, ethanol was the most detected volatile compound in all the samples followed by isoamylalcohol, isobutanol, and 1-hexanol.

Among the acetates and esters, ethyl acetate was found in the highest amounts in all the samples.

Methylpyrazine, 2,3-dimethylpyrazine, and 2-ethyl-5-methylpyrazine were the most detected compounds among the pyrazines.

2-methylbutanal and 3-methylbutanal were the most abundant aldehydes followed by hexanal and benzaldehydes.

Acetoin was the most detected ketone in all the samples.

To better understand the differences among the bread samples, PCA was applied to the volatile compounds detected in the type 0 soft wheat bread (Figure 3 and Table 9) and the whole wheat bread (Figure 4 and Table 10) samples.

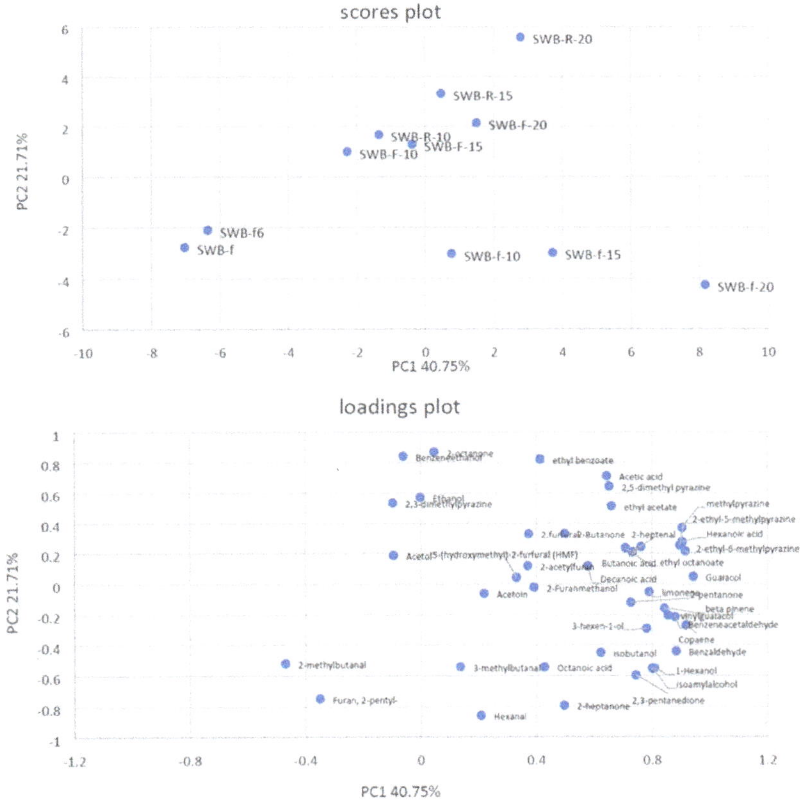

Figure 3. Principal Component Analysis (PCA) of volatile compounds in type 0 soft wheat flour bread (SWB) samples with different concentrations of olive pomace (0%, 10%, 15%, or 20%) stored under different conditions (f, fresh; F, frozen; R, refrigerated).

Table 9. Eigenvalues of the Principal Component Analysis of the volatile organic compound analysis of soft wheat flour bread (SWB) samples. This table reports the eigenvalues of each component of the PCA as well as the percentage of the total variance that is accounted for by each component.

Axis	Eigenvalue	Difference	Proportion (%)	Cumulative (%)
1	17.114715	7.994542	40.75%	40.75%
2	9.120173	1.990407	21.71%	62.46%
3	7.129766	3.920977	16.98%	79.44%
4	3.208789	0.33098	7.64%	87.08%
5	2.877809	2.013748	6.85%	93.93%
6	0.864061	0.210451	2.06%	95.99%
7	0.65361	0.225627	1.56%	97.55%
8	0.427982	0.077306	1.02%	98.56%
9	0.350676	0.098257	0.83%	99.40%
10	0.25242	0.25242	0.60%	100.00%

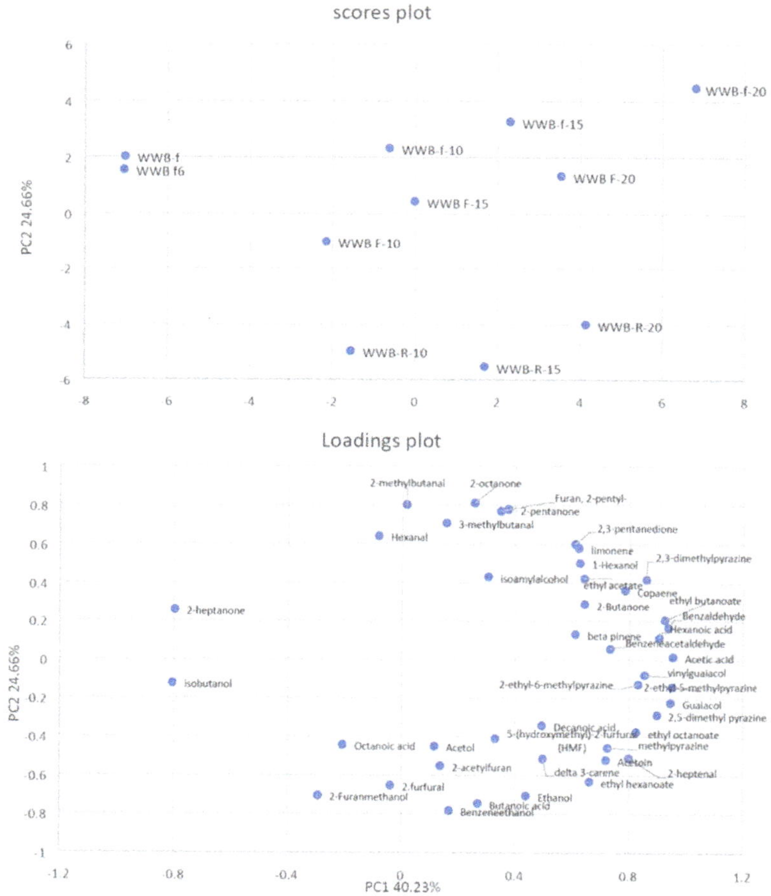

Figure 4. Principal Component Analysis (PCA) of volatile compounds in whole wheat flour bread (WWB) samples with different concentrations of olive pomace (0%, 10%, 15%, or 20%) stored under different conditions (f, fresh; F, frozen; R, refrigerated).

Table 10. Eigenvalues of the Principal Component Analysis of the volatile organic compound analysis of whole wheat flour bread (WWB) samples. This table reports the eigenvalues of each component of the PCA as well as the percentage of the total variance that is accounted for by each component.

Axis	Eigen Value	Difference	Proportion (%)	Cumulative (%)
1	17.29995	6.696625	40.23%	40.23%
2	10.603324	5.611939	24.66%	64.89%
3	4.991385	0.937016	11.61%	76.50%
4	4.054369	1.117212	9.43%	85.93%
5	2.937157	1.670838	6.83%	92.76%
6	1.266318	0.531867	2.94%	95.70%
7	0.734452	0.244383	1.71%	97.41%
8	0.490069	0.152634	1.14%	98.55%
9	0.337434	0.051891	0.78%	99.34%
10	0.285543	0.285543	0.66%	100.00%

Regarding the bread samples produced with type 0 soft wheat flour, the two PCs explained 62.46% of the total variance in the data. The samples were located in three different zones of the plot plane. Regarding the score plot, a clear separation between the SWB-f and SWB-f_6 bread samples produced without olive pomace (negatively associated with PC1) and all the other bread samples produced with different concentrations of olive pomace (positively associated with PC1) was evident.

On the other hand, some differences were also found between samples SWBf-10, 15, and 20 produced with fresh olive pomace (negatively associated with PC2) and with refrigerated and frozen olive pomace (positively associated with PC2).

In particular, the bread samples produced with fresh olive pomace—irrespective of the pomace concentration used—were characterized by terpenoids, phenols, ketones, and aldehydes, whereas the bread samples produced with the refrigerated and frozen olive pomace were characterized by the occurrence of alcohols (mainly ethanol), acids (mainly acetic acid), esters, and acetate.

Figure 4 shows the PCA plot obtained by analyzing the volatile compounds from the bread samples produced with whole wheat flour. The two PCs explained 64.89% of the total variance in the data. The samples were located in three different zones of the plot plane. Also in these samples, regarding the score plot, those produced without olive pomace (negatively associated with PC1) differed from all those produced with different concentrations of olive pomace (positively associated with PC1).

On the other hand, marked differences were also found between the samples produced with fresh olive pomace (positively associated with PC2) and those produced with refrigerated olive pomace (negatively associated with PC2). The bread samples produced with frozen olive pomace plotted close to those produced using fresh olive pomace.

In more detail, the bread samples produced with fresh olive pomace—irrespective of the pomace concentration used—were characterized by terpenoids, ketones, and aldehydes, whereas the bread samples produced with refrigerated olive pomace were characterized by alcohols (mainly ethanol), acids, esters, and acetate. The bread samples produced using frozen olive pomace had a volatile profile that was similar to that of the bread produced with fresh olive pomace.

4. Discussion

The use of olive pomace in bread formulation could represent an innovative and low-cost strategy to produce healthy and value-added food products.

However, two important considerations regarding this ingredient must be made. First, the amount of olive pomace can strongly affect the techno-functional and volatile features of bread; second, the seasonality of olive pomace can reduce the exploitability of this ingredient by the food industry. Hence, the added amount of olive pomace should be carefully investigated in order to allow the food industry to produce marketable bread.

Moreover, effective preservation methods for olive pomace should be investigated to provide a continuous supply to the food industry.

Based on the abovementioned considerations, in this study, olive pomace maintained under different storage conditions was tested for bread-making in proportions up to 20%, providing the food industry with sound results to inform product development and quality assessment.

As for the microbial loads detected in the fresh olive pomace, a progressive and remarkable reduction in the counts of mesophilic aerobes and lactic acid bacteria was observed in the samples stored at +4 °C, whereas a lower reduction was observed in those stored under frozen conditions. Since the analyzed olive pomace samples were obtained after no microbial stabilization (e.g., heat treatment), the presence of mesophilic aerobes is the result of the environmental (e.g., dust or soil) contamination of the milled olives. Of note, the antioxidant compounds detected in olive pomace can strongly modulate the viability of microorganisms, thus producing a negative selective pressure towards certain taxa (e.g., Enterobacteriaceae) or a positive enhancement of others (e.g., lactic acid bacteria and yeasts) [30,31]. Moreover, it is supposed that storage under frozen conditions also guarantees a higher microbial survival time, thus explaining the higher counts after 6 months at −20 °C.

The results obtained from the analysis of the total polyphenolic content of olive pomace at t_0 were in accordance with the concentrations of polyphenols found in fresh olives by Piscopo et al. [32], who indicated a TPC of 13.64 ± 0.64 mg GAE g^{-1} for fresh fruits. In addition, the antioxidant activity of the olive pomace samples was determined by means of three different assays (DPPH, ABT, and FRAP). The values obtained for the initial pomace are higher than those reported for green and black olives by Pellegrini et al. [33]. These authors [33] state an antioxidant capacity of 10.43 ± 7–14.73 ± 3 mmol Trolox kg^{-1} as determined using the DPPH test, and 24.59 ± 6–39.99 ± 4 mmol Fe^{2+} kg^{-1} as determined using the FRAP assay for green and black olives, respectively, whereas the samples analyzed herein yielded values of 16.21 ± 0.41 mg TE g^{-1} when DPPH reagents were used (corresponding to 64.68 ± 1.62 mmol Trolox kg^{-1}) and 15.95 ± 0.19 mg TE g^{-1} when applying the FRAP method (corresponding to 63.72 ± 0.75 mmol Trolox kg^{-1}). The significant and positive Pearson correlation coefficient between the TPC concentrations and AOC values demonstrates that, as expected, the radical scavenger activity (as determined using the DPPH and ABTS tests) and the ferric-reducing ability (as evaluated using the FRAP assay) of olive pomace can be ascribed to polyphenols.

The results obtained for the refrigerated and frozen pomace samples were compared to those obtained for fresh olive pomace and, at the same time, an additional comparison between the samples stored at 4 °C and −20 °C was performed. As a general trend, both the refrigerated and frozen samples maintained their initial characteristics in terms of TPC and antioxidant capacity over time, especially when comparing the t_0 results with those after the longest storage time. In addition, a *t*-test for independent samples was used to evaluate the differences among the two storage methods. Comparing the results obtained for the refrigerated and frozen pomace samples for the four applied tests, it was possible to observe that the frozen material had a statistically higher TPC and AOC than the refrigerated material. Hence, based on these results, storage at −20 °C seems to represent the best strategy for preserving the initial polyphenolic and antioxidant contents of olive pomace.

The addition of olive pomace to bread led to an increase in its total polyphenolic content, irrespective of the flour used, and this behavior was particularly evident in the bread samples containing 20% olive pomace. These results are in accordance with data obtained by Marinopoulou et al. [34], who studied the phenolic concentrations in bread fortified with green and black olive pulp. Indeed, Marinopoulou et al. [34] observed an increase in phenolic amounts in products to which the highest percentage of olive pulp was added. The addition of a naturally polyphenol-rich matrix in wheat bread recipes may indeed lead to a greater quantity of these compounds in the end product. This is reported also for bread fortified with grape-seed extract [35] as well as for crackers prepared with the addition of microalgae [36]. Concerning the antioxidant properties of the bread samples, similar behavior was observed; high antioxidant capacity values were obtained for the bread samples containing olive pomace. These results are in accordance with data obtained in a previous study in which the influence of adding chestnut flour to a bread recipe was tested, leading to an increase in the antioxidant capacity of the bread [22]. Furthermore, Marinopoulou et al. [34] identified that bread containing green and black olive pulp has higher antioxidant activity than bread without olive supplementation, as is the case for the samples studied herein. Thus, the results of the present study support the exploitation of an olive-based substrate to improve the nutritional quality of bread, both in soft and whole wheat types.

Bread samples incorporating olive pomace exhibited a notable fiber content. Of note, the fiber content of the experimental bread samples containing 20% olive pomace and soft wheat flour was above the level recommended by the European Food Safety Authority (EFSA) for foods classified as "high in fiber", which corresponds to 6 g of dietary fiber per 100 g of product [37]. This level of fiber was also generally observed in the bread samples made with whole wheat flour, irrespective of the amount of olive pomace. This result is likely due to the presence of wheat bran in these latter bread samples. It is important to note that the EFSA considers a daily intake of 25 g of dietary fiber to be sufficient for normal laxation in adults as well as for reducing the risk of coronary heart disease and type 2 diabetes and for aiding weight maintenance [37]. Based on these considerations, the consumption of bread enriched with olive pomace could effectively contribute to reaching such a daily threshold.

Based on the results, the specific volume of the bread loaves was not influenced by the addition of olive pomace, irrespective of the wheat flour (type 0 soft or whole) used for bread-making or the storage conditions of the olive pomace. These results are in accordance with those obtained by Cedola et al. [15], who did not observe appreciable differences in the rheological features of bread loaves produced with the addition of 10% olive pomace compared with control loaves.

As expected, the bread loaves studied in the present research significantly differed in their crust and crumb color attributes according to the type of wheat flour (type 0 soft or whole) and the amount of olive pomace used in the bread-making. Notwithstanding, the effect of olive pomace addition was less evident in the bread samples containing whole wheat flour, likely due to the dark appearance of loaves with the presence of wheat bran. The results obtained in the present study are in accordance with those obtained by Cedola et al. [15], who reported that bread samples containing olive pomace were darker in color compared to control loaves without this by-product.

Regarding the different color descriptors, lightness can vary from 0 (black) to 100 (white); hence, the progressive reduction of lightness detected in the samples containing olive pomace reflects the increasing quantities of pigments derived from the olives. As for the a^* parameter, this axis represents the green–red opponent colors, with values < 0 more green and values > 0 more red. In the present study, all the samples had values in the red hue, except for the control bread produced with 100% type 0 soft wheat flour, which yielded an a^* value of 0, thus confirming the strong effect of the addition of olive pomace on bread color. Concerning the b^* parameter, this axis represents the blue–yellow opponents, with values < 0 more blue and those > 0 more yellow. In the present study, the b^* values of all samples were in the yellow hue range, again reflecting a strong effect of the addition of olive pomace, with the bread produced with 100% type 0 wheat flour resulting in the lowest average yellow levels.

As for the bread-making trials, the microbial counts performed on the doughs show that the effect of olive pomace addition varied according to the added amounts; however, no effect on the leavening (fermentation) ability of the yeast (*S. cerevisiae*) was observed between the samples, as was evidenced by Foti et al. [30]. Of note, the high eumycetes counts detected in all of the analyzed bread doughs attest to the viability of the baker's yeast

(*S. cerevisiae*) used as a leavening agent. Although statistically significant differences were observed among a few dough samples, the numerical differences have no significance from a biological point of view, as the counts of eumycetes at the end of leavening assured proper dough development in all trials. Hence, it is likely that the quantity of phytochemicals carried by the olive pomace did not inhibit yeast performance.

As for spore-forming bacteria, this group of microorganisms includes those that can be the causative agents of bread spoilage (e.g., *Bacillus subtilis*) as well as potential human pathogens (e.g., *Bacillus cereus*, sulfite-reducing clostridia, etc.) [38]. Of note, the absence of these bacteria in the experimental bread samples attests that the baking process was properly performed.

The results obtained in this study show that the addition of olive pomace significantly influences the volatile profile of bread, thus representing an advancement in knowledge. In fact, the bread samples obtained with the addition of olive pomace were characterized by higher amounts of terpenoids, mainly beta pinene and copaene, and phenols such as vinylguaiacol and guaiacol than those without pomace. These compounds are likely derived from the raw olive pomace.

The bread samples produced with olive pomace were also characterized by the highest amounts of furans as 2-furfural and 2-furanmethanol, which confer the aromas of toasted caramel and nuts to bread [39], and the highest amounts of pyrazine. As also highlighted by de Gennaro et al. [40], pyrazines are usually formed by the interaction between the products of Maillard reactions and Strecker degradation and, together with furans, significantly contribute to the flavor of baked products.

The results concerning hexanal are very interesting. Indeed, this compound is a representative marker of oxidative rancidity and could be used as an alternative to traditional oxidation indicators (e.g., acidity or peroxide values) [17]. As ascertained by different authors, the lipid fraction of semolina—the main ingredient of bakery products—is very susceptible to lipoxygenase activity, leading to hydroperoxide production [41]. Hydroperoxides are highly unstable and are converted into volatile compounds, such as hexanal, which are responsible for rancid off-flavors [42]. In fact, the presence of hexanal has already been reported in several cereal-based foods, including pasta, bread, and biscuits. In the present study, hexanal was found in high amounts in the samples without olive pomace, indicating that olive pomace has a protective effect against rancidity; in all the bread samples produced without the addition of olive pomace, the amount of hexanal was higher compared to those containing olive pomace.

The bread samples produced with olive pomace were also characterized by the presence of high levels of acids (acetic, butanoic, and hexanoic acid), likely derived from the fresh olive pomace.

Furthermore, the final volatile profile of the bread samples was also influenced by its storage conditions. The results herein highlight that bread produced with frozen olive pomace shows a volatile profile that is more similar to bread made with fresh olive pomace compared to bread made with refrigerated olive pomace. Hence, it is likely that refrigeration temperatures are not able to slow down the biochemical or enzymatic activities that ultimately affect the finished product.

5. Conclusions

In the present study, the suitability of olive pomace for the production of high-value bread was ascertained. It is noteworthy that the storage conditions of the tested olive pomace (fresh, refrigerated, or frozen) did not affect the TPC and antioxidant capacity of the resulting bread, which represents an advance in knowledge regarding the potential use of this olive by-product. The addition of olive pomace, in proportions up to 20%, allowed for the production of bread with increased TPC and with no remarkable influence on the specific volume of the final product. Of note, the olive pomace added to the bread doughs strongly characterized the volatilome component of the loaves, which contained high amounts of terpenoids. The overall results contribute to opening new income opportunities for farmers and the food industry based on more sustainable agriculture in the European Union. Such opportunities aim to reduce food waste and create added value, as expressly stated in the European Green Deal for improving the well-being and health of citizens and future generations. The present research could serve as best practice in the preparation of high-value bread containing olive pomace. Further research is needed to assess the bio-accessibility and functionality of bread bioactive compounds. Finally, consumer tests should be performed in order to evaluate consumer acceptance of this novel high-value bread.

Author Contributions: Conceptualization, L.A.; data curation, F.C., A.R. and L.A.; formal analysis, F.C., L.B., A.R., M.C., F.B., T.D.R., L.D.V., N.C., V.M., C.G. and G.R.; funding acquisition, L.A.; investigation, F.C., L.B., F.B., T.D.R., L.D.V., N.C., V.M., C.C. and G.R.; methodology, F.C. and A.O.; project administration, L.A.; resources, F.C. and L.A.; supervision, A.O. and L.A.; validation, M.C.; visualization, F.C.; writing—original draft, F.C., A.R., M.C. and A.O.; writing—review and editing, A.O. and L.A. All authors have read and agreed to the published version of the manuscript.

Funding: This study was funded by the Marche Region under the "Programma di Sviluppo Rurale (PSR) 2014–2020—Bando Misura 16—Cooperazione, Misura 16.1—Sostegno per la costituzione e la gestione dei gruppi operativi del PEI in materia di produttività e sostenibilità dell'agricoltura"—Action 2 "Fase di gestione del G.O. e realizzazione del Piano di Attività", titled "Valorizzazione dei sottoprodotti della lavorazione di olive Piantone di Mogliano anche attraverso l'implementazione di nuovi metodi di essiccazione a raggi IR", acronym V.A.S.O. (ID 27813).

Institutional Review Board Statement: Not applicable.

Informed Consent Statement: Not applicable.

Data Availability Statement: The original contributions presented in the study are included in the article, further inquiries can be directed to the corresponding author.

Conflicts of Interest: The authors declare no conflicts of interest.

References

1. Nunes, M.A.; Pimentel, F.B.; Costa, A.S.G.; Alves, R.C.; Oliveira, M.B.P.P. Olive by-products for functional and food applications: Challenging opportunities to face environmental constraints. *Innov. Food Sci. Emerg. Technol.* **2016**, *35*, 139–148. [CrossRef]
2. Tura, D.; Gigliotti, C.; Pedò, S.; Failla, O.; Bassi, D.; Serraiocco, A. Influence of cultivar and site of cultivation on levels of lipophilic and hydrophilic antioxidants in virgin olive oils (*Olea europea* L.) and correlations with oxidative stability. *Sci. Hortic.* **2007**, *112*, 108–119. [CrossRef]
3. Batuecas, E.; Tommasi, T.; Battista, F.; Negro, V.; Sonetti, G.; Viotti, P.; Fino, D.; Mancini, G. Life Cycle Assessment of waste disposal from olive oil production: Anaerobic digestion and conventional disposal on soil. *J. Environ. Manag.* **2019**, *237*, 94–102. [CrossRef] [PubMed]
4. Carmona, I.; Aguirre, I.; Griffith, D.M.; García-Borrego, A. Towards a circular economy in virgin olive oil production: Valorization of the olive mill waste (OMW) "alpeorujo" through polyphenol recovery with natural deep eutectic solvents (NADESs) and vermicomposting. *Sci. Total Environ.* **2023**, *872*, 162198. [CrossRef] [PubMed]
5. Goula, A.M.; Lazarides, H.N. Integrated processes can turn industrial food waste into valuable food by-products and/or ingredients: The cases of olive mill and pomegranate wastes. *J. Food Eng.* **2015**, *167*, 45–50. [CrossRef]
6. Harrami, M.; El Fami, N.; Moussadik, A.; Khachani, N.; Taibi, M.; Diouri, A. Elaboration and characterization of composite clays based on "Coal waste-Olive pomace" mixtures. *Mat. Today-Proc.* **2022**, *58*, 1573–1577. [CrossRef]
7. Lila, K.; Belaadi, S.; Solimando, R.; Ralida Zirour, F. Valorisation of organic waste: Use of olive kernels and pomace for cement manufacture. *J. Clean. Prod.* **2020**, *277*, 123703. [CrossRef]
8. Ruschioni, S.; Loreto, N.; Foligni, R.; Mannozzi, C.; Raffaelli, N.; Zamporlini, F.; Pasquini, M.; Roncolini, A.; Cardinali, F.; Osimani, A.; et al. Addition of olive pomace to feeding substrate affects growth performance and nutritional value of mealworm (*Tenebrio molitor* L.) larvae. *Foods* **2020**, *9*, 317. [CrossRef] [PubMed]
9. Roselló-Soto, E.; Koubaa, M.; Moubarik, A.; Lopes, R.P.; Saraiva, J.A.; Boussetta, N.; Grimi, N.; Barba, F.J. Emerging opportunities for the effective valorization of wastes and by-products generated during olive oil production process: Non-conventional methods for the recovery of high-added value compounds. *Trend Food Sci. Technol.* **2015**, *45*, 296–310. [CrossRef]
10. Ying, D.; Hlaing, M.M.; Lerisson, J.; Pitts, K.; Cheng, L.; Sanguansri, L.; Augustin, M.A. Physical properties and FTIR analysis of rice-oat flour and maize-oat flour based extruded food products containing olive pomace. *Food Res. Int.* **2017**, *100*, 665–673. [CrossRef]

11. Balli, D.; Cecchi, L.; Innocenti, M.; Bellumori, M.; Mulinacci, N. Food by-products valorisation: Grape pomace and olive pomace (pâté) as sources of phenolic compounds and fiber for enrichment of tagliatelle pasta. *Food Chem.* **2021**, *355*, 129642. [CrossRef]
12. Simonato, B.; Trevisan, S.; Tolve, R.; Favati, F.; Pasini, G. Pasta fortification with olive pomace: Effects on the technological characteristics and nutritional properties. *LWT* **2019**, *114*, 108368. [CrossRef]
13. Ribeiro, T.B.; Bonifácio-Lopes, T.; Morais, P.; Miranda, A.; Nunes, J.; Vicente, A.A.; Pintado, M. Incorporation of olive pomace ingredients into yoghurts as a source of fibre and hydroxytyrosol: Antioxidant activity and stability throughout gastrointestinal digestion. *J. Food Eng.* **2021**, *297*, 110476. [CrossRef]
14. Cecchi, L.; Schuster, N.; Flynn, D.; Bechtel, R.; Bellumori, M.; Innocenti, M.; Mulinacci, N.; Guinard, J.X. Sensory profiling and consumer acceptance of pasta, bread, and granola bar fortified with dried olive pomace (pâté): A byproduct from virgin olive oil production. *J. Food Sci.* **2019**, *84*, 2995–3008. [CrossRef]
15. Cedola, A.; Cardinali, A.; D'Antuono, I.; Conte, A.; Del Nobile, M.A. Cereal foods fortified with by-products from the olive oil industry. *Food Biosci.* **2020**, *33*, 100490. [CrossRef]
16. Di Nunzio, M.; Picone, G.; Pasini, F.; Chiarello, E.; Caboni, M.F.; Capozzi, F.; Gianotti, A.; Bordoni, A. Olive oil by-product as functional ingredient in bakery products. Influence of processing and evaluation of biological effects. *Food Res. Int.* **2020**, *131*, 108940. [CrossRef] [PubMed]
17. Durante, M.; Bleve, G.; Selvaggini, R.; Veneziani, G.; Servili, M.; Mita, G. Bioactive compounds and stability of a typical Italian bakery products "*Taralli*" enriched with fermented olive paste. *Molecules* **2019**, *24*, 3258. [CrossRef] [PubMed]
18. Miranda-Ramos, K.C.; Sanz-Ponce, N.; Haros, C.M. Evaluation of technological and nutritional quality of bread enriched with amaranth flour. *LWT* **2019**, *114*, 108418. [CrossRef]
19. Taccari, M.; Aquilanti, L.; Polverigiani, S.; Osimani, A.; Garofalo, C.; Milanović, V.; Clementi, F. Microbial diversity of type I sourdoughs prepared and back-slopped with wholemeal and refined soft (*Triticum aestivum*) wheat flours. *J. Food Sci.* **2016**, *81*, M1996–M2005. [CrossRef] [PubMed]
20. Saka, M.; Özkaya, B.; Saka, İ. The effect of bread-making methods on functional and quality characteristics of oat bran blended bread. *Int. J. Gastr. Food Sci.* **2021**, *26*, 100439. [CrossRef]
21. Venturi, M.; Cappelli, A.; Pini, N.; Galli, V.; Lupori, L.; Granchi, L.; Cini, E. Effects of kneading machine type and total element revolutions on dough rheoLogy and bread characteristics: A focus on straight dough and indirect (biga) methods. *LWT* **2022**, *153*, 112500. [CrossRef]
22. Dall'Asta, C.; Cirlini, M.; Morini, E.; Rinaldi, M.; Ganino, T.; Chiavaro, E. Effect of chestnut flour supplementation on physico-chemical properties and volatiles in bread making. *LWT* **2013**, *53*, 233–239. [CrossRef]
23. Martin-Diana, A.B.; Izquierdo, N.; Albertos, I.; Sanchez, M.S.; Herrero, A.; Sanz, M.A.; Rico, D. Valorization of Carob's Germ and Seed Peel as Natural Antioxidant Ingredients in Gluten-Free Crackers. *J. Food Process. Preserv.* **2017**, *41*, e12770. [CrossRef]
24. Jemai, H.; Bouaziz, M.; Sayadi, S. Phenolic composition, sugar contents and antioxidant activity of Tunisian sweet olive cultivar with regard to fruit ripening. *J. Agric. Food Chem.* **2009**, *57*, 2961–2968. [CrossRef] [PubMed]
25. Pulido, R.; Bravo, L.; Saura-Calixto, F. Antioxidant activity of dietary polyphenols as determined by a modified ferric reducing/antioxidant power assay. *J. Agric. Food Chem.* **2000**, *48*, 3396–3402. [CrossRef] [PubMed]
26. Osimani, A.; Belleggia, L.; Botta, C.; Ferrocino, I.; Milanović, V.; Cardinali, F.; Haouet, M.N.; Garofalo, C.; Mozzon, M.; Foligni, R.; et al. Journey to the morpho-textural traits, microbiota, and volatilome of Ciauscolo PGI salami. *Food Biosci.* **2023**, *53*, 102582. [CrossRef]
27. Dreher, J.; Konig, M.; Herrmann, K.; Terjung, N.; Gibis, M.; Weiss, J. Varying the amount of solid fat in animal fat mimetics for plant-based salami analogues influences texture, appearance and sensory characteristics. *LWT* **2021**, *143*, 111140. [CrossRef]
28. Cardinali, F.; Garofalo, C.; Reale, A.; Boscaino, F.; Osimani, A.; Milanovic, V.; Taccari, M.; Aquilanti, L. Liquid sourdough from stone-ground soft wheat (*Triticum aestivum*) flour: Development and exploitation in the breadmaking process. *Food Res. Int.* **2022**, *161*, 111796. [CrossRef]
29. Van Den Dool, H.; Kratz, P.D. A Generalization of the retention index system including linear temperature programmed gas-liquid partition chromatography. *J. Chromatogr. A* **1963**, *11*, 463–471. [CrossRef]
30. Foti, P.; Russo, N.; Randazzo, C.L.; Choupina, A.B.; Pino, A.; Caggia, C.; Romeo, F.V. Profiling of phenol content and microbial community dynamics during pâté olive cake fermentation. *Food Biosci.* **2023**, *52*, 102358. [CrossRef]
31. Zhou, D.; Zhong, J.; Huang, Y.; Cheng, Y. Effect of free and bound polyphenols from Rosa roxburghii Tratt distiller's grains on moderating fecal microbiota. *Food Chem. X* **2023**, *19*, 100747. [CrossRef] [PubMed]
32. Piscopo, A.; De Bruno, A.; Zappia, A.; Poiana, M. Antioxidant activity of dried green olives (Carolea cv.). *LWT* **2014**, *58*, 49–54. [CrossRef]
33. Pellegrini, N.; Serafini, M.; Colombi, B.; Del Rio, D.; Salvatore, S.; Bianchi, M.; Brighenti, F. Total antioxidant capacity of plant foods, beverages and oils consumed in Italy assessed by three different in vitro assays. *J. Nutr.* **2003**, *133*, 2812–2819. [CrossRef] [PubMed]
34. Marinopoulou, A.; Papageorgiou, M.; Irakli, M.; Gerasopoulos, D. Effect of Olive Pulp Enrichment on Physicochemical and Antioxidant Properties of Wheat Bread. *Int. J. Food Stud.* **2020**, *9*, 178–192. [CrossRef]
35. Peng, X.; Ma, J.; Cheng, K.W.; Jiang, Y.; Chen, F.; Wang, M. The effects of grape seed extract fortification on the antioxidant activity and quality attributes of bread. *Food Chem.* **2010**, *119*, 49–53. [CrossRef]

36. Batista, A.P.; Niccolai, A.; Bursic, I.; Sousa, I.; Raymundo, A.; Rodolfi, L.; Biondi, N.; Tredici, M.R. Microalgae as Functional Ingredients in Savory Food Products: Application to Wheat Crackers. *Foods* **2019**, *8*, 611. [CrossRef]
37. EFSA Panel on Dietetic Products, Nutrition, and Allergies (NDA). Scientific opinion on dietary reference values for carbohydrates and dietary fibre. *EFSA J.* **2010**, *8*, 1462. [CrossRef]
38. Pereira, A.P.M.; Stradiotto, G.C.; Freire, L.; Alvarenga, V.O.; Crucello, A.; Morassi, L.L.P.; Silva, F.P.; Sant'Ana, A.S. Occurrence and enumeration of rope-producing spore forming bacteria in flour and their spoilage potential in different bread formulations. *LWT* **2020**, *133*, 110108. [CrossRef]
39. O'Shea, N.; Kilcawley, K.N.; Gallagher, E. Aromatic composition and physicochemical characteristics of crackers containing barley fractions. *Cereal Chem.* **2017**, *94*, 611–618. [CrossRef]
40. de Gennaro, G.; Difonzo, G.; Summo, C.; Pasqualone, A.; Caponio, F. Olive cake powder as functional ingredient to improve the quality of gluten-free breadsticks. *Foods* **2022**, *11*, 552. [CrossRef] [PubMed]
41. Pasqualone, A.; Bianco, A.M.; Paradiso, V.M.; Summo, C.; Gambacorta, G.; Caponio, F.; Blando, A. Production and characterization of functional biscuits obtained from purple wheat. *Food Chem.* **2015**, *180*, 64–70. [CrossRef] [PubMed]
42. Choe, E.; Min, D.B. Mechanisms and factors for edible oil oxidation. *Compr. Rev. Food Sci. Food Saf.* **2006**, *5*, 169–186. [CrossRef]

foods

MDPI

Review

Approaches to Enhance Sugar Content in Foods: Is the Date Palm Fruit a Natural Alternative to Sweeteners?

Estrella Sayas-Barberá [1], Concepción Paredes [1], Manuel Salgado-Ramos [2], Noelia Pallarés [2], Emilia Ferrer [2], Casilda Navarro-Rodríguez de Vera [1,*] and José Ángel Pérez-Álvarez [1]

1 Instituto de Investigación en Innovación Agroalimentaria y Agroambiental (CIAGRO-UMH), Miguel Hernández University, EPS-Orihuela, Ctra. Beniel km 3.2, 03312 Orihuela, Alicante, Spain; estrella.sayas@umh.es (E.S.-B.); c.paredes@umh.es (C.P.); ja.perez@umh.es (J.Á.P.-Á.)

2 Nutrition and Food Science Area, Preventive Medicine and Public Health, Food Science, Toxicology and Forensic Medicine Department, Faculty of Pharmacy, Universitat de València, 46100 Burjassot, València, Spain; manuel.salgado@uclm.es (M.S.-R.); noelia.pallares@uv.es (N.P.); emilia.ferrer@uv.es (E.F.)

* Correspondence: casilda.navarro@umh.es; Tel.: +34-966-749736

Abstract: The current levels of added sugars in processed foods impact dental health and contribute to a range of chronic non-communicable diseases, such as overweight, obesity, metabolic syndrome, type 2 diabetes, and cardiovascular diseases. This review presents sugars and sweeteners used in food processing, the current possibility to replace added sugars, and highlights the benefits of using dates as a new natural, nutritious and healthy alternative to synthetic and non-nutritive sweeteners. In the context of environmental sustainability, palm groves afford a propitious habitat for a diverse array of animal species and assume a pivotal social role by contributing to the provisioning of sustenance and livelihoods for local communities. The available literature shows the date as an alternative to added sugars due to its composition in macro and micronutrients, especially in bioactive components (fiber, polyphenols and minerals). Therefore, dates are presented as a health promoter and a preventative for certain diseases with the consequent added value. The use of damaged or unmarketable dates, due to its limited shelf life, can reduce losses and improve the sustainability of date palm cultivation. This review shows the potential use dates, date by-products and second quality dates as sugar substitutes in the production of sweet and healthier foods, in line with broader sustainability objectives and circular economy principles.

Keywords: *Phoenix dactylifera*; dates; sugar; sweeteners; sugar reduction; alternative sweetener; sustainability

1. Introduction

Presently, sugar is a main contributor to the onset of obesity and diabetes, which may be attributed to the elevated intake of added sugar in the processing of beverages, dairy products, desserts, cookies, candies, jams, among others [1,2]. The implications of excessive of added sugars in processed foods involve an excessive energy consumption, an impact on dental caries, and an increased prevalence of some chronic noncommunicable diseases (overweight and obesity, metabolic syndrome, type 2 diabetes and cardiovascular diseases) [3–5]. These disorders have manifested as significant public health challenges, prompting a call for the reduction of sugar consumption to enhance the nutritional profile of foods in alignment with public health recommendations [3,6,7]. The World Health Organization (WHO) promotes the preparation of guidelines to limit sugar consumption, defining the recommended threshold to be below 10% of the total energy intake in the diet and ideally less than 5% for optimal health benefits [6]. WHO's sugar guidance adopts the concept of free sugars (FS), encompassing monosaccharides and disaccharides added to foods and beverages by manufacturers, cooks, or consumers, as well as natural sugars present in honey, syrups, fruit juices, and fruit juice concentrates [6]. The FS concept is considered more suitable than total or added sugars in this context.

The technological contribution of sugars must be duly considered in the strategies of sugar reduction or substitution due to key role of sugars in food processing, contributing to sensory quality, textural properties and shelf life. The modification or reduction of sugar content represents an important challenge for the food industry, potentially compromising the aforementioned functions [1].

Sugar, being a multifunctional ingredient, holds significant relevance in processed products. It imparts a sweet taste and mouthfeel in solid products and beverages, contributes to textural properties, participates in the Maillard reaction, resulting in brown crust color and appropriate aroma [1]. Additionally, sugar decreases water activity (Aw) in solid products, affects the freezing point, acts as a bulking and preserving agent, extends product shelf-life, and promotes lightness [8]. Among its roles, more relevant is conferring sweet taste to foods. Therefore, when sugar is substituted in a food product, maintaining the flavor, texture, and shelf life of the original product becomes necessary.

Sucrose is acknowledged as the reference sugar for sweetness and serves as a comparative standard for evaluating the sensory and technological attributes of potential alternative sweeteners [9,10]. Moreover, sucrose serves as a moisture retainer, thereby contributing to the extension of shelf life. Additionally, sucrose exerts a profound influence on the structure, appearance, and texture of numerous food, such as baked goods and chocolate, owing to its hygroscopic and crystallization properties [10,11].

The inherent hygroscopicity of sucrose plays a significant role in development the formation of a delicate texture, a heightened porous structure, and the expansion of baked products. Concurrently, the crystallization process of sucrose intricately participates in bestowing crispness and generating a crackling surface in biscuits and cookies [10]. Various researchers have documented the influence of sugar on the sensory and physical attributes of confectionery products, specifically cakes and cake-like items [12,13].

Moreover, sugar plays a pivotal role in binding moisture, and the moisture content varies across different sugar types. For instance, liquid sugars exhibit a higher moisture content compared to brown sugar, and brown sugar, in turn, contains more moisture than crystalline white sugar [14,15]. The impact of sugar in beverages is substantial, manifesting a multifaceted influence encompassing the provision of sweetness, flavor enhancement, enhanced palatability, increased viscosity, texture augmentation, and coloration. Simultaneously, sugar serves as a preservative by reducing water activity [16].

The physical attributes that sucrose possesses regulate fundamental processes that influence food texture, such as rheology, phase transitions of biopolymers and the distribution of water in the different phases of the food and texture-associated sensory attributes [17]. Their results indicated that the functional role of sugar, coupled with its functions as a plasticizer and humectant, significantly influences the rheology of biscuits. Consequently, these aspects can be strategically harnessed for the reformulation of biscuits to replicate the characteristics of their original counterparts [17].

Nevertheless, it is imperative to acknowledge that any reduction, elimination, or substitution of sucrose in food products may induce safety and quality-related undesirable effects [10]. Reducing the added sugar content of processed products to levels that do not compromise the properties and sensory characteristics of the final product poses a challenge for the food industry [1].

This review describes sugars and sweeteners used in food manufacturing. It considers the current potential for substitution of added sugars and highlights the advantages of incorporating dates as a natural, nutritious and sustainable alternative to artificial and non-nutritive sweeteners. This is in addition to the objective of promoting healthy eating behavior and the engagement of the agri-food industry to diversify our food systems towards sustainable production.

2. New Trends and Strategies for Sugar Reduction in Food and Beverages

Considering that processed foods constitute the primary source of FS intake, there exists substantial pressure on the food and beverage industry to engage in reformulation initiatives. The high demand to diminish added sugar content in processed products emerges as a principal concern for public health, necessitating governmental involvement through initiatives such as advertising regulation, regulatory policies, and taxation, among other measures [5]. Of notable concern is the report that the intake of added sugars in the diets of children and youth (aged 4 to 18 years) accounts for approximately 15% of their total dietary energy intake [18,19]. While sugars inherent in fruits and vegetables are enveloped within a natural tissue matrix, added-sugars and sugars found in syrups, honey, and fruit juices lack are free. Consequently, this last category is more readily available for digestive processes [5,18].

Several governments have instituted programs aimed at curbing sugar consumption. These initiatives encompass restrictions on the advertising of sugar-rich foods, salt, etc., the imposition of higher taxes on sugary beverages, and limitations on the availability of unhealthy products in vending machines [1,20].

One of the strategies employed to achieve sugar reduction in both solid and liquid food products involves the partial or complete substitution of sugar with a blend of diverse sweeteners. This approach aims to mitigate the risk of compromising the ultimate sensory characteristics of the products, striving to attain a taste and flavor profile comparable to that of sucrose [8]. Another initiative, in other countries, is the gradual reduction of sugar in liquid and solid foods, encouraging consumers to gradually adapt to the change in sweetness [21].

Beverages and food reformulation focused on reducing the sugar content of processed products is viewed as potentially beneficial in reducing sugar intake and improving health. These changes in the final product composition can compromise the final sensory characteristics in both solid and liquid foods, which implies the need for significant improvements in the sensory quality of sugar-reduced products [5,8,22].

Various methodologies for reducing sugar intake in food products center on alterations in food formulation or product re-design. These approaches comprehend direct sugar reduction, as well as the partial or complete substitution of sugar with sweeteners and bulking agents. Additionally, strategies involve cross-modal interactions or flavors with modifying properties (FMP), modifying the structure of sugar, and establishing a heterogeneous distribution of sugar within the food matrix [5,23].

The primary objective of these strategies is to effect sugar reduction while preserving the essential characteristics of the product, such as sweetness, color, and texture. Recent studies indicate that the simultaneous implementation of diverse sugar reduction strategies proves to be more efficacious than relying on a singular approach [16]. This underscores the complexity of reducing sugar content in food products and, at the same time, assuring consumer acceptance and satisfaction.

2.1. *Changes in the Food Formulation*

This strategy is intended to improve one or more properties of the product, including sensory attributes, safety, nutritional quality, among others [2]. The focus of this paper is specifically directed towards the reduction of added sugar to yield healthier products.

This strategy widely used in the development of low-sugar products involves the reduction of sugar content or the complete or partial substitution of sugar with low-caloric carbohydrates, non-nutritive sweeteners, or bulking agents. This approach is applicable to both solid foods and beverages, and is aligned with the target of improving the health profile of food products, particularly associated with excessive sugar consumption.

2.1.1. Gradual Reduction of the Sugar Content in Foods

This reformulation strategy focused on gradually reducing the sugar content in food products, considering that consumers must accept the changes in sensory profile. Directly reducing the amount of sugar to lower the sweetness intensity of a product could affect product acceptance. In the case of beverages, direct sugar reduction could lead to a decrease in consumer palatability and acceptability, it has been studied that a gradual reduction is more effective than the stepwise reduction strategy because of its lesser impact on consumer sensory perception others [24,25]. Therefore, gradually reducing sugar in beverages can familiarize consumers to lower concentrations of sugar, reduce their preference for sweetness, and maintain consumer satisfaction with sugar-reduced products [26]. However, the impact of this strategy is limited, and the effect would be observed in the long term [16].

Several studies concluded that consumers did not perceive sugar reductions in the range of 6 to 11%, and it has even been shown that reductions of up to 20 to 30% do not seem to induce a major change in overall taste in both adults and children [26]. Reducing sugar in consecutive steps so that consumers do not notice any change in the sensory characteristics of the products is an effective strategy. The gradual reduction of sugar could promote the change of sweet preferences within a food category in favor of products with a lower sugar content [5].

2.1.2. Partial or Total Replacement by Sweetener

This strategy is focused on the replacement of sugar by sweetener, nutritive sweeteners (NS) and non-nutritive sweeteners (NNS) depending on whether they contain calories, or natural and artificial sweeteners depending on origin. They maintain the sweetness of products without negative health effects (obesity, type II diabetes, and cardiovascular diseases) [16]. Nowadays, consumers can find a wide range of commercial processed products that contain at least one non-nutritive sweetener [5].

Since the 1980s, the use of non-nutritive sweeteners and low-calorie carbohydrates (prebiotic fibers) has been used to avoid the negative impacts (obesity, type II diabetes and cardiovascular disease) of high sugar intake, maintains the sweetness of products [27]. The discovery of sweeteners represented a key step for the innovation in the food technology (sweet products with non-caloric intake) [1]. To avoid undesirable effects of sugar reduction or elimination are needed to study the minimum sugar level and/or optimal sugar substitutes to maintain the basic functionality and final product quality.

NS, NNS, polyols, low-calorie carbohydrates (oligofructose, maltodextrin and polydextrose) and bulking agents are used to replace (partially or totally) the sucrose. The sugars commonly used in the food industry are: sucralose, maltitol, stevioside, sorbitol, isomalt, aspartame, erythrito, etc.) and bulking agents (unulin, maltodextrine, polydextrose, oligofructose, syrup, etc.), but they must be adapted to the food product, legislative standards and consumer preference [10]. The NS commonly used include: fructose, glucose, lactose and polyols. While NNS can include natural compounds (stevia, thaumatin and monk fruit), or synthetic compounds (saccharin, aspartame, sucralose, etc.), they are very sweet and tasty, but contain few calories [19].

Polyols (as erythritol, isomaltitol, lactitol, maltitol, sorbitol, mannitol, and xylitol), are food additives, and present insignificant caloric contribution, but a high sweetening capacity, being used in low quantities in food products [19]. In general, they are not cariogenic and do not cause glycemic response, thus being extensively used in hypocaloric diets, for diabetes patients and other specific cases where caloric intake must be controlled [28].

The inclination for a combination of non-nutritive sugar substitutes with sugar alcohols to produce a low-calorie bakery product has increased, with artificial sweeteners such as aspartame and sucralose providing sweetness and sugar alcohols providing the bulking properties.

In the last decades, the use of synthetic sweeteners (e.g., aspartame, sucralose, cyclamate, acesulfame and saccharin) has been commonin the industry to obtain low-calorie foods and beverages, due to their high sweetness, low cost, and calorie-free benefits [28].

In the case of beverages, the reduction of sucrose and its substitution with sweeteners, whether artificial or natural, can lead to a reduction in viscosity and have an adverse effect on the sensory and temporal profiles (manifesting as sensations like dry mouth and decreased viscosity) [16,23]. To avoid these undesirable effects associated with the incorporation of sweeteners into low-calorie beverages, carbohydrate gums and other food additives are introduced.

Sucralose is the most widely used sweetener in the beverage. The aspartame has been shown to exhibit stability in acidic liquid, and instability to heat and alkali; sucralose rapidly reaches peak sweetness and develops an undesirable residual sweetness [29].

New non-nutritive synthetic sweeteners in beverages have been reported such as neotame, advantame and alitame. Neotame has shown to be a good alternative to aspartame with a high sweetening power (7000 to 13,000 times sweeter than sucrose) and successful replacement has been reported up to 30% with no negative taste effects [16].

In the case of beverages, the sweetener aspects that should be considered to be selected are: specific solubility, pH stability, sweetness characteristics, processing temperature stability, temporal sensory profile and undesirable flavors [16]. Nevertheless, controversy exists about the safety over the use of artificial sweeteners in foods and beverages [5,15,30].

Artificial sweeteners share the same palatability as natural sugars, but the metabolic routes are different. The scientific literature shows artificial sweeteners are metabolized differently than natural sweeteners and may not all have the same metabolic impact, which can affect from the composition of the gut microbiota to the degree of digested and absorbed [30]. Artificial sweetener intake could affect body weight and glucose homeostasis through physiological mechanisms involving the gut microbiota, reward-system, adipogenesis, insulin secretory capacity, intestinal glucose absorption, and insulin resistance [30].

It is considered that reducing sugar through sweeteners and not sugar intensity does not reduce the intake of intensely sweet foods and, therefore, does not lead to the development of a preference for products with lower sweetness intensity. This approach has been associated with potential adverse health effects linked to the intake of sweeteners [31]. The activation of sweet receptors without the intake of sugar can lead to metabolic dysregulation [32]. This could be due to the decoupling of sweet taste from energy intake (learned relationship between sweet taste and post-ingestion responses), which may lead to the subsequent development of glucose intolerance. The metabolic response to carbohydrate intake depends on the relationship between energy intake and sweet taste [32].

Consideration of these impacts and the concern of consumers about artificial sweeteners is an important factor among other strategies than the use of artificial sweeteners as a means to reduce or replace sugar.

Natural sweeteners are widely used by the beverage industry because of the demand for clean label products. Stevia, a plant-derived sweetener, is present in some low-calorie drinks due to its good organoleptic properties, despite its bitterness, licorice taste and after-taste [16]. Bulking agents and flavorings can also be used in stevia beverage formulations to avoid these negative impacts on taste [33].

Another natural sweetener with future application is mogroside, a mixture of curcurbitane-type triterpenoid saponin, which is characterized by lower peak sweetness, a longer sweetness duration and after-taste (bitter, chemical and metallic taste) compared with sucrose. Sugar alcohols or polyols are good substitutes for sucrose, because they are minimally metabolized in the body, for example rythritol, calorie-free polyol, has been widely used in sugar-free carbonated beverages [33].

Finally, it is worth mentioning the other sugars (e.g., D-fructose and L-arabinose) and sweet proteins (miraculin, monellin, thaumatin, mabinlins, pentadin, curculin, and brazzein), may have wide potential application in the beverage industry as natural and clean sweeteners [33].

Natural sugars (unrefined sugar) are preferred because they have a high nutritional value due to their high concentration of healthy compounds (bioactive compounds, minerals, fibers, antioxidants, and phytochemicals), which balance the negative effects of refined sugar. Therefore, removing refined sugar or at least reducing its consumption should be promoted as a healthier option in food choices [34].

Natural sweetening agents (honey, xylitol, erythritol, maltose, maltodextrin, stevia, molasses, maple syrup, coconut sugar, agave nectar and date sugar) are presented as an important alternative to sugars, more attractive to consumers and a commercial opportunity for the food industry [34].

Natural sweeteners include the traditional sweeteners, natural sources of sugar. Traditional sweeteners are classed as NS and are obtained from bees (e.g., honey), plant and tree sap (e.g., maple syrup, agave nectar), fruits (e.g., carob syrup), seeds, roots (e.g., Yakon syrup) and leaves (e.g., stevia) and consumed within their natural matrix with minimal pre-processing. They are mainly composed of sugar of sucrose, fructose and glucose (at least 50% of plant-derived syrups and honey), small amounts of polyols, also contain additional nutritive compounds as proteins (< 1.4%), lipids (< 0.5%), dietary fiber (< 3%), and phytochemicals and small amounts of minerals (< 2%) and vitamins (< 0.02%), such as polyphenols [19]. Several positive impacts have been attributed to natural sweeteners, such as improving metabolic health, preventing weight gain and lowering blood glucose. These impacts could be because biomolecules with nutritional and health benefits (e.g., vitamins, phytohormones and minerals) present in NS appear to have the capacity to modify other physiological factors [27,34].

Some studies suggest the presence of phytochemical compounds in traditional sweeteners (honey and agave nectar) could contribute to a reduction in the glycaemic potency compared with glucose syrup and sucrose [19,27]. Phenolic compounds have a range of properties that could have a potential impact on nutrition and health (e.g., reducing the risk of cardiovascular disease, preventing neurodegenerative conditions and type 2 diabetes), as well as anti-oxidant properties. Therefore, by using fruits as a component for a sugar substitute, it will decrease the amount of daily sugar intake. Fruits have a sweet taste, and different studies have reported they are excellent sugar replacers. Ibrahim et al. [11] reformulated dark chocolate using palm sugar and dates as sugar replacers, which was well accepted by the sensory panelist.

Blending sweeteners can be a good alternative to avoid the negative impacts of certain sweeteners, obtaining a better result than any sweetener alone. A suitable combination of sweeteners can provide synergistic impacts on sweetness, stability and enhancement of flavor and temporary taste characteristics [35]. A suitable alternative blends NNS substitutes (providing sweetness) with sugar alcohols (providing bulking properties) to obtain a good low-calorie bakery product [15].

Sweetness enhancers, called positive allosteric modulators (PAMs), are compounds that have no sweet taste, however they increase sweetness intensity when used with a sweetener due to their synergist effect. According to DuBois and Prakash [36], PAMs are not able to activate the sweetener receptors, but their binding mode allows the sweetener to bind to the receptor. PAMs and sweetness are used together to reduce sugar due to their synergist effect and can potentiate the sweetness characteristic of sweetener. Furthermore, PAMs present the advantage that they do not cause bitterness, metallic taste or temporary sensory profile [25]. It has been reported that the combined use of PAM and nutritive sweetener allows for obtaining the same sweet taste using less sweetener, while the PAMs would be reducing off-flavors of non-nutritive sweeteners [37].

2.2. *Flavors with Modifying Properties (FMPs)*

This strategy presents a new type of flavorings, called flavorings with modifying properties (FMP) as an alternative to NNS. FMPs are able to activate responses in human taste receptors to increase sweetness, reduce saltiness and mask bitterness. Some FMPs interact with human sweet taste receptors and improve the sweetness perception [38].

FMPs have the property that they do not have a sweet or salty taste on their own but affect the taste or smell of other flavorings and food additives, and help maintain or improve the flavor profile of solid and liquid food. They facilitate the food industry to develop products more in line with regulatory requirements. It allows for the restriction of the use of sugars (e.g., sucrose, glucose, fructose), sweeteners (e.g., aspartame), and salt (sodium chloride) and masks some undesirable tastes such as the bitterness (e.g., potassium chloride), the persistent taste of certain sweeteners (e.g., stevia fractions), or soften the taste and astringency of vegetable proteins (e.g., soy, peas) [39].

2.3. Multisensory Interactions

Taste perception is a multi-modal sensory information integration process in the central nervous system that is dependent on temperature, medium, physical condition, age, and the information from other stimuli (smell, touch, hearing and sight) [16]. During the food intake, consumers used all their senses (multi-sensory integration) to identify the food product characteristics. Multisensory interactions are based on the role of food oral processing and microstructure and they are linked with sensory perception and texture [25,40], their understanding could be used to reformulate and develop healthier food products.

Several studies have examined interactions between food-intrinsic and extrinsic factors in relation to sweetness to avoid a decrease in consumer satisfaction for the reduced-sugar products [25]. Nevertheless, it is difficult to determine the sweetness-enhancing multisensory factors as individual consumers have different responses to sugar-reduced foods and beverages [16,25].

This strategy focusing on multisensory interactions is based on the enhanced perception of sweetness intensity produced by aroma, color and other stimuli [23]. The use of multisensory interactions is another useful method for sugar reduction. Although the reduction of sugar content is limited, it has the advantage that sweeteners are not used. Multisensory integration can be studied from different perspectives: odor–taste, color–taste and sound– and tactile–taste interactions.

2.3.1. Odor–Taste

Several studies on "sweetness enhancement" have reported volatile compounds that increased the sweet intensity and sweetness perception of foods, independently of glucose, fructose and sucrose content. It has been observed that the effect of odor on taste can be predicted from the sensory characteristics of the odor (e.g., degree of sweetness of an odor) [41]. In studies on different beverage and various odors, Bertelsen et al. [41] concluded that the sweetest odor "caramel" has been found to increase the sweetness of sucrose, while the odor with the least "sweet" odor suppresses the sweetness of sucrose. It should be noted that the effect of odor on taste can be predicted from the sensory characteristics of the odor, such as the degree of sweetness of an odor [42].

Furthermore, the enhancing effect on sweetness can be further increased as well as the optimization of the time phase-shift between odor and taste pulses [42]. When odorant (isoamyl acetate) and tastant (sucrose) pulses were presented out-of-phase, the sweetness intensity was enhanced by more than 35%, compared to a continuous sucrose reference of the same net sucrose concentration [42]. In addition, the sweetness-enhancing effect can be further increased by optimizing the time phase-shift between odor and taste perception [42]. When the odor and taste perception are presented the time phase-shift, the sweetness intensity increases by more than 35% compared to the same sucrose concentration [16,42].

Other authors concluded that the odor–taste interaction was an important strategy for sugar reduction in yogurt and preserving the desirable sensory characteristics [23].

It is known that the odor–taste interaction is produced via stimulating the olfactory receptors in the nasal cavity and the taste receptors in the tongue and mouth, and odor and taste stimulation activate cortical areas of the brain [42].

2.3.2. Color–Taste Interactions

Color of food and beverages can affect the taste perception (e.g., sweetness threshold and intensity). Some authors showed the sucrose solutions containing red colorants obtained lower sweetness threshold [43], and red drinks were the sweetest, followed by blue and purple [44]. Different studies have shown that colors related with the natural ripening of fruits, such as red and yellow, are well suited to modulate the perception of sweetness, while colors opposite to the ripening of the fruit (e.g., green) may decrease sweetness perception; this can be due an association between specific colors and specific product attributes in the mind [45]. This could be due to the expectations that a color generates in the consumer, which affects the perception of sweetness [46,47]. Furthermore, other studies presented different results and reported that the impact of color on the perception of sweet taste was affected by the age of the population (adults or children) [47].

In addition, the package color of the beverages could, in the same way, affect the perception of sweetness, as it could involve the expectations of the product. More knowledge on this subject is necessary to confirm the impact of product or packing color on taste perception [16,28].

2.3.3. Sound–Taste and Tactile–Taste Interactions

Different studies have reported multisensory integration between taste and sound [16]. Although more knowledge is still needed on sound and touch–taste interactions in sweetness perception, it has been reported that beverages tend to be perceived as more enjoyable and sweeter when the beverages are presented in their own containers [48].

2.4. Sugar Structure Modification

Another strategy is focused on the reduction of the size of sugar particles to increase the surface/volume ratio per amount of sugar consumed, thus increasing the perception of sweet for the same amount of sugar when compared to larger sizes (20–100 μm). Using a spray dryer or nano spray dryer are techniques used to obtain these micro- and nanoparticles (350–500 nm). This approach presents some complications due to fact that the sucrose glass-transition temperature is close to 62 °C, making this material sticky [1]. An alternative to this strategy has been reported based on the addition of carriers (as inulin, pectin, lecithin, etc.) to the spray drying feed solution to increase the glass-transition temperature and avoid the stickiness problem. There are few studies on this subject [1,49].

The encapsulation of sweeteners has been used to enhance sweetener perception of alternative sweeteners. This technology allows the creation of structures or matrices with a controlled release of sweetness in food products and reduce the intense flavor of these alternatives, e.g., acesulfame-K, aspartame, thaumatin, xylitol, that need to improve their stability and sensory properties in food products [1].

2.5. Heterogenous Distribution

This strategy is based on the stimulation of taste receptors, liquid release, particle size and viscosity of foods to improve the sweetness in solid and liquid foods. The sugar content may also be reduced through the optimized stimulation of taste receptors to improve sweetness perception (heterogeneous distribution method), facilitated by liquid release from solid foods, the particle size and viscosity in solid and liquid food [1,23].

Richardson et al. [15] considered a new strategy of sugar reduction in confectionery-type products based on the sugar particle size, and they showed that different sugar particle size ranges (924–1877 μm, 627–1214 μm and 459–972 μm) have a significant impact on the physical and sensory properties. The studies concluded that small sugar particles (228 to 377 μm and 459–972 μm) showed an increase in the perceived intensity of sweetness in chocolate brownies, biscuits, so it could be used as a viable, economic and technological strategy to reduce sugar in baked products [14,15].

2.6. Encapsulation for Enhanced Sweet Perception

The encapsulation method is another novel strategy for the improvement of sweetener perception, based on the creation of structures that allow a controlled release of sweetness in food products and mask the intense flavor of these alternative compounds [1]. Different patents have been reported to encapsulate sweeteners and evaluate their effect to improve their application [50]. Alternative sweeteners such as acesulfame-k, aspartame, thaumatin and xylitol are suitable for use in the encapsulation strategy, as this methodology allows for the improvement in their stability and sensory properties in food products [51,52].

3. Nutritive, Non-Nutritive and Traditional Sweeteners

As mentioned, the high intake of sugar not only relates to a high accumulation of body fat, but also can concurrently increase the risk of other adverse health conditions, such as type 2 diabetes or cardiovascular diseases. With this regard, and considering the demand of the population for healthier and more natural foods, over the past several decades the food sector has been focused on sugar substitution in different foodstuffs, thus replacing the traditional sweeteners like glucose or sucrose. Overall, healthier products would be obtained, with the desired health benefits to consumers, and satisfying at the same time their demand for sweetness [53].

In line with the above, artificial, non-nutritive sources are nowadays the most common alternatives [54–56]. However, despite being considered as calorie-free sources, they can sometimes involve metabolic diseases, such as type 2 diabetes or cardiovascular diseases as well [57,58]. For that reason, there is a continuous need for natural alternatives which can be extracted from natural sources [27,53]. Therefore, the classification for sweeteners can be organized considering both the origin of these healthy products (synthetic, natural), or their nutritive power (calorie-load or calorie-free) [59]. The main sweeteners and specific substances employed worldwide are detailed below.

3.1. *Synthetic Sweeteners*

Artificial sweeteners or sugar substitutes are synthetic substances employed to replace sugar during the sweetening process of several products, such as sweets, preserves, dairy products and beverages. The molecules in artificial sweeteners include principally sulfa, dipeptide and sucrose derivatives. These compounds provide a sweet taste without increasing caloric intake and blood sugar levels. Therefore, due to their high efficiency (30–13,000 times the sweetening power of sucrose), a small amount of these compounds provides high sweetness without a caloric intake increase [60,61]. Nonetheless, the amount of use must be safe to guarantee consumers' health [60]. Some of them including aspartame, neotame, saccharin, acesulfame-k, sucralose and advantame, which have been approved as food additives by the Food and Drug Administration (FDA) [27,59].

Synthetic sugar substitutes must have a sucrose-like taste quality and demonstrate safety, with non-toxicity and cariogenic capacity and no effects on blood glucose or insulin. However, some studies suggest that these non-caloric sweeteners might cause an ambiguous psychobiological signal that confuses the body's regulatory mechanisms [61].

Nowadays, the major part of sweeteners available on the market are synthetic compounds. However, they must grant legislative approval and the regulatory requirements of each country. These compounds have undergone a wide safety evaluation process by international and national regulatory food safety authorities, such as the FAO/WHO Joint Expert Committee on Food Additives (JECFA), the US Food and Drug Administration (FDA) or the European Food Safety Authority (EFSA) [62]. Moreover, these authorities continuously review and evaluate any new safety information related to them. In this sense, some problems have been attributed to some of these compounds in terms of their stability, cost, quality of taste and safety [61]. However, over the past few years, because of health concerns, consumers are demanding more natural and healthy foods that are produced in a sustainable way. Thus, food manufacturers are looking for natural and functional sweeteners to be applied in foods [60].

3.1.1. Aspartame

Aspartame (E-951) was discovered in 1965 and was the first sweetener approved by the FDA. Its flavor characteristics are acceptable. This artificial sweetener reduces food intake and may assist with weight control. Some studies evidence that their consumption does not influence blood pressure, glucose and lipid profiles, being a safe option for type 2 diabetics. However, its consumption is still controversial since some studies have associated it with adverse health effects, such as interference of neuronal cell function, hepatotoxicity, kidney disfunction and oxidative stress in blood cells. Moreover, its usage is not recommended for people who suffer from phenylketonuria, since they cannot metabolize phenylalanine, a compound involved in the aspartame synthesis [27,63]. To ensure its safety, the European Commission (EC) has established an acceptable daily intake (ADI) of 40 mg/kg bw/day [64].

3.1.2. Neotame and Advantame

Neotame (E-961) and advantame (E-969) constitute derivatives of aspartame. Neotame is an isomer of aspartame, while advantame is an N-substituted derivative of aspartame and vanillin. Both artificial sweeteners are sweeter than aspartame, and show approximately 13,000 and 20,000 times the sweetening power of common sugar, respectively. Neotame was approved by the FDA in 2002, while advantame was approved in 2014. Similar to aspartame, neotame and advantame present no adverse effects on the human metabolism, constituting a safe option for type 2 diabetes patients [27].

3.1.3. Sucralose

Sucralose (E-955) is also a non-caloric sweetener, that does not break down in the body. Its sweetening potential is approximately 600 times higher than sugar. It constitutes a good option for many industrial applications due its stability at different pH conditions and temperatures. Some studies suggested that this artificial sweetener could interfere with digestive processes, increasing glucose and insulin levels in the body and resulting in weight gain and diabetes risk. Nonetheless, studies related with absorption, distribution, metabolism and excretion pointed out that sucralose is mainly eliminated through fecal excretion, without being absorbed or digested in the organism [27]. The EU Scientific Committee on Food (SCF) established an ADI of 15 mg/kg bw/day) [65].

3.1.4. Saccharin

Saccharin (E-954) was discovered in 1878 and is the oldest and most studied of all sweeteners. This compound presents good technological characteristics, such as stability at low pH and high temperatures. Moreover, it is not metabolized in the gastrointestinal tract and does not alter insulin levels [27,63].

This compound has been involved in some human health concerns, since the USA FDA considered its prohibition as a consequence of some studies that related it with bladder cancer in rats. However, in subsequent studies the relationship between saccharin and bladder cancer has not been demonstrated in humans. The EC has established an ADI of 5 mg/kg bw/day [66].

3.1.5. Acesulfame-K

Acesulfame-K (E-950) was discovered in 1967 and is one of the most common low calorie artificial sweeteners. It constitutes a thermostable component, with good properties to be manipulated. This compound presents around 120 times higher sweetening potential than sugar although it bears a bitter taste, therefore it is usually employed in combination with other sweeteners. It cannot be metabolized without increasing the caloric intake [67].

Acetoacetamide, a breakdown product of acesulfame-K, might be toxic for humans at high concentrations and its consumption has been related by some studies with genotoxicity and the inhibition of glucose fermentation by intestinal bacteria. However, mores studies including bioassays are necessary to clarify this issue [27,63]. The ADI established by EC for acesulfame-K is 9 mg/kg bw/day [67].

Nowadays, there is a need to find healthier substitutes to refined sugar. In this sense, an ideal alternative sweetener should a guarantee lower calorie content and helps to prevent dental decay and diabetes. Moreover, it should present some characteristics to be incorporated into the food during its manufacturing, such as water-solubility, stability in acidic and basic pH, and being metabolized in a wide range of temperatures. It also should demonstrate its non-toxicity [53].

3.2. Nutritive Natural Substitutes

In regard to the above, natural products are normally appealing to the population since they attribute their consumption to health benefits. Thus, as previously mentioned, the production of natural substances by the food sector becomes a valuable hotspot. Concretely, it is well-known that nutritive natural sweeteners involve low glycemic potency and fructose content. In addition, a wide range of bioactive compounds, such as polyphenols or vitamins, can be found in these sources, as well as others as minerals or phytohormones. Overall, these substances are considered as safe, and also bring nutrition to the human body, with positive effects in terms of improving metabolic health, preventing weight gain or lowering blood glucose [19].

Many reports in the literature have already described the main natural, nutritive substances for sweetening [27,53,68]. Among them, honey, molasses (viscous substances come from the refining process of sugar cane), maple syrup, agave nectar or coconut sugar are worthy of note. Briefly, the main properties, composition, and applications of these substances are summarized in Table 1.

Table 1. Brief overview of the main natural-nutritive sweeteners. Information recovered from [27,68] and the side references found in these works.

Sweetener	Composition	Properties	Applicability
Honey	60–85% sugars (mainly glucose and fructose, and 1–15% sucrose), 12–23% water, minerals, vitamins, bioactive compounds (phenolics)	Glycemic index: 54–59. Antioxidant, antimicrobial, anti-inflammatory.	Wide human consumption, nutritional and therapeutic applications. Drug in-home treatment for infections and burns
Molasses	30–40% sucrose, 4–9% glucose, 5–12% fructose 17–25% water, phenolic compounds, and traces of amino acids and vitamins	Glycemic index: 55. Humectant and colligative. Auto-immune, auto-inflammatory and antioxidant.	Food processing (e.g., masker undesired flavor). Coloring agent to improve visual presentation in baked foods
Maple syrup	60–65% sucrose, and other sugars as xylose, glucose or arabinose. Organic acids, amino acids minerals and phenolic compounds	Glycemic index: 54–65. Antioxidant, anti-mutagenic and antiproliferative in human cancer. Benefits in type 2 diabetes and Alzheimer's diseases	Condiment for bakery products, in craft soda, or formulator in some beverage and snacks
Agave nectar	Approximately 90% fructose. Inulin and polyphenols	Low-glycemic index (17–27). Suitable for obesity and diabetes prevention. Boosted metabolic system, anti-obesity, anti-aging, chemoprotective and immunomodulatory effects	Sugar substitute in a wide range of foodstuffs: cheese, cookies, bread, cereal bar snacks, chocolate, guava purees, ice cream, sport drinks or yogurts
Coconut sugar	75% sucrose, < 25% fructose. High vitamin and minerals content	Glycemic index: 35–40	Additive in cakes, cookies, parfaits or sauces, sprinkled on top of granola
Palm sugar	91% sugar and 6% reduced sugars. Significant concentration of vitamins, minerals and phenolics	Glycemic index: 70 Cytoprotective activity against NIH3T3 fibroblast cells and cell proliferation (antioxidative agent)	Food additive in sweet soy sauce, desserts or beverages
Sorghum syrup	69% sugar (11% glucose, 6% fructose). Phenolic compounds, carotenoids, proteins and vitamins	Anticancer and anti-obesity effects. Preventive for cardiovascular diseases	Beverages and food industries

Concurrently, the power of polyols or some kind of oligosaccharides should be also mentioned in this section. The former are commonly used in the food sector since they also entail a distinguished sweetness (50–100% sucrose [19]), and are naturally present in fruits, vegetables, or natural fermented foods. The most common are xylitol, mannitol, sorbitol, or erythritol, all of them with reduced calorie power [69]. Further, some specific oligosaccharides are gaining attention. In this context, 2′-fucosyllactose (2′-FL) or trehalose

(TRH) are noticeable, since they can bring both energy and potential benefits, for instance, improved immunity, blood sugar regulation and weight loss [59]. Trehalose (2′-FL) is commonly found in milk, whereas TRH is widely present in plants, bacteria, or fungi. They have been authorized as sweeteners and additives for commercial issues.

3.3. Non-Nutritive, Natural Sweeteners

Despite the positive properties of the above natural sweeteners, most of them involve a high-calorie content. Regarding that, and also the population concerned about overweight, diabetes, and their health-associated concerns, the development of alternative natural-calorie-free sources has become crucial. In this context, steviol diterpene glycosides recovered from Stevia Rebaudiana are worthy of mention. These substances normally exhibit a high sweetness power, concretely between 40 and 450 times stronger than sucrose, and represent a proportion ranging from 4 to 20% of the dry stevia leaves. In addition, they showed valuable health benefits, as reported [27,59]. Concurrently, the presence of biologically active compounds, such as polyphenols, should be noted in stevia [70]. Among them, flavanols, flavones and tannins are remarkable, as well as hydroxybenzoic and hydroxycinnamic acids. Therefore, in addition to its low-caloric content and the high sweetness intensity by the glycosides structures, Stevia Rebaudiana also entail pharmaceutical and medicinal applications, for instance, anti-cancer, antioxidative, or anti-inflammatory effects [71], possibly making them the most potential natural sweetener.

Briefly, the most valuable steviol glycosides are stevioside and rebaudioside A [70,71], with beneficial health effects, as mentioned. Additionally, others such as rebaudiosides B, D and M should be mentioned [59,71].

Finally, apart from these steviol derivatives, glycyrrhizin is considered another glycoside potential sweetener [59]. This pentacyclic triterpenoid is a recognized bioactive ingredient of licorice, and exhibits around 170-fold higher sweetness compared to sucrose. It presents a wide pharmacological activity such as anti-inflammatory, antitumor or hepatoprotective agent, among others [72].

3.4. New Natural, Healthy Alternatives to Sweeten: Date Fruit

The soluble date sugar extracted from date fruit would be a suitable alternative to refined sugar, with a lower glycemic index than sucrose.

The composition of date fruits rich in carbohydrates (70–80%), most of them in the form of sucrose, fructose and glucose, and in other phytochemicals make them an ideal source for the production of natural sugar. Furthermore, date fruits contain a good amount of dietary fiber ranging from 6.5 to 11.5% (of which 6–16% is soluble), which can help to meet the requirements of a balanced diet [73,74].

Date fruits also show beneficial properties, such as antitumor, anticancer, antioxidant, anti-mutagenic, anti-inflammatory, gastroprotective, hepatoprotective and nephroprotective effects [75–77]. Moreover, date fruits have demonstrated antibacterial activity attributed to bioactive compounds like phenolic molecules [78].

Date syrup constitutes the principal derived date product, and its usage is contemplated as one of the oldest practices in the production of sweeteners. It is employed in the food industry to be incorporated to foodstuffs such as jams, marmalades, concentrated beverages, chocolates, ice cream, confectioneries, and honey. The syrups obtained from date palm present high amounts of sugars, minerals (potassium, iron, magnesium and calcium), vitamins (B1 thiamine, B2 riboflavin, nicotinic acid, A and C) and a distinguished antioxidant activity, mainly related to their high content in phenolic compounds. Moreover, date syrup is rich in unsaturated fatty acids (such as oleic, linoleic, palmitoleic and linolenic acids) [27].

Currently, liquid date sugar is obtained from date fruits by ultrasound-assisted extraction (temperature of 60 °C, extraction time of 30 min, and liquid to solid ratio of 7.6 mL/ga and L/S ratio) [74]. However, some authors proposed alternatives to conventional extraction such as enzymes (pectinase and cellulase) and ultrasonically-assisted methods to extract syrup date, with high efficiency at shorter extraction times [79,80].

4. Sustainability and Valorization of Date Palm

Date palm (Arecaceae family; *Phoenix* genus and *P. dactylifera* species) is considered one of the most ancient cultivated trees in the world and is mainly cultivated in arid and semi-arid areas of southern Europe, North African and southern Central Asian countries [81]. Date palm cultivation has increased over the last decade (Figure 1) [82], with world date production of about 9.7 × 106 tonnes in 2021, where Egypt, Saudi Arabia, Iran and Algeria are the main producers (Figure 2) [81]. In the European Union, the largest palm groves are in Spain, located in Elche and Orihuela (Alicante province, southeast of Spain). Both palm groves are protected spaces because they are considered unique cultural and historical landscapes of great value [83,84].

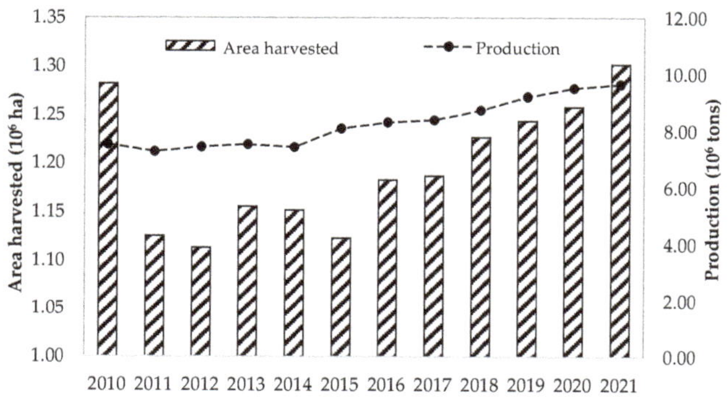

Figure 1. World evolution of the date palm area harvested and the date production in the period 2010–2021 [82].

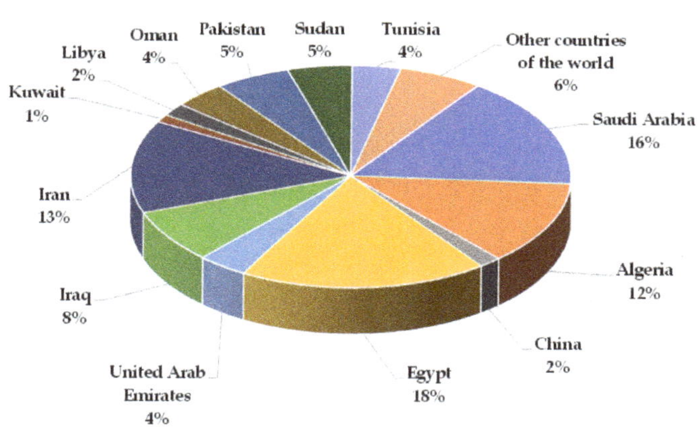

Figure 2. World distribution of date production in 2021 [82].

The date palm excels in challenging environments, thriving in dry climates with low rainfall, high evapotranspiration, and salinity tolerance. It withstands temperatures from 18 °C to 50 °C and short frost periods as low as −5 °C [85,86]. High humidity promotes phytopathogen proliferation, inflorescence rotting, and the production of soft, sticky fruits [81,86]. Hot and dry winds reduce receptivity, and strong winds can disperse pollen, break fruit stalks, and damage developing fruits [81,87]. Date palms thrive in sandy to sandy-loam soils [81,88]. They are highly tolerant to salinity (up to 12 dS/m electrical conductivity), but their production decline starts at 4 dS/m [82,89]. Excessive soil salts result from scarce rainfall and the overexploitation of saline aquifers for irrigation [90].

Climate change threatens food security, necessitating diversified food systems. Promoting climate-resistant foods such as the date palm aligns with the UN's Zero Hunger goal, as the date palm seems less affected by climate change [86,91]. Palm groves combat desertification, create a microclimate, preserving agrobiodiversity and support various animal species, benefiting local communities [92–95]. Intensive date palm cultivation brings environmental challenges, e.g., leading to soil salinization [96–101] or reduced livestock presence [102].

Date Uses and Valorization of By-Products of the Date Palm

Date is a fruit with high nutritional value and is very healthy, its main constituents being carbohydrates and dietary fiber, as well as minerals (especially potassium), vitamins, antioxidant phenolic compounds and carotenoids and to a lesser extent, it also contains proteins and lipids [103]. For this reason, the main use of date is its fresh consumption, especially the highest quality dates (first- and second-grade dates). Lower quality dates are classified as third-grade and cull date. The former is processed and cull dates are destined for animal feed [104]. The following products are obtained from the processed dates:

– Date syrup: this date by-product is obtained by hot aqueous extraction (60 °C) of the date juice and subsequent vacuum evaporation of the extract obtained. Date syrup is used as an ingredient in the preparation of bakery products, ice creams, jams, beverages, etc. [105]. This product has also been used as a sweetener to replace sugar in the preparation of different desserts [106–108] and non-alcoholic beer [109]. In addition, it is added to prebiotic milk and yogurt to improve its organoleptic properties [110,111]. Date syrup can also be used as a carbon source for bacteria in various fermentation processes where the following products are obtained: alcohol, date wine, antibiotics, organic acids, bakery yeast and unicellular proteins [104].

– Date paste: this product is obtained from the grinding of ripe pitted and skinless dates, which have been cooked in hot water or steamed [104]. Date paste has been added to meat products to improve their textural properties, reduce their fat content and increase their concentration of dietary fiber [112], as well as to reduce the oxidation of pigments and lipids during storage of pork liver pâté [113]. This by-product is also used to prepare jam and candies due to its high sugar content [105].

– Date pit: this part of the date fruit can be used for animal feed or added in powder form to different foods to increase its dietary fiber content [105]. Also, the date pit can be treated by pyrolysis to obtain bio-oil and biochar [104]. This biochar has shown very good properties for the removal of organic and inorganic contaminants present in wastewater and drinking water [105]. In addition, oil can be extracted from the date pit with different uses in the food industry, such as cooking or frying oil and for the preparation of margarines and mayonnaises, and the presence of a wide variety of phytochemicals in this oil means that it is also used for the formulation of cosmetic and pharmaceutical products [114]. Date pit oil has also been employed for the production of bio-diesel [115] and as feedstock for the production of polyhydroxyalkanoates with use in the synthesis of biodegradable plastics [116].

On the other hand, the fresh date and methanolic and aqueous extracts of date have been traditionally used for medicinal purposes for the treatment and prevention of different diseases [117].

Also, the cultivation of the date palm generates large amounts of agricultural residues from the pruning operations of the leaves with signs of senescence and the bunches with the harvested dates. These residues can be used to make paper, produce particleboard composites, obtain energy (through thermal treatment by pyrolysis and combustion or through anaerobic digestion) and manufacture composites of natural fiber for use in the automotive industry [85,118]. Likewise, date palm biomass residues have been co-composted with residues from different origins to produce biofertilizers compost [119].

5. Nutritional and Functional Properties of Dates

The diversity of date palm cultivars offers various choices, and their adaptability to different climates contributes to their global popularity. The escalating interest in date fruits and their derivatives is attributed to their role as a highly nutritious and plentiful fruit, and as a cost-effective source of numerous macro- and micronutrients, such as minerals, vitamins, antioxidants, and dietary fibers as well as secondary metabolites essential for human health. The carbohydrates, primarily sugars, constitute the majority of date fruit composition [103,120–122].

Date fruits consist of two main parts: the edible flesh (pulp), representing 85–95% of the total weight, and the seeds (or pits), comprising 5–15% and serving as a notable byproduct in date palm processing. The nutritional composition of date pits, rich in protein, fat, and dietary fiber, has sparked interest in novel functional food applications [122].

With an energy value ranging from 300 to 350 kcal/100 g, date fruits exhibit varying carbohydrate compositions influenced by cultivar types and ripening stages. Date pulps contain easily digestible sugars, primarily glucose, fructose, mannose, maltose, and sucrose constituting over 80% of dry matter [123]. The sugar composition varies, with sucrose predominant in dry dates, while soft dates are characterized by glucose and fructose. Additionally, the dietary fiber content in date pulp varies widely, including insoluble cellulose, hemicelluloses, pectin, hydrocolloids, and lignin [103,120–122,124,125].

In addition to carbohydrates, dates emerge as an exceptional source of proteins with a protein proportion of 2.5–6.5 g/100 g, fats, dietary fibers, and a spectrum of essential minerals and vitamins (rich in B-vitamins) [124]. Dates contain more than twenty different amino acids, which is uncommon in fruits [123].

Dates are established as a superior dietary fiber source compared to cereals. Additionally, dates contain health-promoting β-glucan, that shows potential anticancer properties [123]. Date seeds, with higher dietary fiber content than date flesh, present an opportunity for excellent sources of dietary fiber in food processing [122]. Protein content in date pulp ranges from 1.2% to 6.5%, while date seeds contain 5.1–7% protein, including essential amino acids such as glutamic acid, aspartic acid, and arginine [122]. Dates exhibit low fat content, mainly concentrated in the skin. Date pits, on the other hand, have a significantly higher oil content, making them a potential source of edible oil rich in unsaturated fatty acids [121,122,126].

Noteworthy nutrients include potassium, vital for a healthy nervous system and overall balance, phosphorus collaborating with calcium for bone strength and growth, magnesium, copper, zinc and selenium crucial for cell growth and repair, and iron essential for red blood cell production, facilitating nutrient transport to cells throughout the body, are also found in dates. The low sodium content in dates aligns with recommended daily intake levels [103,120]. Date seeds also contain various dietary minerals, further enhancing their nutritional profile [121–123,125].

Date pulp and seeds are rich in biologically active molecules, with variations based on the cultivar of origin [127], specifically polyphenols, mainly flavonoids [128], carotenoids and phytosterols [121], highlighting their nutritional quality. These phytochemical compounds underscore the antioxidant, anti-diabetic, anti-obesity [129], hepatoprotective [130] and neuroprotective actions [131] and anti-lipidemic properties of date fruits, contributing to their overall health benefits in human consumption [103,120,121,124,132,133]. The study carried out by Alsukaibi et al., [134] indicated the presence of various components in date fruits, responsible for cytotoxicity against cancer cells. Dominant phenolic compounds, such as q-coumaric, ferulic, and vanillic acids, were identified. Antimicrobial assays demonstrated notable biological activities, for second-grade dates. Significantly, these extracts displayed extensive antimicrobial activity against various pathogens [103,120,123]. Alsukaibi et al., [134] found that date kernel (seed) is a natural source of polyphenols that have potential antibacterial activity.

Polyphenolic compounds, particularly phenolic acids and flavonoids, represent primary secondary metabolites in plants. Date fruits emerge as a noteworthy source of these compounds, surpassing other fruits, and can be found in both the pulp and seeds. The concentration and diversity of these phytochemicals generally prevail in the pulp compared to the seeds. The concentration of polyphenolic compounds is contingent upon factors like cultivar, ripening stage, and environmental conditions. Analyses of phenolic acids in date fruit pulp from various studies reveal variations in composition and concentration, with gallic acid frequently standing out [122].

In parallel, date fruit seeds contribute to the pool of phenolic acid compounds [114, 135]. Notably, gallic acid and syringic acid are major compounds in date seed extracts from different cultivars. The concentration of these compounds varies among cultivars. Additionally, studies on date fruit seeds from various regions report the presence of phenolic compounds like ferulic acid, vanillic acid, and p-coumaric acid, showcasing the diversity in phenolic acid composition [122].

Regarding flavonoid content, analyses reveal quercetin as a primary component in date fruit pulp, while date seeds exhibit flavonoids such as rutin, quercetin, and luteolin. The predominant flavonoids in date seeds include catechin, epicatechin, quercetin, and quercetin hexoxide. Rutin is identified as a major flavonoid in date seeds from specific cultivars. Overall, it is crucial to note that the concentration of flavonoids in date seeds tends to be lower than in the pulp [122].

The findings propose that second-grade dates hold substantial promise as efficient, safe, and cost-effective natural antioxidant compounds. This potential creates new possibilities for their utilization in the functional food and nutraceutical industries, highlighting the diverse benefits of dates beyond their nutritional content [103,120].

6. Food Applications of Date as a Sweetener

Elevated sugar consumption has been associated with negative health effects, including dental caries, type 2 diabetes, and cardiovascular diseases, particularly among the demographic of children and adolescents [136,137]. A noteworthy proportion of current consumers is actively seeking healthier alternatives within their lifestyle, emphasizing a change toward a more health-conscious diet. This includes an effort to reduce sugar intake and substitute refined sugar with naturally sourced sugars. Consequently, there exists a considerable interest in the development of food products incorporating natural and healthier sugars or sweeteners derived from natural sources [27]. Research has demonstrated that alternatives to sugars from natural sources, as is the case of the date palm fruit, contains significant levels of bioactive compounds, such as antioxidants, minerals, fibers, and other phytochemicals.

Due to the healthy and medicinal properties associated with the consumption of dates and its products, based on the nutritional and bioactive composition (rich in dietary fiber, minerals, carotenoids, vitamins and phenolic compounds), this fruit is desirable to incorporate into the diet [123,138–140]. Furthermore, this fruit can be regarded as an emerging and potential candidate as alternative for substituting refined sugar in the processing of solid, semi-solid, and liquid food products [138,141]. It should be noted that food matrices are an optimal carrier to facilitate the availability of the biomolecules present in date fruit [139]. This approach not only enhances health benefits and adds value but also contributes to the revalorization of date products and by-products, in that way promoting circular economy principles within the food industry [142].

Dates, ready-to-eat date products, and date-derived products such as syrup, juices, spreads, paste, and liquid sugar [143] possess the potential to function as sweeteners while providing essential vitamins, minerals, phytochemicals, antioxidants, and other health-promoting compounds. These properties contrast with those of refined sugars, which are characterized by empty calories. Additionally, dates exhibit versatile applications beyond their role as sweetening agents, extending to functions as coloring and flavoring agents [144]. Consequently, dates may be utilized as ingredients in specific foods in which sugar is a fundamental component by providing sweet taste and functional properties. Such applications involve beverages, confectionery, desserts, baked goods and dairy products, as shown in Table 2.

A large number of studies focus on the addition of date and date products (syrup, extract and powder) to dairy products [143–147], dessert [107,148] and beverages [109,149,150], candy [151], biscuits [152–159], bread [160–162], snack bars [163–165] and flakes [166]. This integration represents a viable and effective strategy for the creation of novel functional foods, with an improvement in functional and nutritional properties, and good sensory attributes [142].

Amerinasab et al. [143] incorporated varying concentrations (1 to 9%) of date liquid sugar as a substitute for added sugar in the production of dairy products. Their conclusions indicated that yoghurts containing 6% exhibited optimal pH, total titratable acidity, and color characteristics. These yoghurts also demonstrated elevated firmness and viscosity, reduced syneresis, and received the highest scores in sensory evaluations for texture, aroma, flavor, and overall acceptability. Furthermore, a discernible enhancement in antioxidant activity and phenolic content was observed in these yoghurts.

Abdollahzadeh et al. [146] enhanced the nutritional composition of date-flavored probiotic fermented milk by supplementing it with various combinations of date extract at concentrations of 4%, 8%, and 12%. The study revealed a proportional increase in antioxidant activity. Moreover, the probiotic content, specifically Lactobacillus acidophilus, consistently exceeded 6 log10 units throughout the product's shelf life. The authors concluded that date extract presents itself as a viable candidate for enhancing the nutritional profile of probiotic dairy products.

Other authors have added date syrup as a natural and nutritional additive in yogurt [144,145,147], a fermented milk beverage to [149], to produce healthy and nutritious flavored milk beverage with lower amounts of added sugar thus improving its nutritional properties. Djaoud et al. [107] concluded that the incorporation of date by-products (syrup or/and power date) as substitute sugar, could be an alternative to formulate new dairy dessert. They showed dairy dessert with syrup date exhibited the highest total phenolic content, DPPH inhibition, and reducing power, followed by mixed dairy dessert.

Dates have been studied as natural sweeteners for sugar replacement in chocolate products [11,142,167]. These authors replaced sugar by date syrup or powder, alone or with other sweeteners, as an alternative sweetener in the production of chocolate products, improving the taste and flavor and the healthy and physicochemical properties [11,142,167]. Prebiotic chocolate milk (non-fermentative dairy product) with a high sugar content has been reformulated using date syrup as a natural sweetener and inulin as a prebiotic, resulting in an optimal prebiotic chocolate milk, with the added value of having a natural and cost effective as sugar replacer [110]. Additionally, date seed could be used as a good healthy alternative for cocoa powder in chocolate processing, showing that the chocolate sample manufactured with 4% date seed powder was significantly superior in the degree of taste, aroma, and texture and in bioactive compounds (fiber and phenol content) [168].

Other studies have been directed towards reformulation strategies aimed at reducing or replacing sugar content in low-moisture baked products such as biscuits and bread. Aljutaily et al. [158] demonstrated that biscuits supplemented with 5%, 10%, and 15% date fiber exhibited functional anti-obesity properties in obese albino rats. This suggests a potential biological impact of date palm fruit on body weight control in this particular animal group. It has been presented that date syrup exhibited similar effects to sucrose on thermal properties [152], and this aspect can be potential for optimizing sugar replacement in biscuits and dough by utilizing date syrup and liquid sugar [152,156].

Other studies have focused on the incorporation of date powder and flour [153,154,157,159] in biscuit production, resulting in enhanced nutritional value. However, there is a limitation, with a recommended replacement threshold of 10–20% to avoid adverse effects on sensory analysis and physical characteristics. In addition, other varieties of palm, such as *P. canariensis* [155] have been studied in biscuit production in order to develop a new food application for these fruits. These authors evaluated the addition of date powders as a replacement to wheat flour or sugar, and obtained novel biscuits with higher fiber and polyphenolic content [155].

Several studies have researched the effects of substituting sugar with date products in bread production, with the aim of enhancing its nutritional value [160–162]. These studies have demonstrated that the inclusion of date flour and paste can approach the functionalities of sugar in bread production, contributing to improvements in crust color and flavor. Furthermore, this substitution leads to enhancements in the nutritional profile of the bread, characterized by increased levels of protein, minerals, and fiber. These nutritive improvements are attributed to the supply of bioactive compounds and dietary fiber from dates, with minimal baking losses [162].

Dates and their products have shown a great future commercial opportunity in snacks and fruits bars with improved nutritional value and functional properties with the increase in the date content [162–165,169,170]. The conventional snack bars generally include natural sweeteners such as honey and dried fruits, but they can be replaced by other natural substitutes by date and date products which present optimal technological qualities and lower price [165]. Different studies have shown the potential application of dates, rich in functional and bioactive ingredients such as phenolics and flavonoids, to develop balanced, nutritious, and functional date-based bars [170].

Table 2. A selection of studies on the use of date and date products as sweeteners in food processing and its main results.

Food	Way of Incorporation	Concentration Used	Main Results	References
Fruit yogurts	Date liquid sugar (DLS)	1–9%	Higher phenolic compounds and antioxidant activity Yogurts with 6% DLS had the highest scores	[143]
Flavoring yoghurt	Date syrup	6.0, 8.0 and 10%	Higher acidity, solids, proteins and ash Decreased fat, pH, total bacterial count and increased lactobacilli count The best level of addition was 8%	[144]
Flavored drinking yogurt	Date syrup	5 and 10%	Higher acidity Increased viscosity Sensory characteristics acceptable	[145]
Probiotic fermented milk	Date extract	4, 8 and 12%	Higher antioxidant activity and acidity Count reduction Lower pH and syneresis No negative sensory impact	[146]
Functional yoghurt	Date syrup	5%	Higher the nutritional value Enhanced the quality and overall acceptability	[147].
Fermented milk beverages	Date palm with camels' milk and goats'	10%, 20% and 30%	Improved the composition, viscosity, microbiological quality and acceptable sensory attributes Higher acceptable sensory at 10% and 20%	[149]

Table 2. *Cont.*

Food	Way of Incorporation	Concentration Used	Main Results	References
Dairy Desserts	Date syrup (DS) and dried date powder (DP)	16% with the rates: DP/DS = 2; DP/DS = 1 and DP/DS = 0.5)	Enhanced the final product texture. Improved antioxidant activities	[148]
Dairy dessert	Date syrup (DS) date powder (DP)	14% DS and 2% DP	Enhanced the dry matter, lipids, proteins, total phenolic, and antioxidant activity	[107]
Dark chocolate	Dates syrup (70° Brix)	25%	Better physicochemical Well accepted sensory	[11]
Chocolate spread	Date seed powder	2, 4, 10%	Increased crude fiber, total phenol, antioxidant activity and value of L*and h Better in 10% of date seed Decrease in the a*, b* and C	[169]
Chocolate	Date powder	17.94, 19.86 and 25.16%	Improved the taste and flavor of the product	[167]
Prebiotic chocolate milk	Date syrup	4 and 10%	Increased the total solids 10% of date syrup was selected as the optimum	[110]
Biscuits	Date syrup	10, 20, 30, 40, 50, and 60%	Decrease hardness. Lower fracturability Darker cookies	[152]
Biscuits	Date power	5, 10, 20 and 40%.	Increased carbohydrates, crude fibers, ash, crude fat, moisture and protein Decreased physical characteristics of cookies The best was at substitution of 10%	[153].
Biscuits	Date palm flours	15, 17.5, 20, 22.5, 25, 30%	Higher in crispiness Lower the spread ratio Increased fiber content	[154]
Biscuits	Date powder from *P. canariensis*	5%, 7%, 9%, and 11%	Increased in hardness, polyphenol and fiber content, and antioxidant activity The maximum acceptable was 9% and 7%, Two-fold fiber and four-fold polyphenolic content	[155]
Biscuits and Dough	Date syrup and date liquid sugar	Sucrose was replaced at 0, 20, 40, 60, 80 and 100%	Increased pH, cohesiveness and decreased softness and adhesiveness in dough Lower pH and higher ash, moisture, density, antioxidant, mineral content texture and darker color in biscuits,	[156]
Biscuits	Date powders	20 and 30%	Increased in moisture content, starch, ash and fiber content 20% the best in sensory quality	[157]
Biscuits	Date fiber	5, 10 and 15%	Significant positive effects Functional anti-obesity properties resulting in body weight. Lower levels of glucose, and cholesterol in rats	[158]

Table 2. *Cont.*

Food	Way of Incorporation	Concentration Used	Main Results	References
Biscuits	Date powder (+chickpea)	10, 20, 30 and 40%	Higher ash, far, fiber, fat and protein Lower carbohydrate Higher spread factor and spread ratio Decreased overall acceptability	[159]
Bread	Date palm fruit pulp	Replacement at 0, 25, 50, 75, and 100% of sugar	Increased the nutritional value (higher protein, fiber and ash content, and decrease in the level carbohydrate content)	[160]
Bread	Date palm fruit flour	Replacement at 0, 50 and 100% of sugar	Higher essential nutrients with many potential health benefits (increased protein, fiber, ash, vitamin and minerals)	[161]
Fortified bread	Date paste	15, 25, 35%	Improved the nutrient composition, storage stability, physical and sensory properties of bread	[162]
Cereal flakes	Date syrup	25, 50, 75 and 100	Acceptable to consumers, Improved nutrient values and potential health benefits	[166]
Candy	Date palm 10%	0–10%	Improved in the nutritional properties, the functional, phytochemical, and antioxidant properties and decreasing starch content	[151]
Snack Date bar	Date and date syrup	30 and 60% date and 20% syrup	Higher fracturability, fiber, ash, Ca, K, Mg, Fe with 60% date	[136]
Snack bar	Date paste	40, 50, 60 and 70%	Higher fiber. Improved the technological qualities 50% date paste were the formulation with the best sensory characteristics	[165]
Date bars	Date paste from immature fruits	100%	High organoleptic acceptability as well as microbial safety up to 30 days at room temperature and 50 days under refrigeration	[162]
Date-based bars	Date paste and date syrup	50% date paste and 6.5 date syrup	Higher ash, crude fiber, Ca, Cu, Fe, Zn, Mn, and Se, Lysine, Methionine, Histidine, Threonine, Phenylalanine, Isoleucine, and Cystine Better sensory evaluation	[169]
Original beer (nonalcohol)	Bleached date syrup	25, 50, 75 and 100%	The sample with 50% date syrup stands to be acceptable having maintained a Improved the physical characteristics	[109]
Fermented whey beverage	Date syrup	10, 12.5 and 15%	12.5% higher physicochemical, microbial and sensory properties	[150]

Color Indices (L*, a*, b*).

7. Conclusions

Rising cases of obesity and diabetes, coupled with the cardiometabolic risks linked to high sugar consumption, pose a major challenge to the food industry. To address this problem, there is an urgent need for the industry to advocate and facilitate improvements in the nutritional composition of processed products, particularly in terms of sugar type and content. Natural sources of sugar offer not only sweetness but also additional nutritional value, which can protect against certain diseases rather than simply providing empty calories. One promising approach is to replace added sugar with natural alternatives, such as dates, thus introducing a novel strategy to develop healthier foods by providing the food product with the macro- and micronutrients of dates in addition to sweetness.

Products such as fruit bars, dairy products and bread can be reformulated with dates, eliminating the need for additional sugar. This transformation makes them alternative consumption options in both high and low season, enriched by incorporating an undervalued, locally or regionally sourced product into their composition. From a nutritional point of view, this substitution of empty sugars by the sweet fruits of the date palm, which offer high levels of bioactive components such as fiber, polyphenols and minerals such as potassium, thus conferring important health benefits on consumers. This review suggests utilizing damaged dates and by-products as sugar substitutes in sweet food processing, offering health benefits and supporting sustainable practices. This approach efficiently uses discarded resources, aligning with circular economy principles for environmentally friendly production.

Author Contributions: Conceptualization, E.S.-B. and J.Á.P.-Á.; methodology, E.S.-B. and C.N.-R.d.V.; investigation, E.S.-B., C.P., M.S.-R., N.P., E.F., J.Á.P.-Á. and C.N.-R.d.V.; writing—original draft preparation, E.S.-B., C.P., M.S.-R., N.P., E.F. and C.N.-R.d.V.; writing—review and editing, E.S.-B. and C.N.-R.d.V.; funding acquisition, J.Á.P.-Á. All authors have read and agreed to the published version of the manuscript.

Funding: This work is part of the AGROALNEXT-059 Programme financed by MCI with Nextgeneration EU funds (PRTR.C17.I1) and the Generalitat Valenciana, with the title: "Integral valorisation of traditional food resources of the Valencian Community. Development of new products for agricultural and food use based on dates from Elche".

Institutional Review Board Statement: Not applicable.

Informed Consent Statement: Not applicable.

Data Availability Statement: Not applicable.

Acknowledgments: Manuel Salgado-Ramos wishes to thank the post-PhD program from the Universidad de Castilla-La Mancha, for the requalification of the Spanish University System from the Ministry of Universities of the Government of Spain, modality "Margarita Salas—Complementaria" (MS2022) financed by the European Union, Next Generation EU.

Conflicts of Interest: The authors declare no conflict of interest.

References

1. Gomes, A.; Bourbon, A.I.; Peixoto, A.R.; Silva, A.S.; Tasso, A.; Almeida, C.; Nobre, C.; Nunes, C.; Sánchez, C.; Gonçalves, D.A.; et al. Strategies for the reduction of sugar in food products. In *Food Structure Engineering and Design for Improved Nutrition, Health and Well-Being*; Parente Ribeiro Cerqueira, M.A., Pastrana Castro, L.M., Eds.; Academic Press: Cambridge, MA, USA, 2023; pp. 219–241.
2. Onyeaka, H.; Nwaiwu, O.; Obileke, K.; Miri, T.; Al-Sharify, Z.T. Global nutritional challenges of reformulated food: A review. *Food Sci. Nutr.* **2023**, *11*, 2483–2499. [CrossRef] [PubMed]
3. Taskinen, M.-R.; Packard, C.J.; Borén, J. Dietary Fructose and the Metabolic Syndrome. *Nutrients* **2019**, *11*, 1987. [CrossRef] [PubMed]
4. Qi, X.; Tester, R.F. Lactose, maltose, and sucrose in health and disease. *Mol. Nutr. Food Res.* **2020**, *64*, 1901082. [CrossRef] [PubMed]
5. Deliza, R.; Lima, M.F.; Ares, G. Rethinking sugar reduction in processed foods. *Curr. Opin. Food Sci.* **2021**, *40*, 58–66. [CrossRef]
6. World Health Organization (WHO). *Guideline: Sugars Intake for Adults and Children*; WHO Press: Geneva, Switzerland, 2015.
7. Gil, A.; Urrialde, R.; Varela-Moreiras, G. Position statement on the definition of added sugars and their declaration on the labelling of foodstuffs in Spain. *Nutr. Hosp.* **2021**, *38*, 645–660. [PubMed]
8. Di Monaco, R.; Miele, N.A.; Cabisidan, E.K.; Cavella, S. Strategies to reduce sugars in food. *Curr. Opin. Food Sci.* **2018**, *19*, 92–97. [CrossRef]

9. Carocho, M.; Morales, P.; Ferreira, I.C. Sweeteners as food additives in the XXI century: A review of what is known, and what is to come. *Food Chem. Toxicol.* **2017**, *107*, 302–317. [CrossRef]
10. Luo, X.; Arcot, J.; Gill, T.; Louie, J.C.; Rangan, A. A review of food reformulation of baked products to reduce added sugar intake. *Trends Food Sci. Technol.* **2019**, *86*, 412–425. [CrossRef]
11. Ibrahim, S.F.; Dalek, N.E.M.; Raffie, Q.F.M.; Ain, M.F. Quantification of physicochemical and microstructure properties of dark chocolate incorporated with palm sugar and dates as alternative sweetener. *Mater. Today Proc.* **2020**, *31*, 366–371. [CrossRef]
12. Martínez-Cervera, S.; Sanz, T.; Salvador, A.; Fiszman, S.M. Rheological, textural and sensorial properties of low-sucrose muffins reformulated with sucralose/polydextrose. *LWT* **2012**, *45*, 213–220. [CrossRef]
13. Godefroidt, T.; Ooms, N.; Pareyt, B.; Brijs, K.; Delcour, J.A. Ingredient functionality during foam-type cake making: A review. *Compr. Rev. Food Sci. Food Saf.* **2019**, *18*, 1550–1562. [CrossRef] [PubMed]
14. Tyuftin, A.A.; Richardson, A.M.; O'Sullivan, M.G.; Kilcawley, K.N.; Gallagher, E.; Kerry, J.P. The sensory and physical properties of Shortbread biscuits cooked using different sucrose granule size fractions. *J. Food Sci.* **2021**, *86*, 705–714. [CrossRef] [PubMed]
15. Richardson, A.M.; Tyuftin, A.A.; Kilcawley, K.N.; Gallagher, E.; O'Sullivan, M.G.; Kerry, J.P. The impact of sugar particle size manipulation on the physical and sensory properties of chocolate brownies. *LWT* **2018**, *95*, 51–57. [CrossRef]
16. Chen, L.; Wu, W.; Zhang, N.; Bak, K.H.; Zhang, Y.; Fu, Y. Sugar reduction in beverages: Current trends and new perspectives from sensory and health viewpoints. *Food Res. Int.* **2022**, *162*, 112076. [CrossRef] [PubMed]
17. van der Sman, R.G.M.; Jurgens, A.; Smith, A.; Renzetti, S. Universal strategy for sugar replacement in foods? *Food Hydrocoll.* **2022**, *133*, 107966. [CrossRef]
18. Mela, D.J. A proposed simple method for objectively quantifying free sugars in foods and beverages. *Eur. J. Clin. Nutr.* **2020**, *74*, 1366–1368. [CrossRef]
19. Edwards, C.H.; Rossi, M.; Corpe, C.P.; Butterworth, P.J.; Ellis, P.R. The role of sugars and sweeteners in food, diet and health: Alternatives for the future. *Trends Food Sci. Technol.* **2016**, *56*, 158–166. [CrossRef]
20. Prada, M.; Saraiva, M.; Garrido, M.V.; Rodrigues, D.L.; Lopes, D. Knowledge about sugar sources and sugar intake guidelines in Portuguese consumers. *Nutrients* **2020**, *12*, 3888. [CrossRef]
21. MacGregor, G.A.; Hashem, K.M. Action on sugar—Lessons from UK salt reduction programme. *Lancet* **2014**, *383*, 929–931. [CrossRef]
22. Hashem, K.M.; He, F.J.; MacGregor, G.A. Effects of product reformulation on sugar intake and health—A systematic review and meta-analysis. *Nutr. Rev.* **2019**, *77*, 181–196. [CrossRef]
23. Hutchings, S.C.; Low, J.Y.; Keast, R.S. Sugar reduction without compromising sensory perception. An impossible dream? *Crit. Rev. Food Sci. Nutr.* **2019**, *59*, 2287–2307. [CrossRef] [PubMed]
24. Lima, M.; Ares, G.; Deliza, R. Comparison of two sugar reduction strategies with children: Case study with grape nectars. *Food Qual. Prefer.* **2019**, *71*, 163–167. [CrossRef]
25. Wang, Z.X.; Gmitter, F.G.; Grosser, J.W.; Wang, Y. Natural sweeteners and sweetness-enhancing compounds identified in citrus using an efficient metabolomics-based screening strategy. *J. Agric. Food Chem.* **2022**, *70*, 10593–10603. [CrossRef] [PubMed]
26. Velázquez, A.L.; Vidal, L.; Alcaire, F.; Varela, P.; Ares, G. Significant sugar-reduction in dairy products targeted at children is possible without affecting hedonic perception. *Int. Dairy J.* **2021**, *114*, 104937. [CrossRef]
27. Castro-Muñoz, R.; Correa-Delgado, M.; Córdova-Almeida, R.; Lara-Nava, D.; Chávez-Muñoz, M.; Velásquez-Chávez, V.F.; Hernández-Torres, C.E.; Gontarek-Castro, E.; Ahmad, M.Z. Natural sweeteners: Sources, extraction and current uses in foods and food industries. *Food Chem.* **2022**, *370*, 130991. [CrossRef] [PubMed]
28. Carvalho, F.M.; Spence, C. Cup colour influences consumers' expectations and experience on tasting specialty coffee. *Food Qual. Prefer.* **2019**, *75*, 157–169. [CrossRef]
29. Wan, Z.; Khubber, S.; Dwivedi, M.; Misra, N.N. Strategies for lowering the added sugar in yogurts. *Food Chem.* **2021**, *344*, 128573. [CrossRef]
30. Pang, M.D.; Goossens, G.H.; Blaak, E.E. The impact of artificial sweeteners on body weight control and glucose homeostasis. *Front. Nutr.* **2021**, *7*, 598340. [CrossRef]
31. Mora, M.R.; Dando, R. The sensory properties and metabolic impact of natural and synthetic sweeteners. *Compr. Rev. Food Sci. Food Saf.* **2021**, *20*, 1554–1583. [CrossRef]
32. Rother, K.I.; Conway, E.M.; Sylvetsky, A.C. How non-nutritive sweeteners influence hormones and health. *Trends Endocrinol. Metab.* **2018**, *29*, 455–467. [CrossRef]
33. Tireki, S. A review on packed non-alcoholic beverages: Ingredients, production, trends and future opportunities for functional product development. *Trends Food Sci. Technol.* **2021**, *112*, 442–454. [CrossRef]
34. Valle, M.; St-Pierre, P.; Pilon, G.; Marette, A. Differential effects of chronic ingestion of refined sugars versus natural sweeteners on insulin resistance and hepatic steatosis in a rat model of diet-induced obesity. *Nutrients* **2020**, *12*, 2292. [CrossRef] [PubMed]
35. Oktavirina, V.; Prabawati, N.B.; Fathimah, R.N.; Palma, M.; Kurnia, K.A.; Darmawan, N.; Yulianto, B.; Setyaningsih, W. Analytical methods for determination of non-nutritive sweeteners in foodstuffs. *Molecules* **2021**, *26*, 3135. [CrossRef] [PubMed]
36. DuBois, G.E.; Prakash, I. Non-caloric sweeteners, sweetness modulators, andsweetener enhancers. *Annu. Rev. Food Sci. Technol.* **2012**, *3*, 353–380. [CrossRef] [PubMed]
37. Servant, G.; Kenakin, T.; Zhang, L.N.; Williams, M.; Servant, N. The function and allosteric control of the human sweet taste receptor. *Adv. Pharmacol.* **2020**, *88*, 59–82. [PubMed]

38. Harman, C.L.; Hallagan, J.B. Sensory testing for flavorings with modifying properties. *Food Technol.* **2013**, *67*, 44–47.
39. Guentert, M.A. Flavorings with Modifying Properties. *J. Agric. Food Chem.* **2018**, *66*, 3735–3736. [CrossRef] [PubMed]
40. Stieger, M.; van de Velde, F. Microstructure, texture and oral processing: New ways to reduce sugar and salt in foods. *Curr. Opin. Colloid Interface Sci.* **2013**, *18*, 334–348. [CrossRef]
41. Bertelsen, A.S.; Mielby, L.A.; Byrne, D.V.; Kidmose, U. Ternary cross-modal interactions between sweetness, aroma, and viscosity in different beverage matrices. *Foods* **2020**, *9*, 395. [CrossRef]
42. Burseg, K.M.M.; Camacho, S.; Knoop, J.; Bult, J.H.F. Sweet taste intensity is enhanced by temporal fluctuation of aroma and taste, and depends on phase shift. *Physiol. Behav.* **2010**, *101*, 726–730. [CrossRef]
43. Sugrue, M.; Dando, R. Cross-modal influence of colour from product and packaging alters perceived flavour of cider. *J. Inst. Brew.* **2018**, *124*, 254–260. [CrossRef]
44. Velasco, C.; Michel, C.; Youssef, J.; Gamez, X.; Cheok, A.D.; Spence, C. Colour–taste correspondences: Designing food experiences to meet expectations or to surprise. *Int. J. Food Stud.* **2016**, *1*, 83–102. [CrossRef] [PubMed]
45. Simmonds, G.; Spence, C. Thinking inside the box: How seeing products on, or through, the packaging influences consumer perceptions and purchase behaviour. *Food Qual. Prefer.* **2017**, *62*, 340–351. [CrossRef]
46. Higgins, M.J.; Hayes, J.E. Learned color taste associations in a repeated brief exposure paradigm. *Food Qual. Prefer.* **2019**, *71*, 354–365. [CrossRef]
47. Spence, C. On the relationship (s) between color and taste/flavor. *Exp. Psychol.* **2019**, *66*, 99–111. [CrossRef] [PubMed]
48. Cavazzana, A.; Larsson, M.; Hoffmann, E.; Hummel, T.; Haehner, A. The vessel's shape influences the smell and taste of cola. *Food Qual. Prefer.* **2017**, *59*, 8–13. [CrossRef]
49. Abdel-Mageed, H.M.; Fouad, S.A.; Teaima, M.H.; Abdel-Aty, A.M.; Fahmy, A.S.; Shaker, D.S.; Mohamed, S.A. Optimization of nano spray drying parameters for production of α-amylase nanopowder for biotheraputic applications using factorial design. *Dry. Technol.* **2019**, *37*, 2152–2160. [CrossRef]
50. Favaro-Trindade, C.S.; Rocha-Selmi, G.A.; dos Santos, M.G. Microencapsulation of sweeteners. In *Microencapsulation and Microspheres for Food Applications*; Sagis, L.M.C., Ed.; Elsevier: Amsterdam, The Netherlands, 2015; pp. 333–349.
51. Cherukuri, S.; Mansukhani, G. Sweetener Delivery Systems Containing Polyvinyl Acetate. U.S. Patent 4816265, 28 March 1989.
52. Rocha-Selmi, G.A.; Bozza, F.T.; Thomazini, M.; Bolini, H.M.A.; Fávaro-Trindade, C.S. Microencapsulation of aspartame by double emulsion followed by complex coacervation to provide protection and prolong sweetness. *Food Chem.* **2013**, *139*, 72–78. [CrossRef]
53. Eggleston, G.; Aita, G.; Triplett, A. Circular sustainability of sugarcane: Natural, nutritious, and functional unrefined sweeteners that meet new consumer demands. *Sugar Technol.* **2021**, *23*, 964–973. [CrossRef]
54. Sousa Lima, R.; Cazelatto de Medeiros, A.; André Bolini, H.M. Sucrose replacement: A sensory profile and time-intensity analysis of a tamarind functional beverage with artificial and natural non-nutritive sweeteners. *J. Sci. Food Agric.* **2021**, *101*, 593–602. [CrossRef]
55. Farhat, G.; Dewison, F.; Stevenson, L. Knowledge and perceptions of non-nutritive sweeteners within the UK adult population. *Nutrients* **2021**, *13*, 444. [CrossRef]
56. O, B.Y.S.; Coyle, D.H.; Dunford, E.K.; Wu, J.H.Y.; Louie, J.C.Y. The use of non-nutritive and low-calorie sweeteners in 19,915 local and imported pre-packaged foods in Hong Kong. *Nutrients* **2021**, *13*, 1861. [CrossRef] [PubMed]
57. Walbolt, J.; Koh, Y. Non-nutritive sweeteners and their associations with obesity and type 2 diabetes. *J. Obes. Metab. Syndr.* **2020**, *29*, 114–123. [CrossRef] [PubMed]
58. Manavalan, D.; Shubrook, C.; Young, C.F. Consumption of non-nutritive sweeteners and risk for type 2 diabetes: What do we know, and not? *Curr. Diab. Rep.* **2021**, *21*, 53. [CrossRef] [PubMed]
59. Xu, Y.; Wu, Y.; Liu, Y.; Li, J.; Du, G.; Chen, J.; Lv, X.; Liu, L. Sustainable bioproduction of natural sugar substitutes: Strategies and challenges. *Trends Food Sci. Technol.* **2022**, *129*, 512–527. [CrossRef]
60. Zhang, G.; Zhang, L.; Ahmad, I.; Zhang, J.; Zhang, A.; Tang, W.; Ding, Y.; Lyu, F. Recent advance in technological innovations of sugar-reduced products. *Crit. Rev. Food Sci. Nutr.* **2022**, 1–15. [CrossRef] [PubMed]
61. Carniel Beltrami, M.; Döring, T.; De Dea Lindner, J. Sweeteners and sweet taste enhancers in the food industry. *Food Sci. Technol.* **2018**, *38*, 181–187. [CrossRef]
62. Ashwell, M.; Gibson, S.; Bellisle, F.; Buttriss, J.; Drewnowski, A.; Fantino, M.; Gallagher, A.M.; De Graaf, K.; Goscinny, S.; Hardman, C.A.; et al. Expert consensus on low-calorie sweeteners: Facts, research gaps and suggested actions. *Nutr. Res. Rev.* **2020**, *33*, 145–154. [CrossRef] [PubMed]
63. Basílio, M.; Silva, L.J.; Pereira, A.M.; Pena, A.; Lino, C.M. Artificial sweeteners in non-alcoholic beverages: Occurrence and exposure estimation of the Portuguese population. *Food Addit. Contam. Part A* **2020**, *37*, 2040–2050. [CrossRef]
64. EFSA Panel on Food Additives and Nutrient Sources added to Food (ANS). Scientific Opinion on the re-evaluation of aspartame (E 951) as a food additive. *EFSA J.* **2013**, *11*, 3496.
65. SCF (Scientific Committee on Food). Opinion of the Scientific Committee on Food on Sucralose. 2000. Available online: http://ec.europa.eu/food/fs/sc/scf/out68_en.pdf (accessed on 10 September 2023).
66. European Commission (EC). Opinion on Saccharin and Its Sodium, Potassium and Calcium Salts of the Scientific Committee for Food (SCF) Brussels. 1995. Available online: https://ec.europa.eu/food/sites/food/files/safety/docs/sci-com_scf_7_out26_en.pdf (accessed on 10 September 2023).

67. European Commission (EC). Re-Evaluation of Acesulfame K with Reference to the Previous Scientific Committee on Food Opinion of 1991. 2000. Available online: https://ec.europa.eu/food/sites/food/files/safety/docs/sci-com_scf_out52_en.pdf (accessed on 10 September 2023).
68. Anwar, S.; Syed, Q.A.; Munawar, F.; Arshad, M.; Ahmad, W.; Rehman, M.A.; Arshad, M.K. Inclusive Overview of Sweeteners Trends: Nutritional Safety and Commercialization. *ACS Food Sci. Technol.* **2023**, *3*, 245–258. [CrossRef]
69. Park, Y.C.; Oh, E.J.; Jo, J.H.; Jin, Y.S.; Seo, J.H. Recent advances in biological production of sugar alcohols. *Curr. Opin. Biotechnol.* **2016**, *37*, 105–113. [CrossRef] [PubMed]
70. Bursać Kovačević, D.; Barba, F.J.; Granato, D.; Galanakis, C.M.; Herceg, Z.; Dragović-Uzelac, V.; Putnik, P. Pressurized hot water extraction (PHWE) for the green recovery of bioactive compounds and steviol glycosides from Stevia Rebaudiana Bertoni leaves. *Food Chem.* **2018**, *254*, 150–157. [CrossRef] [PubMed]
71. Bursać Kovačević, D.; Maras, M.; Barba, F.J.; Granato, D.; Roohinejad, S.; Mallikarjunan, K.; Montesano, D.; Lorenzo, J.M.; Putnik, P. Innovative technologies for the recovery of phytochemicals from Stevia Rebaudiana Bertoni leaves: A review. *Food Chem.* **2018**, *268*, 513–521. [CrossRef] [PubMed]
72. Zhao, Y.; Lv, B.; Feng, X.; Li, C. Perspective on biotransformation and de Novo biosynthesis of licorice constituents. *J. Agric. Food Chem.* **2017**, *65*, 11147–11156. [CrossRef]
73. Al-Farsi, M.; Alasalvar, C.; Al-Abid, M.; Al-Shoaily, K.; Al-Amry, M.; Al-Rawahy, F. Compositional and functional characteristics of dates, syrups, and their by-products. *Food Chem.* **2007**, *104*, 943–947. [CrossRef]
74. AlYammahi, J.; Hai, A.; Krishnamoorthy, R.; Arumugham, T.; Hasan, S.W.; Banat, F. Ultrasound-assisted extraction of highly nutritious date sugar from date palm (*Phoenix dactylifera*) fruit powder: Parametric optimization and kinetic modeling. *Ultrason. Sonochem.* **2022**, *88*, 106107. [CrossRef]
75. Ben Thabet, I.; Besbes, S.; Masmoudi, M.; Attia, H.; Deroanne, C.; Blecker, C. Compositional, physical, antioxidant and sensory characteristics of novel syrup from date palm (*Phoenix dactylifera* L.). *Food Sci. Technol. Int.* **2009**, *15*, 583–590. [CrossRef]
76. Tang, Z.-X.; Shi, L.-E.; Aleid, S.M. Date fruit: Chemical composition, nutritional and medicinal values, products. *J. Sci. Food Agric.* **2013**, *93*, 2351–2361. [CrossRef]
77. Julai, K.; Sridonpai, P.; Ngampeerapong, C.; Tongdonpo, K.; Suttisansanee, U.; Kriengsinyos, W.; On-Nom, N.; Tangsuphoom, N. Effects of extraction and evaporation methods on physico-chemical, functional, and nutritional properties of syrups from Barhi dates (*Phoenix dactylifera* L.). *Foods* **2023**, *12*, 1268. [CrossRef]
78. Taleb, H.; Maddocks, S.E.; Morris, R.K.; Kanekanian, A.D. The antibacterial activity of date syrup polyphenols against *S. aureus* and *E. coli*. *Front. Microbiol.* **2016**, *7*, 198. [CrossRef]
79. Entezari, M.H.; Nazari, S.H.; Haddad Khodaparast, M.H. The direct effect of ultrasound on the extraction of date syrup and its micro-organisms. *Ultrason. Sonochem.* **2004**, *11*, 379–384. [CrossRef] [PubMed]
80. El-Sharnouby, G.A.; Aleid, S.M.; Al-Otaibi, M.M. Liquid sugar extraction from date palm (*Phoenix dactylifera* L.) fruits. *J. Food Process. Technol.* **2014**, *5*, 1–5.
81. Alotaibi, K.D.; Alharbi, H.A.; Yaish, M.W.; Ahmed, I.; Alharbi, S.A.; Alotaibi, F.; Kuzyakov, Y. Date palm cultivation: A review of soil and environmental conditions and future challenges. *Land Degrad. Dev.* **2023**, *34*, 2431–2444. [CrossRef]
82. FAOSTAT. Available online: https://www.fao.org/faostat/en/#data/QCL (accessed on 31 March 2023).
83. Ferry, M.; Gómez, S.; Jiménez, E.; Navarro, J.; Ruipérez, E.; Vilella-Esplá, J. The date palm grove of Elche, Spain: Research for the sustainable preservation of a world heritage site. *Palm* **2002**, *46*, 139–148.
84. Ley 16/1985, de 25 de junio, del Patrimonio Histórico Español. *Boletín Of. Estado* **1985**, *155*, 20342–20352.
85. Hanieh, A.A.; Hasan, A.; Assi, M. Date palm trees supply chain and sustainable model. *J. Clean. Prod.* **2020**, *258*, 120951. [CrossRef]
86. Krueger, R.R. Date palm (*Phoenix dactylifera* L.) biology and utilization. In *The Date Palm Genome*; Al-Khayri, J.M., Jain, S.M., Johnson, D.V., Eds.; Springer: Cham, Switzerland, 2021; Volume 1, pp. 3–28.
87. Zaid, A.; De Wet, P.F. Climatic requirements of date palm. In *Date Palm Cultivation*; FAO Plant Production and Protection Paper 156. Rev. 1; Zaid, A., Arias-Jimenez, E.J., Eds.; FAO: Rome, Italy, 2002; pp. 57–72.
88. Liebenberg, P.J.; Zaid, A. Date palm irrigation. In *Date Palm Cultivation*; FAO Plant Production and Protection Paper 156. Rev. 1; Zaid, A., Arias-Jimenez, E.J., Eds.; FAO: Rome, Italy, 2002; pp. 131–144.
89. Maas, E.V.; Grattan, S.R. Crop yields as affected by salinity. In *Agricultural Drainage*; Skaggs, R.W., van Schilfgaarde, J., Eds.; American Society of Agronomy, Crop Science Society of America, and Soil Science Society of America: Madison, WI, USA, 1999; Volume 38, pp. 55–108.
90. Hazzouri, K.M.; Flowers, J.M.; Nelson, D.; Lemansour, A.; Masmoudi, K.; Amiri, K.M.A. Prospects for the study and improvement of abiotic stress tolerance in date palms in the post-genomics era. *Front. Plant Sci.* **2020**, *11*, 293. [CrossRef]
91. FAO. 5 Facts about Dates That Make Them an Important Food of Our Future. Available online: https://www.fao.org/fao-stories/article/en/c/1251859/ (accessed on 19 April 2023).
92. Al-Khayri, J.M.; Jain, S.M.; Johnson, D.V. *Date Palm Genetic Resources and Utilization. Africa and the Americas, Volume 1*; Springer: Dordrecht, The Netherlands, 2015.
93. Al-Khayri, J.M.; Jain, S.M.; Johnson, D.V. *Date Palm Genetic Resources and Utilization. Asia and Europe, Volume 2*; Springer: Dordrecht, The Netherlands, 2015.

94. Mihi, A.; Tarai, N.; Chenchouni, H. Can palm date plantations and oasification be used as a proxy to fight sustainably against desertification and sand encroachment in hot drylands? *Ecol. Indic.* **2019**, *105*, 365–375. [CrossRef]
95. Santoro, A. Traditional oases in Northern Africa as multifunctional agroforestry systems: A systematic literature review of the provided Ecosystem Services and of the main vulnerabilities. *Agrofor. Syst.* **2023**, *97*, 81–96. [CrossRef]
96. Allam, A.; Cheloufi, H. Biodiversity of fruit species in the valley of Oued Righ: The case of the area of Touggourt (Algeria). *Fruits* **2013**, *68*, 33–37. [CrossRef]
97. Hamza, F.; Hanane, S. The effect of microhabitat features, anthropogenic pressure and spatial structure on bird diversity in southern Tunisian agroecosystems. *Ann. Appl. Biol.* **2021**, *179*, 195–206. [CrossRef]
98. Loumassine, H.E.; Bonnot, N.; Allegrini, B.; Bendjeddou, M.L.; Bounaceur, F.; Aulagnier, S. How arid environments affect spatial and temporal activity of bats. *J. Arid Environ.* **2020**, *180*, 104206. [CrossRef]
99. Derouiche, L.; Bouhadad, R.; Fernandes, C. Mitochondrial DNA and morphological analysis of hedgehogs (Eulipotyphla: Erinaceidae) in Algeria. *Biochem. Syst. Ecol.* **2016**, *64*, 57–64. [CrossRef]
100. Ayeb, N.; Majdoub, B.; Dbara, M.; Fguiri, I.; Khorchani, S.; Hammadi, M.; Khorchani, T. Quality and fatty acid profile of milk of indigenous dairy goats fed from oasis resources in Tunisian arid areas. *Anim. Prod. Sci.* **2020**, *60*, 2044–2049. [CrossRef]
101. Belhadj Elmehdi, E.; Remini, B.; Rezzoug, C.; Hamoudi, S. Study of ancestral irrigation systems in the oasis of Taghit in the South West of Algeria. *J. Water Land. Dev.* **2020**, *44*, 13–18.
102. El Janati, M.; Akkal-Corfini, N.; Bouaziz, A.; Oukarroum, A.; Robin, P.; Sabri, A.; Thomas, Z. Benefits of circular agriculture for cropping systems and soil fertility in oases. *Sustainability* **2021**, *13*, 4713. [CrossRef]
103. Benmeziane-Derradji, F. Nutritional value, phytochemical composition, and biological activities of Middle Eastern and North African date fruit: An overview. *Euro-Mediterr. J. Environ. Integr.* **2019**, *4*, 39. [CrossRef]
104. Oladzad, S.; Fallah, N.; Mahboubi, A.; Afsham, N.; Taherzadeh, M.J. Date fruit processing waste and approaches to its valorization: A review. *Bioresour. Technol.* **2021**, *340*, 125625. [CrossRef]
105. Najjar, Z.; Stathopoulos, C.; Chockchaisawasdee, S. Utilization of Date By-Products in the Food Industry. Emir. *J. Food Agric.* **2020**, *32*, 808–815.
106. Manickvasagan, A.; Chandini, S.; Al Attabi, Z. Effect of sugar replacement with date paste and date syrup on texture and sensory quality of kesari (traditional Indian dessert). *J. Agric. Mar. Sci.* **2018**, *22*, 67–74. [CrossRef]
107. Djaoud, K.; Boulekbache-Makhlouf, L.; Yahia, M.; Mansouri, H.; Mansouri, N.; Madani, K.; Romero, A. Dairy dessert processing: Effect of sugar substitution by date syrup and powder on its quality characteristics. *J. Food Process Preserv.* **2020**, *44*, e14414. [CrossRef]
108. Lajnef, I.; Khemiri, S.; Ben Yahmed, N.; Chouaibi, M.; Smaali, I. Straightforward extraction of date palm syrup from *Phoenix dactylifera* L. byproducts: Application as sucrose substitute in sponge cake formulation. *J. Food Meas. Charact.* **2021**, *15*, 3942–3952. [CrossRef]
109. Ghafari, Z.; Hojjatoleslamy, M.; Shokrani, R.; Shariaty, M.A. Use of date syrup as a sweetener in non alcoholic beer: Sensory and rheological assessment. *J. Microbiol. Biotechnol. Food Sci.* **2013**, *3*, 182–184.
110. Kazemalilou, S.; Alizadeh, A. Optimization of sugar replacement with date syrup in prebiotic chocolate milk using response surface methodology. *Korean J. Food Sci. Anim. Resour.* **2017**, *37*, 449–455. [CrossRef] [PubMed]
111. El-Nagga, E.A.; Abd El-Tawab, Y.A. Compositional characteristics of date syrup extracted by different methods in some fermented dairy products. *Ann. Agric. Sci.* **2012**, *57*, 29–36. [CrossRef]
112. Sánchez-Zapata, E.; Fernández-López, J.; Peñaranda, M.; Fuentes-Zaragoza, E.; Sendra, E.; Sayas, E.; Pérez-Alvarez, J.A. Technological properties of date paste obtained from date byproducts and its effect on the quality of a cooked meat product. *Food Res. Int.* **2011**, *44*, 2401–2407. [CrossRef]
113. Martín-Sánchez, A.M.; Ciro-Gómez, G.; Sayas, E.; Vilella-Esplá, J.; Ben-Abda, J.; Pérez-Álvarez, J.A. Date palm by-products as a new ingredient for the meat industry: Application to pork liver pâté. *Meat Sci.* **2013**, *93*, 880–887. [CrossRef] [PubMed]
114. Mrabet, A.; Jiménez-Araujo, A.; Guillén-Bejarano, R.; Rodríguez-Arcos, R.; Sindic, M. Date Seeds: A Promising Source of Oil with Functional Properties. *Foods* **2020**, *9*, 787. [CrossRef]
115. Al-Muhtaseb, A.; Jamil, F.; AlHaj, L.; Myint, M.T.Z.; Mahmoud, E.; Ahmad, M.; Hasan, A.; Rafiq, S. Biodiesel production over a catalyst prepared from biomass-derived waste date pits. *Biotechnol. Rep.* **2018**, *20*, e00284. [CrossRef]
116. Zainab-L, I.; Uyama, H.; Li, C.; Shen, Y.; Sudesh, K. Production of Polyhydroxyalkanoates From Underutilized Plant Oils by Cupriavidus necator. *Clean Soil Air Water* **2018**, *46*, 1700542. [CrossRef]
117. Kamarubahrin, A.F.; Haris, A. Nutritional and Potential Planting of Date Palm: Review of Recent Trends and Future Prospects in Malaysia. *Int. J. Fruit Sci.* **2020**, *20*, S1097–S1109. [CrossRef]
118. Jonoobi, M.; Shafie, M.; Shirmohammadli, Y.; Ashori, A.; Hosseinabadi, H.Z.; Mekonnen, T. A Review on Date Palm Tree: Properties, Characterization and Its Potential Applications. *J. Renew. Mater.* **2019**, *7*, 1055–1075. [CrossRef]
119. Vico, A.; Pérez-Murcia, M.D.; Bustamante, M.A.; Agulló, E.; Marhuenda-Egea, F.C.; Sáez, J.A.; Paredes, C.; Pérez-Espinosa, A.; Moral, R. Valorization of date palm (*Phoenix dactylifera* L.) pruning biomass by co-composting with urban and agri-food sludge. *J. Environ. Manag.* **2018**, *226*, 408–415. [CrossRef] [PubMed]
120. Farag, K.M. Date Palm: A Wealth of Healthy Food. In *Encyclopedia of Food and Health*; Caballero, B., Finglas, P.M., Toldrá, F., Eds.; Academic Press: Cambridge, MA, USA, 2016; pp. 356–360.

121. Bentrad, N.; Hamida-Ferhat, A. Chapter 22—Date palm fruit (*Phoenix dactylifera*): Nutritional values and potential benefits on health. In *The Mediterranean Diet*, 2nd ed.; Preedy, V.R., Watson, R.R., Eds.; Academic Press: Cambridge, MA, USA, 2020; pp. 239–255.
122. Fernández-López, J.; Viuda-Martos, M.; Sayas-Barberá, E.; Navarro-Rodríguez de Vera, C.; Pérez-Álvarez, J.Á. Biological, Nutritive, Functional and Healthy Potential of Date Palm Fruit (*Phoenix dactylifera* L.): Current Research and Future Prospects. *Agronomy* **2022**, *12*, 876. [CrossRef]
123. Maqsood, S.; Adiamo, O.; Ahmad, M.; Mudgil, P. Bioactive compounds from date fruit and seed as potential nutraceutical and functional food ingredients. *Food Chem.* **2020**, *308*, 125522. [CrossRef] [PubMed]
124. Haris, S.; Alam, M.; Galiwango, E.; Mohamed, M.M.; Kamal-Eldin, A.; Al-Marzouqi, A.H. Characterization analysis of date fruit pomace: An underutilized waste bioresource rich in dietary fiber and phenolic antioxidants. *Waste Manag.* **2023**, *163*, 34–42. [CrossRef] [PubMed]
125. Gnana Rani, V.; Thangamathi, P.; Harihar, B.; Ananth, S.; Suvaithenamudhan, S. Evaluation and assessment of nutritional composition for quality profiling of *Phoenix dactylifera* cultivars using multivariate analytical tools. *Ecol. Genet. Genom.* **2023**, *27*, 100173.
126. Alahyanea, A.; ElQarnifa, S.; Ayour, J.; Elateri, I.; Ouamnina, A.; Ait-Oubahou, A.; Benichou, M.; Abderrazik, M. Date seeds (*Phoenix dactylifera* L.) valorization: Chemical composition of lipid fraction. *Braz. J. Biol.* **2022**, *84*, e260771. [CrossRef]
127. Kchaou, W.; Abbès, F.; Mansour, R.B.; Blecker, C.; Attia, H.; Besbes, S. Phenolic profile, antibacterial and cytotoxic properties of second grade date extract from Tunisian cultivars (*Phoenix dactylifera* L.). *Food Chem.* **2016**, *194*, 1048–1055. [CrossRef]
128. Muñoz-Bas, C.; Muñoz-Tébar, N.; Candela-Salvador, L.; Pérez-Alvarez, J.A.; Lorenzo, J.M.; Viuda-Martos, M.; Fernández-López, J. Quality Characteristics of Fresh Date Palm Fruits of "Medjoul" and "Confitera" cv. from the Southeast of Spain (Elche Palm Grove). *Foods* **2023**, *12*, 2659. [CrossRef] [PubMed]
129. Kamal, H.; Hamdi, M.; Mudgil, P.; Aldhaheri, M.; Affan Baig, M.; Hassan, H.M.; Alamri, A.S.; Galanakis, C.M.; Maqsood, S. Nutraceutical and bioactive potential of high-quality date fruit varieties (*Phoenix dactylifera* L.) as a function of in-vitro simulated gastrointestinal digestion. *J. Pharm. Biomed. Anal.* **2023**, *223*, 115113. [CrossRef] [PubMed]
130. Alqahtani, N.K.; Mohamed, H.A.; Moawad, M.E.; Younis, N.S.; Mohamed, M.E. The Hepatoprotective Effect of Two Date Palm Fruit Cultivars' Extracts: Green Optimization of the Extraction Process. *Foods* **2023**, *12*, 1229. [CrossRef] [PubMed]
131. Hussein, A.M.; Mahmoud, S.A.; Elazab, K.M.; Ahmed, F.; Abouelnaga, A.F.; Abass, M.; Mosa, A.A.H.; Hussein, M.A.M.; Elsayed, M.E.G. Possible Mechanisms of the Neuroprotective Actions of Date Palm Fruits Aqueous Extracts against Valproic Acid-Induced Autism in Rats. *Curr. Issues Mol. Biol.* **2023**, *45*, 1627–1643. [CrossRef] [PubMed]
132. Hamdi, M.; Mostafa, H.; Aldhaheri, M.; Mudgil, P.; Kamal, H.; Alamri, A.S.; Galanakis, C.M.; Maqsood, S. Valorization of different low-grade date (*Phoenix dactylifera* L.) fruit varieties: A study on the bioactive properties of polyphenolic extracts and their stability upon in vitro simulated gastrointestinal digestion. *Plant Physiol. Biochem.* **2023**, *200*, 107764. [CrossRef] [PubMed]
133. Ayyash, M.; Tarique, M.; Alaryani, M.; Al-Sbiei, A.; Masad, R.; Al-Saafeen, B.; Fernández-Cabezudo, M.; al-Ramadi, B.; Kizhakkayil, J.; Kamal-Eldin, A. Bioactive properties and untargeted metabolomics analysis of bioaccessible fractions of non-fermented and fermented date fruit pomace by novel yeast isolates. *Food Chem.* **2022**, *296*, 133666. [CrossRef] [PubMed]
134. Alsukaibi, A.K.D.; Alenezi, K.M.; Haque, A.; Ahmad, I.; Saeed, M.; Verma, M.; Ansari, I.A.; Hsieh, M.F. Chemical, biological and in silico assessment of date (*P. dactylifera* L.) fruits grown in Ha'il region. *Front. Chem.* **2023**, *11*, 1138057. [CrossRef]
135. Bhaskaracharya, R.K.; Bhaskaracharya, A.; Stathopoulos, C. A systematic review of antibacterial activity of polyphenolic extract from date palm (*Phoenix dactylifera* L.) kernel. *Front. Pharmacol.* **2023**, *13*, 1043548. [CrossRef]
136. Alvi, T.; Khan, M.K.I.; Maan, A.A.; Razzaq, Z.U. Date fruit as a promising source of functional carbohydrates and bioactive compounds: A review on its nutraceutical potential. *J. Food Biochem.* **2022**, *46*, e14325. [CrossRef]
137. Caporizzi, R.; Severini, C.; Derossi, A. Study of different technological strategies for sugar reduction in muffin addressed for children. *NFS J.* **2021**, *23*, 44–51. [CrossRef]
138. Arshad, S.; Rehman, T.; Saif, S.; Rajoka, M.S.R.; Ranjha, M.M.A.N.; Hassoun, A.; Cropotova, J.; Trif, M.; Younas, A.; Aadil, R.M. Replacement of refined sugar by natural sweeteners: Focus on potential health benefits. *Heliyon* **2022**, *8*, e10711. [CrossRef] [PubMed]
139. Echegaray, N.; Gullón, B.; Pateiro, M.; Amarowicz, R.; Misihairabgwi, J.M.; Lorenzo, J.M. Date fruit and its by-products as promising source of bioactive components: A Review. *Food Rev. Int.* **2023**, *39*, 1411–1432. [CrossRef]
140. Younas, A.; Naqvi, S.A.; Khan, M.R.; Shabbir, M.A.; Jatoi, M.A.; Anwar, F.; Aadil, R.M. Functional food and nutra-pharmaceutical perspectives of date (*Phoenix dactylifera* L.) fruit. *J. Food Biochem.* **2020**, *44*, e13332. [CrossRef] [PubMed]
141. Abdeen, E.S.M.M. Enhancement of functional properties of dairy products by date fruits. *Egypt. J. Food* **2018**, *46*, 197–206.
142. Muñoz-Tebar, N.; Viuda-Martos, M.; Lorenzo, J.M.; Fernández-López, J.; Pérez-Álvarez, J.A. Strategies for the Valorization of Date Fruit and Its Co-Products: A New Ingredient in the Development of Value-Added Foods. *Foods* **2023**, *12*, 1456. [CrossRef] [PubMed]
143. Amerinasab, A.; Labbafi, M.; Mousavi, M.; Khodaiyan, F. Development of a novel yoghurt based on date liquid sugar: Physico-chemical and sensory characterization. *J. Food Sci. Technol.* **2015**, *52*, 6583–6590. [CrossRef] [PubMed]
144. Moustafa, R.M.A.; Abdelwahed, E.M.; El-Neshwy, A.A.; Taha, S.N. Utilization of date syrup (dips) in production of flavoured yoghurt. *Zagazig J. Food Dairy Res.* **2016**, *43 Pt B*, 2463–2471.
145. Jafarpour, D.; Amirzadeh, A.; Maleki, M.; Mahmoudi, M.R. Comparison of physicochemical properties and general acceptance of flavored drinking yogurt containing date and fig syrups. *Foods Raw Mater.* **2017**, *5*, 36–43. [CrossRef]

146. Abdollahzadeh, S.M.; Zahedani, M.R.; Rahmdel, S.; Hemmati, F.; Mazloomi, S.M. Development of Lactobacillus acidophilus-fermented milk fortified with date extract. *LWT* **2018**, *L98*, 577–582. [CrossRef]
147. Abdel-Ghany, A.S.; Zaki, D.A. Production of novel functional yoghurt fortified with bovine colostrum and date syrup for children. *Alex. Sci. Exch. J.* **2018**, *39*, 651–662. [CrossRef]
148. Jridi, M.; Souissi, N.; Salem, M.B.; Ayadi, M.; Nasri, M.; Azabou, S. Tunisian date (*Phoenix dactylifera* L.) by-products: Characterization and potential effects on sensory, textural and antioxidant properties of dairy desserts. *Food Chem.* **2015**, *188*, 8–15. [CrossRef]
149. Tawfek, M.; Baker, E.; El-Sayed, H. Study Properties of Fermented Camels' and Goats' Milk Beverages Fortified with Date Palm (*Phoenix dactylifera* L.). *Food Nutr. Sci.* **2021**, *12*, 418–428.
150. Gab-Allah, R.H.; Shehta, H.A. A new functional whey beverage, containing calcium and Date syrup (Dibs). *Egypt. J. Nutr.* **2020**, *35*, 53–75.
151. Oluwasina, O.O.; Demehin, B.F.; Awolu, O.O.; Igbe, F.O. Optimization of starch-based candy supplemented with date palm (*Phoenix dactylifera*) and tamarind (*Tamarindus indica* L.). *Arab. J. Chem.* **2020**, *13*, 8039–8050. [CrossRef]
152. Woodbury, T.J.; Lust, A.L.; Mauer, L.J. The effects of commercially available sweeteners (sucrose and sucrose replacers) on wheat starch gelatinization and pasting, and cookie baking. *J. Food Sci.* **2021**, *86*, 687–698. [CrossRef] [PubMed]
153. Amin, A.A.E.N.; Abdel Fattah, A.F.A.K.; El-Sharabasy, S.F. Quality attributes of cookies fortified with date powder. *Arab Univ. J. Agric. Sci.* **2019**, *27*, 2539–2547. [CrossRef]
154. Agu, H.O.; Onuoha, G.O.; Elijah, O.E.; Jideani, V.A. Consumer acceptability of acha and malted Bambara groundnut (BGN) biscuits sweetened with date palm. *Heliyon* **2020**, *6*, e05522. [CrossRef] [PubMed]
155. Turki, M.; Barbosa-Pereira, L.; Bertolino, M.; Essaidi, I.; Ghirardello, D.; Torri, L.; Bouzouita, N.; Zeppa, G. Physico-Chemical Characterization of Tunisian Canary Palm (*Phoenix canariensis* Hort. Ex Chabaud) Dates and Evaluation of Their Addition in Biscuits. *Foods* **2020**, *9*, 695. [CrossRef] [PubMed]
156. Majzoobi, M.; Mansouri, H.; Mesbahi, G.; Farahnaky, A.; Golmakani, M.T. Effects of sucrose substitution with date syrup and date liquid sugar on the physicochemical properties of dough and biscuits. *J. Agric. Sci. Technol.* **2016**, *18*, 643–656.
157. Panhwar, A.A.; Khaskheli, S.G.; Sheikh, S.A.; Soomro, A.H. Physico-chemical and sensorial properties of biscuits supplemented with date powder. In Proceedings of the VII International Date Palm Conference, Abu Dhabi, United Arab Emirates, 14–16 March 2022; Volume 1371, pp. 361–366.
158. Aljutaily, T.; Elbeltagy, A.; Ali, A.A.; Gadallah, M.G.; Khalil, N.A. Anti-Obesity Effects of Formulated Biscuits Supplemented with Date's Fiber; Agro-Waste Products Used as a Potent Functional Food. *Nutrients* **2022**, *14*, 5315. [CrossRef]
159. Dhankhar, J.; Vashistha, N.; Sharma, A. Development of biscuits by partial substitution of refined wheat flour with chickpea flour and date powder. J. Microbiol. Biotechnol. *Food Sci.* **2019**, *8*, 1093–1097.
160. Nwanekezi, E.C.; Ekwe, C.C.; Agbugba, R.U. Effect of substitution of sucrose with date palm (*Phoenix dactylifera*) fruit on quality of bread. *J. Food Process. Technol.* **2015**, *6*, 1.
161. Awofadeju, O.F.J.; Awe, A.B.; Adewumi, O.J.; Adeyemo, E.A. Influence of substituting sucrose with date palm fruit flour (DPFF) on the nutritional and organoleptic properties of bread. *Croat. J. Food Sci. Technol.* **2021**, *13*, 1–6. [CrossRef]
162. Shinde, D.B.; Popale, S.R.; Salunke, S.G.; Kadam, S. Quality characteristics of breads fortified with date (*Phoenix dactylifera* L.) paste. *J. Pharmacogn. Phytochem.* **2019**, *8*, 4489–4492.
163. Hussien, H.A.; Salem, E.M.; Masoud, M.R. Innovation of High Nutritional Value Snack Bars from Dates and Extruded Cereals. *Egypt. J. Agric. Res.* **2018**, *96*, 149–158. [CrossRef]
164. Ibrahim, S.A.; Fidan, H.; Aljaloud, S.O.; Stankov, S.; Ivanov, G. Application of date (*Phoenix dactylifera* L.) fruit in the composition of a novel snack bar. *Foods* **2021**, *10*, 918. [CrossRef] [PubMed]
165. Ibrahim, S.A.; Ayad, A.A.; Williams, L.L.; Ayivi, R.D.; Gyawali, R.; Krastanov, A.; Aljaloud, S.O. Date fruit: A review of the chemical and nutritional compounds, functional effects and food application in nutrition bars for athletes. *Intern. J. Food Sci. Technol.* **2021**, *56*, 1503–1513. [CrossRef]
166. Aljobair, M.O. Characteristics of cereal flakes manufactured using date syrup in place of sugar. *Nutr. Food Sci.* **2018**, *48*, 899–910. [CrossRef]
167. Erukainure, O.L.; Egagah, T.I.; Bolaji, P.T.; Ajiboye, A.J. Development and quality assessment of date chocolate products. *Am. J. Food Technol.* **2010**, *5*, 324–330. [CrossRef]
168. Tlay, R.H.; Al-Baidhani, A.M. The possibility of benefiting from date seed powder in the manufacture of chocolate spread and studying its quality characteristics. *Euphrates J. Agric. Sci.* **2023**, *15*, 294–307.
169. Alfheeaid, H.A.; Barakat, H.; Althwab, S.A.; Musa, K.H.; Malkova, D. Nutritional and Physicochemical Characteristics of Innovative High Energy and Protein Fruit-and Date-Based Bars. *Foods* **2023**, *12*, 2777. [CrossRef]
170. Barakat, H.; Alfheeaid, H.A. Date Palm Fruit (*Phoenix dactylifera*) and Its Promising Potential in Developing Functional Energy Bars: Review of Chemical, Nutritional, Functional, and Sensory Attributes. *Nutrients* **2023**, *15*, 2134. [CrossRef]

MDPI AG
Grosspeteranlage 5
4052 Basel
Switzerland
Tel.: +41 61 683 77 34

Foods Editorial Office
E-mail: foods@mdpi.com
www.mdpi.com/journal/foods

www.ingramcontent.com/pod-product-compliance
Lightning Source LLC
Chambersburg PA
CBHW040745250926
51850CB00007BB/1414
* 9 7 8 3 7 2 5 8 4 7 3 1 0 *